Surveys in Computer Science

Editors:

G. Schlageter
F. Stetter
E. Goto

S. Ceri G. Gottlob L. Tanca

Logic Programming and Databases

With 42 Figures

Springer-Verlag Berlin Heidelberg NewYork
London Paris Tokyo Hong Kong

Stefano Ceri
Dipartimento di Matematica
Università di Modena
Via Campi 213
I-41100 Modena

Georg Gottlob
Institut für Angewandte Informatik
und Systemanalyse
Abteilung für Verteilte Datenbanken
und Expertensysteme
Technische Universität Wien
Paniglgasse 16/181
A-1040 Wien

Letizia Tanca
Dipartimento di Elettronica
Politecnico di Milano
Piazza Leonardo Da Vinci 32
I-20133 Milano

ISBN 3-540-51728-6 Springer-Verlag Berlin Heidelberg New York
ISBN 0-387-51728-6 Springer-Verlag New York Berlin Heidelberg

Library of Congress Cataloging-in-Publication Data.
Ceri, Stefano, 1955–
Logic programming and databases / S. Ceri, G. Gottlob, L. Tanca. p. cm. –
(Surveys in computer science)
Includes bibliographical references.
ISBN 0-387-51728-6 (U.S.)
1. Logic programming. 2. Data base management. I. Gottlob, G. (Georg).
II. Tanca, L. (Letizia). III. Title. IV. Series.   QA76.63.C47   1990   005.74 – dc20
89-28960 CIP

This work is subject to copyright. All rights are reserved, whether the whole or part of the material is concerned, specifically the rights of translation, reprinting, reuse of illustrations, recitation, broadcasting, reproduction on microfilms or in other ways, and storage in data banks. Duplication of this publication or parts thereof is only permitted under the provisions of the German Copyright Law of September 9, 1965, in its version of June 24, 1985, and a copyright fee must always be paid. Violations fall under the prosecution act of the German Copyright Law.

© Springer-Verlag Berlin Heidelberg 1990
Printed in Germany

The use of registered names, trademarks, etc. in this publication does not imply, even in the absence of a specific statement, that such names are exempt from the relevant protective laws and regulations and therefore free for general use.

Printing: Druckhaus Beltz, Hemsbach; bookbinding: J. Schäffer, Grünstadt
2145/3140-543210 – Printed on acid-free paper

Preface

The topic of logic programming and databases has gained increasing interest in recent years. Several events have marked the rapid evolution of this field: the selection, by the Japanese Fifth Generation Project, of *Prolog* and of the relational data model as the basis for the development of new machine architectures; the focusing of research in database theory on logic queries and on recursive query processing; and the pragmatic, application-oriented development of expert database systems and of knowledge-base systems. As a result, an enormous amount of work has been produced in the recent literature, coupled with the spontaneous growth of several advanced projects in this area.

The goal of this book is to present a systematic overview of a rapidly evolving discipline, which is presently not described with the same approach in other books. We intend to introduce students and researchers to this new discipline; thus we use a plain, tutorial style, and complement the description of algorithms with examples and exercises. We attempt to achieve a balance between theoretical foundations and technological issues; thus we present a careful introduction to the new language *Datalog*, but we also focus on the efficient interfacing of logic programming formalisms (such as *Prolog* and *Datalog*) with large databases.

The book is divided into three parts, preceded by two preliminary chapters. Chapter 1 offers an overview of the field. Chapter 2 discusses those aspects of the relational model and *Prolog* which are required for understanding the rest of the book. Of course, Chapter 2 is not a complete tutorial on these fields; it just redefines terminology, notation, and basic concepts in order to keep the book self-contained. However, in order to fully understand the problems, the reader should have some background in these subjects.

Part I is devoted to the coupling of *Prolog* with relational databases. Chapter 3 presents *Prolog* as a query language, applied to the formulation of two classical problems, the *anti-trust* and the *bill-of-materials* problems. Chapter 4 describes the alternative architectures and techniques for coupling a *Prolog* system to a relational database. Chapter 5 presents a review of the major current projects and prototypes for coupling *Prolog* and relational databases.

Part II is devoted to the precise definition of the *Datalog* language. Chapter 6 defines formally the syntax and semantics of *Datalog*. The semantics of *Datalog* is described by a nonprocedural, model-theoretic approach. Chapter 7 presents the proof theory of the language, by introducing an algorithm for evaluating *Datalog* goals, and by showing that the method is sound and complete with respect to the model-theoretic semantics. Chapter 7 also introduces two other paradigms for the evaluation of *Datalog* programs: *fixpoint theory* and *backward chaining*. In particular, resolution and SLD-resolution are defined in the context of *Datalog*.

Part III is devoted to the description of query optimization techniques for *Datalog*. Chapter 8 presents a general classification of the optimization techniques; we distinguish *rewriting methods*, which assume as input a *Datalog* program and produce as output an optimized *Datalog* program, from *evaluation methods*, which assume as input a *Datalog* program and produce the result of the query. Furthermore, we show a simple translation from *Datalog* programs to systems of algebraic equations. This translation enables us to describe a class of algebraic methods for query optimization. These arguments are then studied in the subsequent Chapters 9 and 10. Chapter 9 deals with evaluation methods, and presents both bottom-up and top-down evaluation methods (including the Naive, Semi-naive, and Query-Subquery methods). Chapter 10 deals with rewriting methods, and presents the logical rewriting methods (including the Magic Set, Counting, and the Static Filtering methods) and the algebraic rewriting methods.

Chapter 11 deals with extensions to pure *Datalog*, such as sets and negation. This chapter should be considered as an introduction to the subject, rather than a full treatment. Finally, Chapter 12 presents an overview of the main projects on the integration of logic programming and databases, including *Nail, LDL*, and the Fifth Generation Project.

The book is organized so that the three parts can be read independently by different readers. In fact, the various chapters are rather independent.

This book does not present a full overview of all topics which belong to the field of deductive databases. For instance, it does not deal with incompleteness, disjunctive data, or the validation of integrity constraints. We apologize to those who find their favorite topics missing, in particular the community of logicians who work in the area of deductive databases. In the choice of arguments, we have concentrated our attention on the use of *large* databases; loyal to the tradition of the database community, we are mainly concerned with the efficiency of database access, even when we use logic programming as a query language.

This book is primarily the outcome of research work conducted by the authors in cooperation with other colleagues. We wish to thank Gio Wiederhold for his contribution to the CGW approach, which was developed in the framework of the KBMS project at Stanford University; Stefano Crespi-Reghizzi, Gianfranco Lamperti, Luigi Lavazza, and Roberto Zicari, for their contribution to the development of the algebraic approach to logic queries within the framework of the ALGRES project; Silvia Cozzi, Fabrizio Gozzi, Marco Lugli, and Guido Sanguinetti, who developed the PRIMO system as part of their diploma theses at the University of Modena; and Roberta Cantaroni, Stefania Ferrari, and Franca Garzotto, who have addressed with us problems related to logic databases.

Many colleagues and students have produced useful comments in extending, reviewing, and correcting the manuscript; among them, we wish to thank Maurice Houtsma, who has made a very careful reading, suggesting several corrections and improvements; Hervé Gallaire, Jean-Marie Nicolas, Johann Christoph Freytag, and François Bry, who have provided useful information concerning the entire manuscript and more specifically about the projects developed at ECRC; Shamin Naqvi has also made specific comments about the LDL project developed at MCC; Wolfgang Nejdl has provided us with material about the QSQ method and its modifications. Werner Schimanovich and Alex Leitsch have helped us with encouragement and interesting discussions.

Particular appreciation is given to Gunter Schlageter and to his PhD students and to Renate Pitrik, Wilhelm Rossak, and Robert Truschnegg, whose careful reading and critical review has improved the quality of the book. Remaining errors and omissions are, of course, the responsibility of the authors.

We would like to thank our home institutions for providing support and equipment for editing this manuscript: the University of Modena, the Politecnico di Milano, the Technical University of Wien, and Stanford University. During the preparation of the manuscript, Letizia Tanca was supported by a grant from C.I.L.E.A. The accurate and fast final preparation of this manuscript has been supervised by Dr. Hans Wössner, Ingeborg Mayer, and the copy editor Dr. Gillian Hayes, from Springer-Verlag.

The "Progetto Finalizzato Informatica e Calcolo Parallelo" of the Italian National Research Council, starting in 1989, will provide us with a research environment for realizing many of the ideas presented in this book.

October 1989
Stefano Ceri
Georg Gottlob
Letizia Tanca

Table of Contents

Chapter 1
Logic Programming and Databases: An Overview 1

1.1 Logic Programming as Query Language 2
1.2 Prolog and Datalog 9
1.3 Alternative Architectures 11
1.4 Applications 14
1.5 Bibliographic Notes 14

Chapter 2
A Review of Relational Databases and Prolog 16

2.1 Overview of Relational Databases 16
2.1.1 The Relational Model 16
2.1.2 Relational Languages 18
2.2 Prolog: A Language for Programming in Logic 23
2.3 Bibliographic Notes 26

Part I Coupling Prolog to Relational Databases 27

Chapter 3
Prolog as a Query Language 29

3.1 The Anti-Trust Control Problem 30
3.2 The Bill of Materials Problem 34
3.3 Conclusions 38
3.4 Bibliographic Notes 38
3.5 Exercises 38

Chapter 4
Coupling Prolog Systems to Relational Databases 40

4.1 Architectures for Coupling Prolog
     and Relational Systems 40
4.1.1 Assumptions and Terminology 40
4.1.2 Components of a CPR System 42
4.1.3 Architecture of CPR Systems 45
4.2 Base Conjunctions 47
4.2.1 Determining Base Conjunctions in LCPR Systems . . 49
4.2.2 Improving Base Conjunctions in TCPR Systems . . . 54
4.3 Optimization of the Prolog/Database Interface 57
4.3.1 Caching of Data 58
4.3.2 Caching of Data and Queries 58
4.3.3 Use of Subsumption 59
4.3.4 Caching Queries 60
4.3.5 Parallelism and Pre-fetching in Database Interfaces . 61
4.4 Conclusions 62
4.5 Bibliographic Notes 62
4.6 Exercises 63

Chapter 5
**Overview of Systems for Coupling Prolog
to Relational Databases** 65

5.1 PRO-SQL 65
5.2 EDUCE 67
5.3 ESTEAM 68
5.4 BERMUDA 69
5.5 CGW and PRIMO 70
5.6 QUINTUS-PROLOG 72
5.7 Bibliographic Notes 74

Part II Foundations of Datalog 75

Chapter 6
Syntax and Semantics of Datalog 77

6.1 Basic Definitions and Assumptions 77
6.1.1 Alphabets, Terms, and Clauses 77
6.1.2 Extensional Databases and Datalog Programs 81
6.1.3 Substitutions, Subsumption, and Unification 83

6.2	The Model Theory of Datalog	86
6.2.1	Possible Worlds, Truth, and Herbrand Interpretations	86
6.2.2	The Least Herbrand Model	91
6.3	Conclusions	92
6.4	Bibliographic Notes	92
6.5	Exercises	93

Chapter 7
Proof Theory and Evaluation Paradigms of Datalog 94

7.1	The Proof Theory of Datalog	94
7.1.1	Fact Inference	95
7.1.2	Soundness and Completeness of the Inference Rule EP	98
7.2	Least Fixpoint Iteration	101
7.2.1	Basic Results of Fixpoint Theory	101
7.2.2	Least Fixpoints and Datalog Programs	104
7.3	Backward Chaining and Resolution	107
7.3.1	The Principle of Backward Chaining	107
7.3.2	Resolution	113
7.4	Conclusions	120
7.5	Bibliographic Notes	121
7.6	Exercises	121

Part III Optimization Methods for Datalog .. 123

Chapter 8
Classification of Optimization Methods for Datalog 124

8.1	Criteria for the Classification of Optimization Methods	124
8.1.1	Formalism	124
8.1.2	Search Strategy	125
8.1.3	Objectives of Optimization Methods	126
8.1.4	Type of Information Considered	126
8.2	Classification of Optimization Methods	127
8.3	Translation of Datalog into Relational Algebra	130
8.4	Classification of Datalog Rules	136
8.5	The Expressive Power of Datalog	142
8.6	Bibliographic Notes	143
8.7	Exercises	144

Chapter 9
Evaluation Methods 145

9.1   Bottom-up Evaluation 145
9.1.1   Algebraic Naive Evaluation 145
9.1.2   Semi-naive Evaluation 150
9.1.3   The Method of Henschen and Naqvi 154
9.2   Top-down Evaluation 155
9.2.1   Query-Subquery 155
9.2.2   The RQA/FQI Method 160
9.3   Bibliographic Notes 161
9.4   Exercises 162

Chapter 10
Rewriting Methods 163

10.1   Logical Rewriting Methods 163
10.1.1 Magic Sets 165
10.1.2 The Counting Method 174
10.1.3 The Static Filtering Method 177
10.1.4 Semi-naive Evaluation by Rewriting 183
10.2   Rewriting of Algebraic Systems 185
10.2.1 Reduction to Union-Join Normal Form 185
10.2.2 Determination of Common Subexpressions 187
10.2.3 Query Subsetting and Strong Components 189
10.2.4 Marking of Variables 191
10.2.5 Reduction of Variables 193
10.2.6 Reduction of Constants 193
10.2.7 Summary of the Algebraic Approach 200
10.3   A General View of Optimization 200
10.4   Bibliographic Notes 205
10.5   Exercises 206

Chapter 11
Extensions of Pure Datalog 208

11.1   Using Built-in Predicates in Datalog 210
11.2   Incorporating Negation into Datalog 211
11.2.1 Negation and the Closed World Assumption 212
11.2.2 Stratified Datalog 215
11.2.3 Perfect Models and Local Stratification 224
11.2.4 Inflationary Semantics and Expressive Power 226

11.3	Representation and Manipulation of Complex Objects	228
11.3.1	Basic Features of LDL	229
11.3.2	Semantics of Admissible LDL Programs	235
11.3.3	Data Models for Complex Objects	240
11.4	Conclusions	241
11.5	Bibliographic Notes	241
11.6	Exercises	244

Chapter 12
Overview of Research Prototypes for Integrating Relational Databases and Logic Programming 246

12.1	The LDL Project	247
12.2	The NAIL! Project	251
12.3	The POSTGRES Project	255
12.4	The FIFTH GENERATION Project	257
12.5	The KIWI Project	260
12.6	The ALGRES Project	262
12.7	The PRISMA Project	264
12.8	Bibliographic Notes	265

Bibliography 267

Index 277

Chapter 1
Logic Programming and Databases: An Overview

This book deals with the *integration of logic programming and databases to generate new types of systems,* which extend the frontiers of computer science in an important direction and fulfil the needs of new applications. Several names are used to describe these systems:

a) The term *deductive database* highlights the ability to use a logic programming style for expressing deductions concerning the content of a database.
b) The term *knowledge base management system (KBMS)* highlights the ability to manage (complex) knowledge instead of (simple) data.
c) The term *expert database system* highlights the ability to use expertise in a particular application domain to solve classes of problems, but having access over a large database.

The confluence between logic programming and databases is part of a general trend in computer science, where different fields are explored in order to discover and profit from their common concepts.

Logic programming and databases have evolved in parallel throughout the seventies. *Prolog*, the most popular language for PROgramming in LOGic, was born as a simplification of more general theorem proving techniques to provide efficiency and programmability. Similarly, the relational data model was born as a simplification of complex hierarchical and network models, to enable set-oriented, nonprocedural data manipulation. Throughout the seventies and early eighties, the use of both *Prolog* and relational databases has become widespread, not only in academic or scientific environments, but also in the commercial world.

Important studies on the relationships between logic programming and relational databases have been conducted since the end of the seventies, mostly from a theoretical viewpoint. The success of this confluence has been facilitated by the fact that *Prolog* has been chosen as the programming language paradigm within the Japanese *Fifth Generation Project*. This project aims at the development of the so-called "computers of the next generation", which will be specialized in the execution of Artificial Intelligence applications, hence capable of performing an extremely high number of *deductions per time unit.* The project also includes the use of the relational data model for storing large collections of data.

The reaction to the Japanese *Fifth Generation Project* was an incentive to research in the interface area between logic programming and relational databases. This choice indicated that this area is not just the ground for theoretical investigations, but also has great potential for future applications.

By looking closely at logic programming and at database management, we discover several features in common:

a) *DATABASES*. Logic programming systems manage small, single-user, main-memory databases, which consist of deduction rules and factual information. Database systems deal instead with large, shared, mass-memory data collections, and provide the technology to support efficient retrieval and reliable update of persistent data.

b) *QUERIES*. A query denotes the process through which relevant information is extracted from the database. In logic programming, a query (or *goal*) is answered by building chains of deductions, which combine rules and factual information, in order to *prove* or *refute* the validity of an initial statement. In database systems, a query (expressed through a special-purpose data manipulation language) is processed by determining the most efficient access path in mass memory to large data collections, in order to extract relevant information.

c) *CONSTRAINTS*. Constraints specify correctness conditions for databases. Constraint validation is the process through which the correctness of the database is preserved, by preventing incorrect data being stored in the database. In logic programming, constraints are expressed through general-purpose rules, which are activated whenever the database is modified. In database systems, only a few constraints are typically expressed using the data definition language.

Logic programming offers a greater power for expressing queries and constraints as compared to that offered by data definition and manipulation languages of database systems. Furthermore, query and constraint representation is possible in a homogeneous formalism and their evaluation requires the same inferencing mechanisms, hence enabling more sophisticated reasoning about the database content. On the other hand, logic programming systems do not provide the technology for managing large, shared, persistent, and reliable data collections.

The natural extension of logic programming and of database management consists in building new classes of systems, placed at the intersection between the two fields, based on the use of *logic programming as a query language*. These systems combine a logic programming style for formulating queries and constraints with database technology for efficiency and reliability of mass-memory data storage.

1.1 Logic Programming as Query Language

We give an informal presentation of how logic programming can be used as a query language. We consider a relational database with two relations:

$$PARENT(PARENT, CHILD), \quad \text{and} \quad PERSON(NAME, AGE, SEX).$$

The tuples of the *PARENT* relation contain pairs of individuals in parent-child relationships; the tuples of the *PERSON* relation contain triples whose first,

PARENT	
PARENT	CHILD
john	jeff
jeff	margaret
margaret	annie
john	anthony
anthony	bill
anthony	janet
mary	jeff
claire	bill
janet	paul

PERSON		
NAME	AGE	SEX
paul	7	male
john	78	male
jeff	55	male
margaret	32	female
annie	4	female
anthony	58	male
bill	24	male
janet	27	female
mary	75	female
claire	45	female

Fig. 1.1. Example of relational database

second, and third elements are the person's name, age, and sex, respectively. We assume that each individual in our database has a different name. The content of the database is shown in Fig. 1.1.

We express simple queries to the database using a logic programming language. We use *Prolog* for the time being; we assume the reader has some familiarity with *Prolog*. We use two special *database predicates*, *parent* and *person* with the understanding that the ground clauses (facts) for these predicates are stored in the database. We use standard *Prolog* conventions on upper and lower case letters to denote variables and constants. For instance, the tuple <*john, jeff*> of the database relation PARENT corresponds to the ground clause:

$$parent(john, jeff).$$

The query: *Who are the children of John?* is expressed by the following *Prolog* goal:

$$? - parent(john, X).$$

The answer expected from applying this query to the database is:

$$X = \{jeff, anthony\}.$$

Let us consider now which answer would be given by a *Prolog* interpreter, operating on facts for the two predicates *parent* and *person* corresponding to the database tuples; we assume facts to be asserted in main memory in the order shown above.

The answer is as follows: After executing the goal, the variable X is first set equal to *jeff*; if the user asks for more answers, then the variable X is set equal to *anthony*; if the user asks again for more answers, then the search fails, and the interpreter prompts *no*. Note that *Prolog* returns the result one tuple at a time, instead of returning the set of all result tuples.

The query: *Who are the parents of Jeff?* is expressed as follows:

$$? - parent(X, jeff).$$

The set of all answers is:

$$X = \{john, mary\}.$$

Once again, let us consider the *Prolog* answer: After executing this goal, the variable X is set equal to *john*; if the user asks for more answers, then the variable X is set equal to *mary*; if the user asks again for more answers, then the search fails.

We can also express queries where all arguments of the query predicate are constants. For instance:

$$? - parent(john, jeff).$$

In this case, we expect a positive answer if the tuple <*john, jeff*> belongs to the database, and a negative answer otherwise. In the above case, a *Prolog* system would produce the answer *yes*.

Rules can be used to build an *Intensional Database (IDB)* from the *Extensional Database (EDB)*. The EDB is simply a relational database; in our example it includes the relations PARENT and PERSON. The IDB is built from the EDB by applying rules which define its content, rather than by explicitly storing its tuples. In the following, we build an IDB which includes the relations FATHER, MOTHER, GRANDPARENT, SIBLING, UNCLE, AUNT, ANCESTOR, and COUSIN. Intuitively, all these relationships among persons can be built from the two EDB relations PARENT and PERSON.

We start by defining the relations FATHER and MOTHER, by indicating simply that a father is a male parent and a mother is a female parent:

$father(X, Y) :- person(X, \_, male), parent(X, Y).$
$mother(X, Y) :- person(X, \_, female), parent(X, Y).$

As a result of this definition, we can deduce from our sample EDB the IDB shown in Fig. 1.2.

Note that here we are presenting the tuples of the IDB relations as if they actually existed; in fact, tuples of the IDB are not stored. One can regard the two rules *father* and *mother* above as *view definitions*, i.e., programs stored in the database which enable us to build the tuples of *father* starting from the tuples of *parent* and *person*.

The IDB can be queried as well; we can, for instance, formulate the query: *Who is the mother of Jeff?*, as follows:

$$? - mother(X, jeff).$$

1.1 Logic Programming as Query Language

FATHER	
FATHER	CHILD
john	jeff
jeff	margaret
john	anthony
anthony	bill
anthony	janet

MOTHER	
MOTHER	CHILD
margaret	annie
mary	jeff
claire	bill
janet	paul

Fig. 1.2. The IDB relations FATHER and MOTHER

With a *Prolog* interpreter, after the execution of this query X is set equal to *mary*. Notice that the interpreter does not evaluate the entire IDB relation MOTHER in order to answer the query, but rather it finds just the tuple which contributes to the answer.

We can proceed with the definition of the IDB relations GRANDPARENT, SIBLING, UNCLE, and AUNT, with obvious meanings:

$grandparent(X, Z) :- parent(X, Y), parent(Y, Z).$
$sibling(X, Y) :- parent(Z, X), parent(Z, Y), not(X = Y).$
$uncle(X, Y) :- person(X, \_, male), sibling(X, Z), parent(Z, Y).$
$aunt(X, Y) :- person(X, \_, female), sibling(X, Z), parent(Z, Y).$

Complex queries to the EDB and IDB can be formulated by building new rules which combine EDB and IDB predicates, and then presenting goals for those rules; for instance, *Who is the uncle of a male nephew?* can be formulated as follows:

$query(X) :- uncle(X, Y), person(Y, \_, male).$
$? - query(X).$

More complex IDB relations are built from *recursive rules*, i.e., rules whose head predicate occurs in the rule body (we will define recursive rules more precisely below). Well-known examples of recursive rules are the ANCESTOR relation and the COUSIN relation.

The ANCESTOR relation includes as tuples all ancestor-descendent pairs, starting from parents.

$ancestor(X, Y) :- parent(X, Y).$
$ancestor(X, Y) :- parent(X, Z), ancestor(Z, Y).$

The COUSIN relation includes as tuples either two children of two siblings, or, recursively, two children of two previously determined cousins.

$cousin(X, Y) :- parent(X1, X), parent(Y1, Y), sibling(X1, Y1).$
$cousin(X, Y) :- parent(X1, X), parent(Y1, Y), cousin(X1, Y1).$

The IDB resulting from the two definitions above is shown in Fig. 1.3.

ANCESTOR	
ANCESTOR	DESCENDENT
john	jeff
jeff	margaret
margaret	annie
john	anthony
anthony	bill
anthony	janet
mary	jeff
claire	bill
janet	paul
john	margaret
mary	margaret
jeff	annie
john	bill
john	janet
anthony	paul
john	annie
mary	annie
john	paul

COUSIN	
PERSON1	PERSON2
margaret	bill
margaret	janet
annie	paul

Fig. 1.3. The IDB relations ANCESTOR and COUSIN

This example shows that recursive rules can generate rather large IDB relations. Furthermore, the process of generating IDB tuples is quite complex; for instance, a *Prolog* interpreter operating on the query *ancestor(X, Y)* would generate some of the IDB tuples more than once. Therefore, the efficient computation of recursive rules is quite critical. On the other hand, recursive rules are very important because they enable us to derive useful IDB relations that cannot be expressed otherwise.

Rules can also express integrity constraints. Let us consider the EDB relation PARENT; we would like to express the following constraint:

a) *SelfParent constraint:* a person cannot be his (her) own parent.

The formulation of this constraint in *Prolog* is as follows:

a) $incorrectdb : -parent(X, X)$.

This constraint formulation enables us to inquire about the correctness of the database. For instance, consider the *Prolog* goal:

$? - incorrectdb.$

If no individual X exists satisfying the body of the rule then the answer to this query is *no*. In this case, the database is correct. If, instead, such an individual does exist, then the answer is *yes*.

Let us consider a few more examples of constraints. For instance:

b) $OneMother:$ Each person has just one mother.
c) $PersonParent:$ Each parent is a person.
d) $PersonChild:$ Each child is a person.
e) $AcyclicAncestor:$ A person cannot be his(her) own ancestor.

These constraints are formulated as follows:

b) $incorrectdb : - mother(X, Z), mother(Y, Z), not(X = Y)$.
c) $incorrectdb : - parent(X, \_), not(person(X, \_, \_))$.
d) $incorrectdb : - parent(\_, Y), not(person(Y, \_, \_))$.
e) $incorrectdb : - ancestor(X, X)$.

Note that constraints b) and e) also use in their formulation some IDB relations, while contraints a), c), and d) refer just to EDB relations. The two cases, however, are not structurally different. Moreover, note that constraint b) is a classic functional dependency, while constraints c) and d) express inclusion dependencies (also called referential integrity).

Let us consider a collection of constraints of this nature. Constraint evaluation can be used either to preserve the integrity of an initially correct database, or to determine (and then eliminate) all sources of inconsistency.

Let us consider the former application of constraints, namely, how to preserve the integrity of a correct database. We recall that the content of a database is changed by the effects of the execution of *transactions*. A transaction is an atomic unit of execution, containing several operations which insert new tuples, delete existing tuples, or change the content of some tuples. Atomicity of transactions means that their execution can terminate either with an *abort* or with a *commit*. An abort leaves the initial database state unchanged; a commit leaves the database in a final state in which all operations of the transaction are successfully performed. Thus, to preserve consistency, we should accept the commit of a transaction iff it produces a final database state that does not violate any constraint. Efficient methods have been designed for testing the correctness of the final state of a transaction. These methods assume the database to be initially correct, and test integrity constraints on the part of the database that has been modified by the transaction.

Let us consider, then, the application of constraints to restore a valid database state. The above constraint formulation enables a *yes/no* answer, which is not very helpful for such purposes. However, we might, for instance, restate the constraints as follows:

a) $incorrectdb(selfparent, [X]) :- parent(X, X).$
b) $incorrectdb(onemother, [X, Y, Z]) :-$
$\qquad mother(X, Z), mother(Y, Z), not(X = Y).$
c) $incorrectdb(personparent, [X]) :- parent(X, \_), not(person(X, \_, \_)).$
d) $incorrectdb(personchild, [Y]) :- parent(\_, Y), not(person(Y, \_, \_)).$
e) $incorrectdb(acyclicancestor, [X]) :- ancestor(X, X).$

In this formulation, the head of the rule has two arguments; the first argument contains the constraint name, and the second argument contains the list of variables which violate the constraint. This constraint formulation enables us to inquire about the causes of incorrectness of the database. For instance, consider the *Prolog* goal:

$$? - incorrectdb(X, Y).$$

If there exists no constraint X which is invalidated, then the answer to this query is *no*. In this case, the database is correct. If instead one such constraint exists, then variables X and Y are set equal to the constraint name and the list of values of variables which cause constraint invalidity. For instance, the answer:

$$X = personparent, Y = [Karen]$$

reveals that Karen belongs to the relation PARENT but not to the relation PERSON; this should be fixed by adding a tuple for Karen to the relation PERSON.

However, the answer to the above query might not be sufficient to understand the action required in order to restore the correctness of the database. This happens with rules b) and e), which express a constraint on IDB relations. We have already observed that IDB relations are generally not stored explicitly; they are defined by rules, and their value depends on the EDB relations which appear in these rules. Thus, violations to constraints b) and e) should be compensated by actions applied to the underlying EDB relations.

We conclude this example by showing, in Table 1.1, the correspondence between similar concepts in logic programming and in databases that we have seen so far:

DATABASE CONCEPTS	LOGIC PROGRAMMING CONCEPTS
Relation	Predicate
Attribute	Predicate argument
Tuple	Ground clause (fact)
View	Rule
Query	Goal
Constraint	Goal (returning an expected truth value)

Table 1.1. Correspondence between similar concepts in logic programming and in databases

1.2 Prolog and Datalog

In the previous section, we have shown how a *Prolog* interpreter operates on a database of facts, and we have demonstrated that *Prolog* can act as a powerful database language. The choice of *Prolog* to illustrate the usage of logic programming as a database language is almost mandatory, since *Prolog* is the most popular logic programming language. On the other hand, the use of *Prolog* in this context also has some drawbacks, which have been partially revealed by our example:

1) *Tuple-at-a-time processing.* While we expect that the result of queries over a database be a *set* of tuples, *Prolog* returns individual tuples, one at a time.
2) *Order sensitivity and procedurality.* Processing in *Prolog* is affected by the order of rules or facts in the database and by the order of predicates within the body of the rule. In fact, the *Prolog* programmer uses order sensitivity to build efficient programs, thereby trading the so-called *declarative nature* of logic programming for procedurality. Instead, database languages (such as SQL or relational algebra) are *nonprocedural*: the execution of database queries is insensitive to order of retrieval predicates or of database tuples.
3) *Special predicates.* *Prolog* programmers control the execution of programs through special predicates (used, for instance, for input/output, for debugging, and for affecting backtracking). This is another important loss of the declarative nature of the language, which has no counterpart in database languages.
4) *Function symbols.* *Prolog* has function symbols, which are typically used for building recursive functions and complex data structures; neither of these applications are useful for operating over a *flat* relational database, although they might be useful for operating over *complex database objects.* We will not address this issue in this book.

These reasons motivate the search for an alternative to *Prolog* as a database and logic programming language; such an alternative is the new language *Datalog*.

Datalog is a logic programming language designed for use as a database language. It is nonprocedural, set-oriented, with no order sensitivity, no special predicates, and no function symbols. Thus, *Datalog* achieves the objective of eliminating all four drawbacks of *Prolog* defined above.

Syntactically, *Datalog* is very similar to pure *Prolog*. All *Prolog* rules listed in the previous section for expressing queries and constraints are also valid *Datalog* rules. Their execution produces the *set* of all tuples in the result; for instance, after executing the goal:

$$? - parent(john, X).$$

We obtain:

$$X = \{\ jeff,\ anthony\ \}.$$

As an example of the difference between *Prolog* and *Datalog* in order sensitivity, consider the following two programs:

Program Ancestor1:

$ancestor(X,Y) :- parent(X,Y).$
$ancestor(X,Y) :- parent(X,Z), ancestor(Z,Y).$

Program Ancestor2:

$ancestor(X,Y) :- ancestor(Z,Y), parent(X,Z).$
$ancestor(X,Y) :- parent(X,Y).$

Input goal:

$$? - ancestor(X,Y).$$

Both Ancestor1 and Ancestor2 are syntactically correct programs either in *Prolog* or in *Datalog*. *Datalog* is neither sensitive to the order of rules, nor to the order of predicates within rules; hence it produces the correct expected answer (namely, the set of all ancestor-descendent pairs) in either version. A *Prolog* interpreter, instead, produces the expected behavior in version Ancestor1 (namely, the first ancestor-descendent pair); but it loops forever in version Ancestor2. In fact, the *Prolog* programmer must avoid writing looping programs, while the *Datalog* programmer need not worry about this possibility.

The process that has led to the definition of *Datalog* is described in Fig. 1.4. The picture was shown by Jeff Ullman at Sigmod 1984; it indicates that the evolution from *Prolog* to *Datalog* consists in going from a procedural, record-oriented language to a nonprocedural, set-oriented language; that process was also characteristic of the evolution of database languages, from hierarchic and network databases to relational databases. Even though *Datalog* is a declarative language and its definition is independent of any particular search strategy, *Datalog* goals are usually computed with the breadth-first search strategy, which produces the *set of all answers*, rather than with the depth-first search strategy of *Prolog*, which produces answers with a tuple-at-a-time approach. This is consistent with the set-oriented approach of relational query languages. Furthermore, in *Datalog* the programmer does not need to specify the *procedure* for accessing data, which is left as system responsibility; this is again consistent with relational query languages, which are nonprocedural.

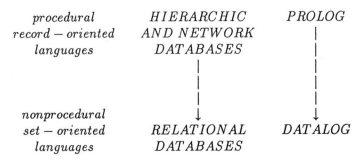

Fig. 1.4. Datalog as an evolution of Prolog

On the other hand, these features limit the power of *Datalog* as a general-purpose programming language. In fact, *Datalog* is obtained by subtracting some features from *Prolog*, but not, for the time being, by adding to it features which belong to classical database languages. Therefore, *Datalog* is mostly considered a good abstraction for illustrating the use of logic programming as a database language, rather than a full-purpose language. We expect, however, that *Datalog* will evolve to incorporate a few other features and will turn into a full-purpose database language in the near future.

In Table 1.2, we summarize the features that characterize *Datalog* in contrast to *Prolog*.

PROLOG	DATALOG
Depth-first search	(usually) Breadth-first search
Tuple-at-a-time	Set-oriented
Order sensitive	No order sensitivity
Special predicates	No special predicates
Function symbols	No function symbols

Table 1.2. Comparison of Prolog and Datalog

1.3 Alternative Architectures

Turning *Prolog* and *Datalog* into database languages requires the development of new systems, which integrate the functionalities of logic programming and database systems. Several alternative architectures have been proposed for this purpose; in this section, we present a first classification of the various approaches.

The first, broader distinction concerns the relationship between logic programming and relational systems.

a) We describe the development of an interface between two separate subsystems, a logic programming system and a database system, as *coupling*. With this approach, we start from two currently available systems, and we couple them so as to provide a single-system image. Both subsystems preserve their individuality; an interface between them provides the procedures required for bringing data from the persistent database system into the main-memory logic programming execution environment in order to evaluate queries or to validate constraints.

b) We describe the development of a single system which provides logic programming on top of a mass-memory database system as *integration*. This approach corresponds to the development of an entirely new class of data structures and algorithms, specifically designed to use logic programming as a database language.

Given the above alternatives, it is reasonable to expect that *Prolog*-based systems will mostly use the coupling approach, and *Datalog*-based systems will mostly use the integration approach. This is due to the present availability of many efficient *Prolog* systems that can be coupled with existing database systems with various degrees of sophistication. In fact, several research prototypes and even a few commercial products that belong to this class are already available. On the other hand, *Datalog* is an evolution of *Prolog* specifically designed to act as a database language; hence it seems convenient to use this new language in the development of radically new integrated systems. This mapping of *Prolog* to coupling and of *Datalog* to integration should not be considered mandatory. Indeed, we should recall that the *Fifth Generation Project* will produce an integrated system based on a parallel version of *Prolog*.

The coupling approach is easier to achieve but also potentially much less efficient than the integration approach. In fact, we cannot expect the same efficiency from the interface required by the coupling approach as from a specifically designed system. Furthermore, the degree of complexity of the interfaces can be very different. At one extreme, the simplest interface between a *Prolog* system and a relational system consists in generating a distinct SQL-like query in correspondence to every attempt at unification of each database predicate. This approach is very simple, but also potentially highly inefficient.

Hence, we expect that coupling will be sufficient for dealing with some applications, while other applications will require integration; further, coupling may be made increasingly efficient by superimposing ad-hoc techniques to the standard interfaces, thus achieving the ability of dealing with several special applications.

Within coupling, we further distinguish two alternative approaches:

a) *Loose coupling*. With this approach, the interaction between the logic programming and database systems takes place independently of the actual inference process. Typically, coupling is performed at compile time (or at program load time, with interpreters), by extracting from the database all the facts that might be required by the program; sometimes, coupling is performed on a rule-by-rule basis, prior to the activation of that rule. Loose coupling is also called *static coupling* because coupling actions are performed independently of the actual pattern of execution of each rule.

b) *Tight coupling.* With this approach, the interaction between the logic programming and database systems is driven by the inference process, by extracting the specific facts required to answer the current goal or subgoal. In this way, coupling is performed whenever the logic programming system needs more data from the database system in order to proceed with its inference. Tight coupling is also called *dynamic coupling* because coupling actions are performed in the frame of the execution of each rule.

It follows from this presentation that loose and tight coupling are very different in complexity, selectivity, memory required, and performance. With loose coupling, we execute fewer queries of the database, because each predicate or rule is separately considered once and for all; while with tight coupling each rule or predicate can be considered several times. However, queries in loosely coupled systems are less selective than queries in tightly coupled systems, because variables are not instantiated (bound to constants) when queries are executed. If coupling is performed at compile or load time, queries are presented a priori, disregarding the actual pattern of execution of the logic program. In fact, it is even possible to load data at compile or load time concerning a rule or predicate that will not be used during the work session.

From these considerations, we deduce that the amount of main memory required for storing data which is to be retrieved by a loosely coupled system is higher than that required by a tightly coupled system. On the other hand, this consideration does not allow us to conclude that the performance of tightly coupled systems is always better; in general, tight coupling requires more frequent interactions with the database, and this means major overhead for the interface, with frequent context switching between the two systems. Thus, a comparison between the two approaches is difficult, and includes a trade-off analysis between memory used and execution time.

In Table 1.3, we summarize the comparison between loose and tight coupling.

LOOSE COUPLING	TIGHT COUPLING
fewer queries	more queries
less instantiated queries	more instantiated queries
all queries applied	relevant queries applied
more memory required	less memory required

Table 1.3. Comparison between loose and tight coupling

1.4 Applications

There are a number of new applications needing integrated systems at the confluence between databases and logic programming; the following is a list of the features of these applications.

a) *Database need.* The application must need to access data stored in a database. This means access to persistent, shared data that are resilent to failures and that can be accessed and updated concurrently by other applications. If we consider most expert systems or knowledge bases presently available, we observe that these systems have access to persistent data files, but that these are locally owned and controlled, with no sharing, concurrency, or reliability requirement with other applications.

b) *Selective access.* We cannot postulate that the entire database will be examined by the application during a work session, or else we should deal with data retrieval loads which exceed those of traditional database applications. Instead, we can postulate that during the work session the application will retrieve only a limited portion of the database, due to its selective access to data.

c) *Limited working set.* As a result of the previous assumption, the *working set* of data, namely the data required in main memory at a given time, is limited. This requirement is particularly important for loosely coupled systems, as loose coupling does not profit from the access selectivity, which is not expressed at compile time.

d) *Demanding database activity.* It is quite important to understand that systems which perform *millions of deductions per second*, as stated in the requirements of the *Fifth Generation Project*, are likely to put quite a heavy demand on the database. For instance, the computation of recursive rules requires a high number of interactions with the database. This feature contrasts with the typical database transactions, which serve *thousands of transactions per second*, each one responsible for a small amount of input/output operations.

Dealing with the above features extends the current spectrum of applicability of expert systems and other artificial intelligence applications; it also solves classical database problems, such as the *bill-of-materials* or the *anti-trust control* problems. These problems will be described in Chap. 3.

1.5 Bibliographic Notes

The relationship between logic programming and databases has been investigated since November 1977, when the conference on *Logic and Databases* took place in Toulouse; this event was followed by other two conferences on *Advances in Database Theory*, held in 1979 and 1982, again centered on this subject. The proceedings of the conferences, edited by Gallaire and Minker [Gall 78] and by Gallaire, Minker, and Nicolas [Gall 81, Gall 84a], contain fundamental papers for the systematization of this field. Perhaps the best synthesis of the results in Logic and Databases achieved before 1984 is contained in a paper, again by Gallaire, Minker, and Nicolas, which appeared in *ACM Computing Surveys* [Gall 84b].

Two events characterize the growth of interest in this field: the selection, by the Japanese *Fifth Generation Project*, of an architecture based on *Prolog* as main programming language and of the relational model for data representation [Itoh 86]; and the growth of interest in the database theory community in logic queries and recursive query processing, marked by the seminal paper of Ullman [Ullm 85a]. Ullman has also presented, at the ACM-SIGMOD conference in 1984, the picture shown in Sect. 1.2 indicating the relationship between *Prolog* and *Datalog*. All recent database conferences (ACM-SIGMOD, ACM-PODS, VLDB, Data Engineering, ICDT, EDBT) have presented one or more sessions on logic programming and databases.

Parallel interest in *Expert Database Systems*, characterized by a more pragmatic, application-oriented approach, has in turn been presented through the new series of Expert Database Systems (EDS) conferences, organized by L. Kerschberg [Kers 84], [Kers 86], and [Kers 88]. The reading of the conference proceedings makes it possible to follow the growth, systematization, and spread of this area.

Specialized workshops on knowledge base management systems and on deductive databases were held in Islamorada [Brod 86], Washington [Mink 88], and Venice [Epsi 86]. Good overview papers describing methods for recursive query processing have been presented by Bancilhon and Ramakrishnan [Banc 86b], by Gardarin and Simon [Gard 87], and by Roelants [Roel 87]. A comparison of ongoing research projects for integrating databases and logic is provided by a special issue of IEEE-Database Engineering, edited by Zaniolo [Zani 87].

Chapter 2
A Review of Relational Databases and Prolog

This chapter provides an overview of the prerequisite notions that are required in order to understand this book. The reader who has a background in these topics can skip this chapter or part of it, perhaps after taking a look at the notation.

The first section gives a description of the fundamental concepts of relational databases and relational algebra. The second section provides a brief introduction to the language *Prolog*. We assume the reader to be familiar with the basic concepts of first-order logic. Other aspects of logic programming will be treated in Chaps. 6 and 7, where we present a complete introduction to the language *Datalog*.

2.1 Overview of Relational Databases

The *Relational Model* was introduced by Codd in 1970 and represents, together with the *Network* and the *Hierarchical* models, one of the three data models currently supported by most *Database Management Systems (DBMS)*. Many of the concepts in this book apply to any database system, but the relational model has been selected for its simplicity and formality.

2.1.1 The Relational Model

In the relational model, data are organized using *relations*, which are defined as follows: Let $D_1, \ldots, D_n$ be sets, called *domains* (not necessarily distinct). Let D be the Cartesian product of $D_1, \ldots, D_n$. A *relation* defined on $D_1, \ldots, D_n$ is any subset R of D; n is the *arity* of R.

We suppose domains to be finite sets of data. Unless stated otherwise, we will therefore assume finite relations. Elements of relations are called *tuples*; let $< d_1, \ldots, d_n >$, with $d_1 \in D_1, \ldots, d_n \in D_n$, be a tuple of R. The number of tuples in R is called the *cardinality* of the relation. Note that a relation is a set, therefore tuples in a relation are distinct, and the order of tuples is irrelevant. The empty set is a particular relation, the *null (or empty) relation*. Relations can also be viewed as *tables*, each tuple being a row of a table. The column names of a relation are also called *attributes*.

The ordered set of all the attributes of a relation R is the *schema* of R. We refer to attributes of relations either by their *name* or by the *position* (column number) that the attribute occupies in the relation schema.

A *key* for a relation R is a subset K of the attributes of R whose values are unique within the relation; it can thus be used to *uniquely identify* the tuples of the relation. A key should also be *minimal*: it should not include attributes which are not strictly required for a unique identification of the tuples of a relation. Given that relations are sets of tuples, a key must exist; at least, the set of all attributes of the relation constitutes the key. Each relation can have multiple keys. The property of being a key is *semantic*, and should hold for any value taken by tuples of a relation.

Example 2.1. Consider a university with the sets of its students, professors, courses, years in which courses are offered, departments in which courses take place, and grades. We call these domains, respectively, *STUDENT, PROFESSOR, COURSE, YEAR, DEPARTMENT*, and *GRADE*; two relations *OFFERING* and *EXAM* on these domains are represented in Fig. 2.1.

OFFERING			
PROFESSOR	COURSE	YEAR	DEPARTMENT
Date	Databases	1987	Computer Science
Saunders	Geometry	1985	Mathematics
Floyd	Automata theory	1987	Computer Science
Brill	Chemistry	1986	Chemistry
Brill	Chemistry	1987	Chemistry

EXAM			
STUDENT	COURSE	YEAR	GRADE
Jones	Automata Theory	1987	A
Smith	Databases	1987	B
Delsey	Chemistry	1986	A
Carey	Chemistry	1986	B
Jones	Geometry	1987	C

Fig. 2.1. Relations *OFFERING* and *EXAM*

The schemas of the relations *OFFERING* and *EXAM* are:

OFFERING (PROFESSOR, COURSE, YEAR, DEPARTMENT).

EXAM (STUDENT, COURSE, YEAR, GRADE).

In the above example, we assume (as a *semantic property* of the underlying university system) that each course is taught each year by a single professor; thus, the key of $OFFERING$ is the pair of attributes $COURSE, YEAR$. Each professor can teach the same course over many years, and multiple courses in the same year; hence, pairs

$$PROFESSOR, COURSE \quad \text{and} \quad PROFESSOR, YEAR$$

are not keys. The pair $STUDENT, COURSE$ is the key of the relation $EXAM$, since we assume that each student takes an exam in a specific course at most once. □

2.1.2 Relational Languages

Let us now review some of the languages for manipulating relations and for retrieving data from the database. A *query*, expressed in a *query language*, specifies a portion of data contained in the database. There are two main formalisms for expressing queries on relations:

Relational Algebra is defined through several operators that apply to relations, and produce other relations; this is a "set-oriented" formalism, because each operator is applied to *whole* relations. The order in which operations of relational algebra are specified within queries also suggests the order in which operations have to be applied to the database. Indeed, many query optimization procedures are based on *equivalence transformations*, which permit relational algebra expressions to be transformed into other expressions that can be processed by the Database Management System (DBMS) in a more efficient way.

Relational Calculus expresses a query by means of a first-order logic formula on the tuples of the database; the result of the query is the set of tuples that satisfy the formula. This is a "tuple-oriented" formalism, since each formula defines the properties that characterize a tuple in the result. Relational calculus is a *nonprocedural* query language, because the query expression in calculus does not suggest a method for computing the query answer.

In the following, we present the operations of *relational algebra* that will be used throughout this book to express queries over a relational database. Each operation of relational algebra takes one or two relations as operand and produces one relation as result. Operations of relational algebra can be composed into arbitrarily complex *algebraic expressions*; these are used to formulate queries.

The five *primitive* (or *basic*) operations of relational algebra are: selection, projection, Cartesian product, union, and difference. This set of operators is *relationally complete*, in the sense that it has the same expressive power as first-order calculus. If we exclude from the above operators the *difference* operator, we obtain a sublanguage called *Positive Relational Algebra (RA^+)*.

- *Selection:* Let F be a valid formula which expresses a selection predicate. Valid formulas in a selection are logical combinations of simple predicates, where each simple predicate is a comparison either between two attributes or between one attribute and a constant value. Then, the result of a selection

$\sigma_F R$ on the relation R is the set of tuples of R for which the formula F is true.
- *Projection:* Let R have arity n. Suppose $i_1, \ldots, i_k$, $k \leq n$, are names (or numbers) of attributes of R. Let D_i be the domain of attribute i. Then, the result of a projection $\Pi_{i_1,\ldots,i_k} R$ is a relation of arity k having $i_1, \ldots, i_k$ as attributes. The tuples of the result are derived from the tuples of the operand relation by suppressing the values of attributes which do not belong to the above list. Note that the cardinality of the result can change due to projection, because duplicate tuples are removed.
- *Cartesian product:* Let R have arity r, and S have arity s; then $R \times S$ is a relation of arity $r+s$, whose schema is formed by the concatenation of schemas of R and S, and whose tuples are formed by all the possible concatenations of tuples of R and tuples of S.
- *Union:* This applies to two relations which have identical schemas, and produces a relation with the same schema. Given relations R and S, $R \cup S$ is the set of tuples belonging to the union (in the set-theoretic sense) of R and S.
- *Difference:* This also applies to relations with identical schemas, and produces a relation of that schema. Given relations R and S, $R - S$ is the set of tuples belonging to the difference (in the set-theoretic sense) of R and S.

The following operations derive from the ones that we have just defined.

- *Join:* Let F be a logical formula which is restricted to being a logical combination of comparisons of attributes of the two operands. The join is defined from Cartesian product and selection:

$$R \bowtie_F S = \sigma_F(R \times S)$$

If the formula F only involves equality predicates, then the join is called an *equijoin*. The *natural join* of R and S ($R \bowtie S$) is properly defined only when attributes are denoted by their name, and corresponds to an equijoin in which all pairs of attributes with the same name are set equal. In the natural join, for each pair of equal columns, one is finally projected out of the result.
- *Semijoin:* Let F be a join predicate. The semijoin of R and S ($R \ltimes_F S$), is defined as:

$$R \ltimes_F S = \Pi_{Schema(R)}(R \bowtie_F S).$$

where $Schema(R)$ denotes the relation schema of R. From this definition, it follows that the result relation produced by a semijoin is contained within the first operand relation. The *natural semijoin* of R and S is:

$$R \ltimes S = \Pi_{Schema(R)}(R \bowtie S).$$

Figure 2.2 shows examples of the use of each of the above operations.

We now turn to queries corresponding to algebraic expressions involving more than one operation, using the two relations $OFFERING$ and $EXAM$ defined in Example 2.1.

Initial relations:

R		
A1	A2	A3
a	b	c
f	d	h
f	e	h

S	
B1	B2
d	e
g	h
f	m

T	
B1	B2
b	e
d	e
g	m

Basic operations:

$\sigma_{A1=f} R$		
A1	A2	A3
f	d	h
f	e	h

$\Pi_{A1,A3} R$	
A1	A3
a	c
f	h

$S \cup T$	
B1	B2
d	e
g	h
f	m
b	e
g	m

$S - T$	
B1	B2
g	h
f	m

$R \times S$				
A1	A2	A3	B1	B2
a	b	c	d	e
a	b	c	g	h
a	b	c	f	m
f	d	h	d	e
f	d	h	g	h
f	d	h	f	m
f	e	h	d	e
f	e	h	g	h
f	e	h	f	m

Derived operations:

$R \bowtie_{A2=B2} S$				
A1	A2	A3	B1	B2
f	e	h	d	e

$R \ltimes_{A3=B2} S$		
A1	A2	A3
f	d	h
f	e	h

Fig. 2.2. Examples of algebraic operations

Example 2.2. Consider the following queries to the database:

a) We want to know all students who received an A in 1987.

$$\Pi_{STUDENT}(\sigma_{GRADE=A \wedge YEAR=1987} EXAM)$$

The answer to this query is a relation having a single tuple:

b) We want to know the name and grade of all students who have taken an exam in one of the courses administered by the Computer Science department.

$$\Pi_{STUDENT,GRADE}(EXAM \bowtie (\sigma_{DEPT.=C.S.}OFFERING))$$

The answer to this query is the following:

STUDENT	GRADE
Jones	A
Smith	B

Notice that the natural join between $EXAM$ and $OFFERING$ corresponds to the formula $COURSE = COURSE \wedge YEAR = YEAR$, involving the two attributes with the same name in the two operand relations.

c) We want to know the students who took a class with Professor Brill. The query is expressed in relational algebra as follows:

$$\Pi_{STUDENT}\sigma_{PROFESSOR=Brill}(EXAM \bowtie OFFERING)$$

Notice that the above expression is equivalent to:

$$\Pi_{STUDENT}(EXAM \bowtie \sigma_{PROFESSOR=Brill}OFFERING)$$

However, the second expression leads to a *more efficient* computation than the first, because the selection condition is directly applied to the relation $OFFERING$, thus reducing its size *before* evaluating the join. This is an example of the application of equivalence transformations of algebraic expressions, to improve their computation. The result of this query is:

STUDENT
Delsey
Carey

□

Relational algebra, however, is not a "user-friendly" language, as understanding the meaning of an expression in relational algebra (or calculus) can be difficult for the average user. Thus, relational algebra is normally used as a language for *internal* representation of queries, while user-oriented query languages have been defined for interacting with the database system. By and large, the most popular relational query language is SQL, which has become a de-facto standard within relational DBMS. We briefly present the most typical statement of the language, the so-called *query block*. An SQL block has the form:

SELECT < attribute_list >
FROM < relations >
WHERE < predicate >

The SQL block has a simple interpretation in relational algebra: it is equivalent to performing a selection operation using the predicate of the $WHERE$ clause, on the Cartesian product of the relations specified by the $FROM$ clause, and then projecting the result on the attributes of the $SELECT$ clause. Obviously, the Cartesian product reduces to a join if the predicate of the $WHERE$ clause includes the join condition; no Cartesian product is required if the $FROM$ clause contains just one relation.

The expressive power of the SQL language comes from the fact that blocks can be *nested* by including entire SQL blocks within the predicate of the $WHERE$ clause. The following is a simple example of block nesting:

> $SELECT\ <attribute\_list>$
> $FROM\ <relations>$
> $WHERE\ <attribute>\ IN$
> $SELECT\ <attribute\_list>$
> $FROM\ <relations>$
> $WHERE\ <predicate>$

As we have noted already, SQL is the most popular relational query language; other query languages are $QUEL$ and QBE.

Example 2.3. Let us translate into SQL the previous algebraic expressions. The query (a) is expressed in SQL as follows:

> $SELECT\ STUDENT$
> $FROM\ EXAM$
> $WHERE\ GRADE = A$
> $AND\ YEAR = 1987$

The query (b) is expressed using nested blocks as follows:

> $SELECT\ STUDENT, GRADE$
> $FROM\ EXAM$
> $WHERE\ <COURSE,\ YEAR> =$
> $SELECT\ COURSE,\ YEAR$
> $FROM\ OFFERING$
> $WHERE\ DEPARTMENT = Computer\ Science$ □

A *view* is a relation which is not stored explicitly in the database, but is defined through a query language expression. This expression may refer to database relations or to other views. Once a view is defined, it can be used for data retrieval exactly as if it were a normal database relation. The DBMS operates a *query composition*, thereby combining the user query with the view definition to generate a new query expression which operates just on database relations.

Example 2.4. An example of a view on the university database, giving math grades of the students, is:

$$MG = \Pi_{STUD.,C.SE,GRADE}(EXAM \bowtie (\sigma_{DEPT.=Math.}OFFERING))$$

This view is equivalent to a relation, having as attributes:

> $STUDENT,\ COURSE,$ and $GRADE$

presenting all grades of students in courses administred by the Mathematics department. Once defined, it is possible to use this view as any other relation, hence making queries using MG. For instance:

$$\Pi_{STUDENT}(\sigma_{COURSE=Geometry \wedge GRADE=C}MG)$$

asks for the names of all students that took C in Geometry. In our case, the answer to this query is:

STUDENT
Jones

Notice that MG is not explicitly stored, but rather its definition is combined with the query in order to reformulate the query in terms of the database relations $OFFERING$ and $EXAM$; the reformulated query is ultimately executed. □

2.2 Prolog: A Language for Programming in Logic

We give here a very brief description of some of the features of the language *Prolog*; for a complete description, we refer the reader to well-known textbooks. The language structure is based on the notions of *atom, variable, functor, predicate, term, literal,* and *clause*, defined below.

An *atom* is a finite string of alphanumeric characters starting with a lower case letter. Examples of atoms are: book, george, 'George', x12op. A *constant* is either an atom or a number (such as 0, 1213, 4).

A *variable* is a finite string of alphanumeric characters starting with an upper case letter. Examples are: Book, X, Var. The special variable "_", called *anonymous variable*, is thought of as standing for any unnamed variable.

A *functor* is characterized by its name (which is an atom) and its *arity*, i.e. the number of its arguments. A functor can be an operator, like "+", or a predicate name, like "≥" or "ancestor". Each *predicate* defines a mapping from the domain D for which the predicate is defined to the set of values $\{true, false\}$. For instance, we can apply "*ancestor*" to the pair $(bill, joe)$, and its value will be *true* if Bill is an ancestor of Joe, and *false* otherwise.

Atoms and variables are *terms*. Structured terms are built recursively by applying functors to other terms. Thus in *Prolog* predicates are also viewed as terms. A *literal* is either a structured term whose outermost symbol is a predicate, or the negation of such a term. Examples of literals are: $likes(george, mary)$, or $member(X, L)$.

A *Prolog clause* is a collection of literals. A *Prolog program* is a collection of clauses. Variables are *local* to clauses, i.e., the scope of each variable is limited to the clause in which it appears. There are three kinds of clauses:

- *Rules:* A rule is a clause consisting of a *head*, which is one positive literal, and a *body*, formed by several literals separated by commas. The head and the body are separated by the symbol ": −". A rule expresses a conditional assertion. For example:

$$likes(george, Y) : -woman(Y), likes(Y, books).$$

says that George likes any woman, provided that she likes books. Notice that variables in the rule head are to be considered as *universally* quantified, while variables in the body are considered *existentially* quantified. The comma between literals in the rule body represents conjunction. A rule body can also be in *disjunctive* form. For example,

$$likes(mary, Y) : -nice(Y); loves(Y, cats).$$

says that Mary likes a person either if this person is nice, or if he/she loves cats.

- *Facts:* A fact is a clause consisting of a unique literal. It is an assertion which is an axiom, i.e. it is true independently of any condition. For instance, "$likes(george, mary)$." says that the fact that George likes Mary is true in the program.
- *Goals*, or headless clauses: these represent the queries. When a goal is presented to a *Prolog engine*, it tries to prove that the goal is true by using the facts and rules that are present in the program. A *Prolog* goal is always preceded by the symbol "?−". For example ? − $likes(george, lucy)$. will cause the *Prolog engine* to try to prove that George likes Lucy, by using facts and rules contained in the program.

A *Prolog Engine* is an inferential machine which operates on a given *Prolog* program, aiming to prove the truth of a given goal. A goal is proven if it *matches* the head of some clause, and each of the literals (if any) of the body of that clause, considered as a goal, is also proven.

In this way, goals are arranged in a *proof tree*. Computation is performed by the *Prolog* interpreter by traversing this tree in a top-down, depth-first fashion. In this context, *matching* of two literals has to be understood informally as the fact that an instantiation of variables of both literals that makes their values equal exists.

The most important features of *Prolog* are *recursion* and *backtracking*:

- *Prolog* allows the definition of recursive predicates. A predicate is recursive if its definition involves the predicate itself. An example is the following program to detect list membership:

$$member(X, [X|\_]).$$
$$member(X, [\_|Y]) \;\; : -member(X, Y).$$

The first argument of the predicate *member* is a variable, the second argument is a list, denoted by enclosing its elements in square brackets ([]). The first fact says that X is member of a list if it is the head of the list. The second clause says that X is member of a list if it is a member of its tail. Notice the use of the anonymous variable "$\_$", and the list concatenation operator "|".

- If, during the exploration of the proof tree, *Prolog* encounters a goal that is not satisfiable, it retraces its own course by going backwards along the last tree branch, and resumes traversal by trying to re-satisfy the goal to the left of the one that just failed. This feature is called *backtracking*.

The aim of backtracking is to enable the exploration of the proof tree with all possible variable instantiations. However, sometimes this makes the computation rather slow. In order to compensate for this and other limitations of *Prolog*, other constructs have been introduced that introduce procedurality. These are extralogical primitives, which express control information about how the search must be carried out.

The most important of these primitives is the *cut*("!"), used to control backtracking. Its effect is to make the variables occurring to the left of the cut in the current rule irrevocably instantiated: variable instantiations made by the *Prolog engine* before reaching the cut cannot be changed any more.

Other extralogical primitives are, for instance, the predicates *retract* and *assert*, which, respectively, withdraw and add clauses to the program, and the predicate *fail*, which forces failure of the rule.

Example 2.5. We now briefly present a few examples of the *Prolog* programming style. More complex examples, showing the power of *Prolog* both as a programming language and a database language, will be given in Chap. 3.

Consider the following definition of factorial:

$fact(0, 1).$
$fact(N, F) : -M \text{ is } N - 1, \ fact(M, G), \ multiply(G, N, F).$

The first clause states that the factorial of 0 is 1. The second clause states that we can compute the factorial F of N from the factorial G of M, where $M = N - 1$, by making a recursive call to compute G, and then by multiplying G by N to compute F.

Suppose now we have the goal $fact(-1, Answer)$. In this case, the program will run forever, trying recursively to find:

$fact(-2, G).$
$fact(-3, G').$
$fact(-4, G'')....\text{etc.}$

The following is a correct version of this program, assigning value 1 to the factorial of every negative number:

$fact(N, 1) : - \ N <= 0.$
$fact(N, F) : - \ not(N <= 0), \ M \text{ is } N - 1, \ fact(M, G), \ multiply(G, N, F).$

Another, more efficient version of the program uses the *cut*:

$fact(N, 1) : -N <= 0, !.$
$fact(N, F) : -M \text{ is } N - 1, \ fact(M, G), \ multiply(G, N, F).$

The *cut* in the first rule blocks the search for solutions if N is negative. The second rule will only be used if the literal "$N <= 0$" is not provable. This version avoids

computing $N <= 0$ twice, but makes individual clauses less understandable. Note that in both the correct versions of the program we were able to specify an *if – then – else*, either by the use of negation, or, alternatively, by the use of *cut*.
□

2.3 Bibliographic Notes

The relational model of data was defined by Codd [Codd 70]. Several textbooks describe database concepts and systems, including a systematic treatment of the relational model and algebra, by Date [Date 83], [Date 86]; Korth and Silberschatz [Kort 86]; Ullman [Ullm 82]; and Wiederhold [Wied 87]. A systematic approach to database theory is presented by Maier [Maie 83]. Aho, Sagiv, and Ullman [Aho 79b] have studied the equivalence properties of algebraic expressions. The query language SQL is presented by Chamberlin et al. [Cham 76]; the transformation from SQL into relational algebra is presented by Ceri and Gottlob [Ceri 85].

The foundations of mathematical logic can be found in fundamental books by Kleene [Klee 67] and Mendelson [Mend 64]. The foundations of theorem proving and of the resolution principle can be found in the papers by Robinson [Robi 65] and [Robi 68] and in the book by Chang and Lee [Chan 73]. The foundations of logic programming can be found in the papers by Van Emden and Kowalski [VanE 76], and in the book by Lloyd [Lloy 87]. A good introduction to the programming language *Prolog* can be found in the paper by Colmerauer [Colm 85] and in the textbooks by Clocksin and Mellish [Cloc 81] and by Sterling and Shapiro [Ster 86].

Part I
Coupling Prolog to Relational Databases

In this part of the book, we are concerned with the *Prolog* language and with its coupling to relational databases. We first demonstrate that *Prolog* is a powerful formalism for expressing queries and computations; this is achieved by presenting two classical problems, the *anti-trust control* and the *bill-of-materials*. They are both inherently recursive, and hence cannot be dealt with through standard query languages.

Then, we define the architectures and mechanisms which are required for coupling a *Prolog engine* to a database system. The choice of relational databases is dictated by convenience, since many of the available interfaces between *Prolog* and databases currently use relational databases, and since the relational model and language are the appropriate formalisms for expressing the properties of interfaces; on the other hand, the treatment of coupling techniques can be extended to nonrelational systems with little effort.

Finally, we overview some of the existing prototypes of *coupled systems*; this state of the art reflects the present situation, where most coupled systems are prototypes developed by industrial or academic research centers.

Chapter 3
Prolog as a Query Language

In Chap. 1 we provided an overview of the main reasons for the integration of database and logic programming technologies, and we established a comparison between the main features of relational databases and logic programming languages. Here we will just recall some important aspects of *Prolog as a query language*; these features essentially enrich the expressive power of classical query languages based on relational algebra or calculus.

- *Intensional database specification:* rules in *Prolog* provide powerful means of definition and evaluation of *derived relations* (also called *views*). The set of derived relations defined through *Prolog* rules is the so-called *Intensional Database (IDB)*. More examples will be given in the next paragraphs.
- *Recursion:* predicates and rules in *Prolog* can be recursively defined. As a consequence, the expressive power of *Prolog* as a query language is strictly greater than the expressive power of conventional query languages.
- *Incomplete knowledge:* It is possible to mention null values in the rules and facts of the program, by means of the anonymous variable "_".
- *Total order in the fact and rule base:* The *Prolog* programmer has the possibility of ordering facts and rules in the most efficient way with respect to computation efficiency. The *Prolog engine* is order sensitive.
- *Inference control: Prolog* allows the programmer to enhance efficiency of programs by introducing some procedural features, such as the *cut* predicate.
- *Negative information: Prolog* permits negative information to be expressed. For instance, literals like:

$$not(likes(george, vanessa)) \quad \text{or} \quad not(member(X, l))$$

are perfectly legal in the right side of a rule.

The first two features are typical of *any* logic programming language, while the remaining ones are typical of *Prolog*. Indeed, we have shown in the previous chapter that *Prolog* enhances pure logic programming by its extralogical features, such as the *cut* predicate. Further, the use of negation in the rule body yields non-Horn clause programs. Extralogical features introduce procedural aspects, for instance, by enabling easy specification of alternative cases in programs (*if − then − else* constructs, see Example 2.5). Procedural aspects make *Prolog* programs quite powerful, but also more liable to unsafeness: the burden of correctness is left on the programmer, and, as we have seen in Chap. 2, an ill-designed *Prolog* program may not even terminate.

30   Chapter 3 Prolog as a Query Language

This chapter presents two significant examples of typical problems that are solved in a natural way by using the full computational power of *Prolog*. These problems are intrinsically recursive; furthermore, they have some procedural features (list processing, use of arithmetic operations, alternative choices) which deviate from a pure Horn logic specification. Both problems assume the existence of a database system which stores potentially large amounts of data.

3.1 The Anti-Trust Control Problem

When a certain company produces goods that are sold in some geographical area of demand for these goods, or *market*, we say that this company *owns directly* a *quota* of that market. A company can also control a market quota by owning a majority share of another company, which in its turn controls a part of the market.

Some countries issue economic policies for the purpose of preventing firms from controlling, either directly or indirectly, more than a fixed market quota. Companies that escape these rules are said to form a *trust*. However, in spite of the existence of anti-trust policies, firms very often escape law control: indirect ownership can involve complex chains of different firms, each controlling the following one in the chain, so that trusts can be very difficult to detect, due to complicated maps of ownership.

Consider, for example, the graph in Fig. 3.1, showing the ownership relationships within an over-simplified commercial system with 11 companies. Each company is represented by the name of a *flower*, and operates in the *toys* market. The number in parentheses next to each company node is the market quota (in percentage) directly owned by that company. A directed edge (an arrow)

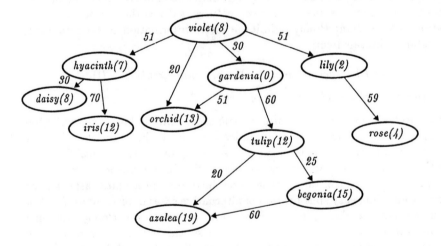

Fig. 3.1. Example of map of ownership

$< A, B >$ in the graph says that A owns shares of B; the edge label represents the percentage of share.

Thus, the company *violet* holds directly a market quota of 8 per cent. Further, it has shares of companies *hyacinth* (51 per cent), *orchid* (20 per cent), *gardenia* (30 per cent) and *lily* (51 per cent). The company *orchid* owns directly 13 per cent of the market for toys, but does not possess shares in any other company. The company *gardenia*, instead, does not hold directly any quota of the market, but has a 60 per cent share of the company *tulip*, which in turn owns directly a 12 per cent market quota, and shares of two other companies, *begonia* (25 per cent), and *azalea* (20 per cent). Indeed, *begonia* itself also has a share of *azalea*, so that *tulip* participates directly as well as indirectly (through *begonia*) in *azalea*'s business.

The information content of a participation graph can be represented through *Prolog* predicates (corresponding to database relations).

The predicate $company(a, b, n)$ indicates that a is a company that owns directly a quota of $n\%$ for the market of goods b. The predicate

$$hasshares(comp_1, comp_2, n)$$

indicates that company $comp_1$ owns $n\%$ of company $comp_2$. Finally, a predicate $trust\_limit(a, b)$ indicates that, for the particular market a, b is the share of market which designates a trust.

If firm A directly owns more than 50% of firm B, we say that A *controls* B. In this case, A has control over all the market quotas owned by B, and over the companies controlled by B. Thus, the problem of determining whether a given company violates the anti-trust control is inherently recursive. We now describe the *Prolog* program which is used to discover both the share of market

---

$controlled(Comp2, Comp1) : -$ $hasshares(Comp1, Comp2, N),$
  $N > 50.$

$market(Comp1, Mrkt, Total) : -$ $company(Comp1, Mrkt, Quota),$
  $findall(Comp2,$
    $controlled(Comp2, Comp1), List),$
  $inquire(List, Mrkt, Quota, Total).$

$inquire([], Mrkt, Quota, Quota).$
$inquire([Head|Tail], Mrkt, Quota, Total) : -$
  $market(Head, Mrkt, Quota2),$
  $Quota3$ is $Quota2 + Quota,$
  $inquire(Tail, Mrkt, Quota3, Total).$

$findtrust(Mrkt, Comp, Total) : -$ $trust\_limit(Mrkt, Threshold),$
  $market(Comp, Mrkt, Total),$
  $Total > Threshold.$

controlled by a given company (through the *market* rule) and the violations to the anti-trust law (through the *findtrust* rule).

The *findtrust* rule can be used to trigger the program, yielding one by one the names and market percentages of companies that have a trust in a certain market *Mrkt*. The rule is quite simple: first, it looks up the threshold value from the base predicate *trust_limit*. Then, it calls the goal *market* to compute the total percentage owned by company *Comp*; finally, it compares this total with the threshold value; the goal succeeds if the anti-trust rule is violated, and fails otherwise.

The goal *market*, once executed, returns the *Total* quota of market owned by company *Comp*1. It first looks up the *Quota* of that market which is directly owned by the company, then finds the *List* of all companies directly controlled by *Comp*1, then examines the situation of each of the companies of the *List*, by calling the goal *inquire*.

List is built by using the built-in predicate *findall*, which finds all companies *Comp*2 that are *controlled* by company *Comp*1. *findall* is a system predicate which always succeeds and is not resatisfiable. Its first argument must be some non-instantiated variable X; the second argument must be a predicate containing the same variable X; the third argument is the list of all elements X that satisfy the predicate. The *Prolog engine* assumes this predicate as a goal, and instances which satisfy the goal are inserted in the list. In our case, the non-instantiated variable is *Comp*2, and the predicate is *controlled*, which is satisfied when the company in the first argument is controlled by the company in the second argument, since the latter owns directly more than 50% of the shares of the former.

Consider the graph of Fig. 3.1. Suppose *Comp*1 = *violet*. Then, *List* = [*hyacinth*, *lily*]. Notice that the list will only contain the names of directly controlled companies, i.e., only those that have distance 1 from *Comp*1.

The two rules that define the predicate *inquire* are the main rules of the program. The second rule scans a list of companies controlled by the company under investigation, traversing the first level of the graph shown in Fig. 3.1. List analysis is done by processing the list's head and by recursively calling the *inquire* rule on the list's tail. Initially, the variable *Quota* contains the market quota directly owned by the company we are examining. By calling recursively the goal *market* within the goal *inquire*, all the companies of the list are investigated in their turn, adding up their shares. The first rule ends recursion by assigning to the last variable the latest quota value that has been computed.

Notice the use of mutual recursion, since the goal *inquire* calls the goal *market*, which in turn will call again the goal *inquire*. At each recursion step, the call to the goal *market* moves the search one step downward in the graph, while the call to *inquire* traverses all the nodes at the same level. Mutual recursion of this program is not strictly required, since rules could be rewritten so as to eliminate the need for it, but this would lead to a more intricate final program.

Finally, let us note that the program is assuming acyclicity of the *hasshare* relationship, thereby assuming that maps of ownership are in fact directed acyclic graphs. In fact, the program is also correct under less restrictive assumptions, namely the acyclicity of the *controlled* relationship, which follows immediately from the former assumption.

Let us now execute the program on an extensional database. The information of Fig. 3.1. is shown by the relations of Fig. 3.2.

HASSHARES		
violet	hyacinth	51
violet	orchid	20
violet	gardenia	30
violet	lily	51
hyacinth	daisy	30
hyacinth	iris	70
gardenia	orchid	51
fiat	opel	2
gardenia	tulip	60
lily	rose	59
tulip	azalea	20
tulip	begonia	25
begonia	azalea	60

COMPANY		
violet	toys	8
hyacinth	toys	7
opel	cars	33
orchid	toys	13
gardenia	toys	0
tulip	toys	12
ford	cars	30
daisy	toys	8
iris	toys	12
lily	toys	2
rose	toys	4
azalea	toys	19
begonia	toys	15
fiat	cars	35

TRUST-LIMIT	
toys	20
cars	38
gasoline	40

Fig. 3.2. Relational representation of the map of ownership in Fig. 3.1

In the following, we simulate a *Prolog* session:

?- *findtrust(toys, gardenia, Total).*
Total = 25.

?- *findtrust(toys, violet, Total).*
Total = 33.

?- *findtrust(toys, tulip, Total).*
no.

?- *findtrust(Market, begonia, Total).*
Market = *toys.*
Total = 34.

3.2 The Bill of Materials Problem

The *bill of materials* of a certain product is a description of all the items that compose it, down to the lowest level of detail, where no further decomposition is possible. A component can appear several times within the bill of materials of a given main product, possibly at different depths in the subcomponent's tree. Manufacturing companies make large use of bills of materials, since they need to store information about components of their products. A (very rough) representation of the components of an internal combustion engine is shown in Fig. 3.3.

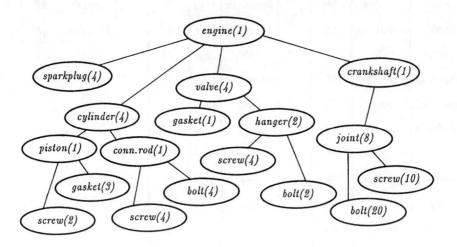

Fig. 3.3. Bill of materials of an internal combustion engine

The engine contains 4 valves, 4 spark plugs, 1 crankshaft, and 4 cylinders. These elements are decomposed in turn. A cylinder contains a piston and a connecting-rod; pistons contain screws and gaskets, and connecting-rods contain screws and bolts. Elements in the graph are labeled by a number expressing the number of occurrences of a subpart within a part.

We present a program that computes, for each node of the bill of materials tree, the list of its descendents and the total number of occurrences of each component.

The predicate $comp(prod_1, prod_2, qty, flag)$ says that $prod_1$ contains qty occurrences of $prod_2$. The flag can assume the value b (basic element) or c (composite element). Basic elements are leaves of the tree, composite elements are internal nodes; the flag is not really needed, but is only used for the sake of efficiency.

The program for finding elementary components is triggered by calling the goal *lookfor*, with the variable *Prod* instantiated to the product we want to examine,

$lookfor(Prod, IL, FL, Factor) : -findall($
$\qquad\qquad\qquad\qquad (Pr1, Qty, Flag),$
$\qquad\qquad\qquad\qquad comp(Prod, Pr1, Qty, Flag),$
$\qquad\qquad\qquad\qquad CL),$
$\qquad\qquad\qquad\qquad inquire(CL, IL, FL, Factor).$

$inquire([\,], List, List, \_).$
$inquire([(Pr1, Qty, Flag)|TC], List, FL, Fact) : -$
$\qquad\qquad\qquad\qquad \%\text{if}\%$
$\qquad\qquad\qquad\qquad ((Flag == c, !,$
$\qquad\qquad\qquad\qquad Fact1 \text{ is } Fact * Qty,$
$\qquad\qquad\qquad\qquad lookfor(Pr1, List, ML, Fact1));$
$\qquad\qquad\qquad\qquad \%\text{elsif}\%$
$\qquad\qquad\qquad\qquad (member((Pr1, \_), List), !,$
$\qquad\qquad\qquad\qquad update(List, ML, Pr1, Qty, Fact));$
$\qquad\qquad\qquad\qquad \%\text{else}\%$
$\qquad\qquad\qquad\qquad (Q1 \text{ is } Qty * Fact,$
$\qquad\qquad\qquad\qquad add((Pr1, Q1), List, ML)),$
$\qquad\qquad\qquad\qquad \%\text{anyway}\%$
$\qquad\qquad\qquad\qquad inquire(TC, ML, FL, Fact).$

$update([(N, Q1)|T], [(N, Q2)|T], N, Q, Fact) : -$
$\qquad\qquad\qquad\qquad Q2 \text{ is } Q1 + Q * Fact.$
$update([H|T], [H|T1], N, Q, Fact) : -update(T, T1, N, Q, Fact).$

$add(E, [\,], [E]).$
$add(E, [H, T], [E|[H|T]]).$

$member(H, [H|\_]).$
$member(E, [\_|T]) : -member(E, T).$

$costs(Prod, Cost) : -lookfor(Prod, [\,], FL, 1),$
$\qquad\qquad\qquad\qquad examine(FL, Cost).$

$examine([\,], 0).$
$examine([(El, Numb)|T], Cost) : -price(El, Cost1),$
$\qquad\qquad\qquad\qquad examine(T, Cost2),$
$\qquad\qquad\qquad\qquad Cost \text{ is } Cost1 * Numb + Cost2.$

the initial list IL instantiated to the empty list ($[\,]$), and $Factor$ instantiated to 1. At a generic level of recursion, the value of $Factor$ gives the number of occurrences of the current value of $Prod$ in its parent component. The program returns into the variable FL the list of components of the given product, each with the corresponding total number of occurrences. For example, if we want the elementary components of the product *engine* in Fig. 3.3., we execute the goal:

$$? - lookfor(engine, [\,], FL, 1).$$

The *lookfor* rule uses the *findall* special predicate, defined in the previous section, in order to find all first-level components $Pr1$ of $Prod$, along with their $Flags$, and with the number Qty of occurrences of $Pr1$ in $Prod$; the triplets $< Pr1, Qty, Flag >$ are inserted into the list CL. Then, the recursive goal *inquire* is called; at the end of the recursive descent, the variable FL of *inquire* will contain the result of the computation as previously defined (namely, the list of pairs containing *all* the components with their occurrences).

The predicate *inquire* is recursive, with the first rule defining a termination condition. The second rule of *inquire* receives as input the list of triplets CL; list analysis is done, as in the anti-trust example, by processing the triple at the list's head and by recursively calling the *inquire* rule on the list's tail. Three cases are distinguished, based on the type of the component in the current triple.

a) If the component $Pr1$ is in turn composite, then the goal *lookfor* is further activated, thus descending one level in the bill of materials tree. The variable $Fact1$ is instantiated to the product of the occurrences Qty of $Pr1$ and the number $Fact$ of occurrences of component $Pr1$ in its parent element.
b) When $Pr1$ is a basic component and has already been found within the component list that is being built, then the number of occurrences of $Prod1$ in the list must be updated. This is done by means of a specialized *update* rule, not further described here.
c) If neither of the above cases hold, then $Pr1$ is a new basic component, which is added to the list; the number of occurrences is given by the product of Qty and the number $Fact$ of occurrences of the component $Pr1$ within its parent element.

Note the use of the cut for an efficient implementation of alternatives: if either of the two first alternatives succeeds, then the search for other solutions is blocked. Again, the program has mutual recursion, as *inquire* is used to traverse the tree staying at the same level, while *lookfor* is called to descend one step within the tree.

The predicate *costs* computes the total cost of a certain element starting from the prices of its elementary components. It calls the goal *lookfor* and then uses the goal *examine* to add up the prices of the components. These are represented by the binary predicate *price(Comp, Price)*, giving the *Price* of each elementary *Component*. In this computation, the prices of intermediate components are not added up.

Let us now execute the program on an extensional database. The information of Fig. 3.3. is represented by the relations of Fig. 3.4. As in the previous section, we simulate a *Prolog* session:

3.2 The Bill of Materials Problem

COMP			
engine	sparkplug	4	b
engine	cylinder	4	c
engine	valve	4	c
engine	crankshaft	1	c
cylinder	piston	1	c
cylinder	connecting-rod	1	c
valve	gasket	1	b
valve	hanger	2	c
crankshaft	joint	8	c
piston	screw	2	b
piston	gasket	3	b
connecting-rod	screw	4	b
connecting-rod	bolt	4	b
hanger	screw	4	b
hanger	bolt	2	b
joint	screw	10	b
joint	bolt	20	b

PRICE	
sparkplug	10
screw	2
gasket	3
bolt	2

Fig. 3.4. Relational representation of the bill of materials of an internal combustion engine

$? - lookfor(engine, [\,], FL, 1).$

$FL = [\ (sparkplug, 4),$
$\qquad\quad (screw, 136),$
$\qquad\quad (gasket, 16),$
$\qquad\quad (bolt, 192)].$

$? - lookfor(cylinder, [\,], FL, 1).$

$FL = [\ (screw, 6),$
$\qquad\quad (gasket, 3),$
$\qquad\quad (bolt, 4)].$

$? - costs(engine, Cost).$

$Cost = 744.$

3.3 Conclusions

The examples provided in this chapter have shown that the full computational power of *Prolog* can enrich the possibilities of traditional query languages. Moreover, we have also seen that *Prolog* combines a description of database retrievals and computations within a single linguistic framework. This provides higher programming readability and uniformity, thus presenting substantial differences from other database interfaces which require two different languages, one for the programming itself and the other for the database calls.

3.4 Bibliographic Notes

A simple introduction to the use of *Prolog* as a query language can be found in papers by Parsaye [Pars 83], Naqvi [Naqv 84], and Zaniolo [Zani 84]. Brodie and Jarke [Brod 84] evaluate logic, logic programming, and *Prolog* with respect to databases, giving a systematic overview of the field. Sciore and Warren [Scio 84] and Warren [Warr 81] address the issue of how to produce efficient implementations of *Prolog engines* operating over large databases, by introducing access methods (such as indexes) or by making effective use of buffering. With their approach, the *Prolog* programmer has great control over these access methods.

The two problems presented in this chapter were worked out by Gozzi and Lugli in their diploma thesis [Gozz 87].

3.5 Exercises

3.1 Modify the anti-trust program of Sect. 3.1 to determine whether a *group* of companies create a trust.

3.2 Modify the anti-trust program of Sect. 3.1 by changing the notion of control, including also indirect shares. In other words, in order to evaluate whether the company $Comp1$ controls the company $Comp2$, add to the direct shares all shares which belong to other companies $Comp3$ which are controlled by $Comp1$.

3.3 Modify the bill of materials program of Sect. 3.2 by also including in the price computation the price of intermediate components.

3.4 Write a *Prolog* program which operates over a large database describing train schedules. Assume the following schema for the *SCHEDULE* relation:

SCHEDULE(TRAIN-NUM,DEP-CITY,DEP-TIME,ARR-CITY,ARR-TIME).

Assume that all trains run daily; with this representation, each tuple corresponds to one route segment. A train going from Milano to Rome with two intermediate stops at Bologna and Firenze is hence represented through the relation we give below.

Write a program for finding the shortest schedule connecting any two given cities, assuming the departure time as given. The program should produce a list of tuples of the same format as the SCHEDULE relation, indicating all route segments required for going from the start city to the destination city, under the assumption that all connections are feasible (namely, the arrival city of segment i is the same as the departure city of segment $i+1$, and the arrival time of segment i is before the departure time of segment $i+1$).

TRAIN-NUM	DEP-CITY	DEP-TIME	ARR-CITY	ARR-TIME
567	Milano	7.50	Bologna	9.30
567	Bologna	9.32	Firenze	10.30
567	Firenze	10.32	Roma	12.15

A simple solution is acceptable; but at the end of your work, you should think about possible improvements which make your program more efficient.

Chapter 4
Coupling Prolog Systems to Relational Databases

In Chap. 3 we have seen that *Prolog* is a very expressive query language, one that can express queries with no counterpart in relational languages such as algebra or calculus. In this chapter, we describe how it is possible to build systems that interface *Prolog* to databases, providing efficient data access.

We consider the following issues:

a) We discuss the *components* and the *alternative architectures* of systems for coupling *Prolog* systems to relational databases.
b) Within the alternative architectures, we discuss methods for *interfacing a database and a Prolog system*; we introduce the concept of *base conjunction* as the unit of interaction between the *Prolog* and the database environments.
c) We then discuss various methods for *optimizing the performance of the interface*. Our major concern is the saving of database access operations. In particular, *caching* is used to store data or intermediate queries in order not to repeat the same activities during the execution of a program.

The concepts and methods presented in this chapter are based on several existing prototypes and systems, including PRO-SQL, EDUCE, the ESTEAM prototype, BERMUDA, PRIMO, and QUINTUS-PROLOG; these are reviewed in Chap. 5.

4.1 Architectures for Coupling Prolog and Relational Systems

We recall from Chap. 1 the general features of systems for **C**oupling **P**rolog to **R**elational databases (we will call these **CPR** systems). With the *coupling* approach, an interface is built between currently available *Prolog* and database systems, which preserve their individuality; the interface provides the procedures required for bringing data from the mass-memory database into the main-memory logic programming execution environment.

4.1.1 Assumptions and Terminology

Before studying CPR systems in greater detail, we introduce some assumptions and terminology.

a) A significant subset of the predicates of the *Prolog* programs considered in this chapter are *database predicates* (**dbps**). Each of these predicates corresponds to a relation stored in secondary storage. **Dbps** are not allowed to appear on the left-hand side of a rule; note that this limitation does not limit the computational power of a *Prolog* program.

b) We assume that the order of facts in the mass-memory database is *not* relevant. However, if facts get stored in the main-memory *Prolog* database, then their order becomes relevant. This preserves the conventional semantics of *Prolog*, in particular with regard to backtracking.

c) We only consider retrievals and not updates, which are under the control of the database system. We show how to satisfy the data needs of a *Prolog* program by generating a suitable set of queries; the serializability of these queries with respect to all the other applications which use the database can be provided by any concurrency control method.

In Fig. 4.1 we show a *Prolog* program which references a **dbp** rel and the corresponding relation REL stored in the database. In the example shown in Fig. 4.1, the **dbp** rel has three *occurrences* (one for each rule of the program). An *adornment* is a mapping that assigns either "bound" or "free" to each argument of a predicate. If an argument is a constant, it is assigned "bound"; if it is a variable, it is assigned "free". The *static adornment* of a **dbp** is the one that can be determined prior to execution; during execution, the adornment changes when variables are bound to constants. The static adornment of $rel(a,Z,X,X,T)$ is *(bound, free, free, free, free)*.

Fragment of *Prolog* program

$p(X) : -rel(a, X, Z, Z, Y), t(Y).$
$q(X) : -rel(a, Z, X, X, T), q(T).$
$t(X) : -rel(b, X, Y, c, f), q(X).$

Database Relation

REL				
A	B	C	D	E
a	b	c	b	c
a	a	c	b	d
c	a	c	b	d
c	a	d	f	d
a	c	c	b	e

Fig. 4.1. Example of interaction between Prolog and a relational database

Whenever two or more occurrences of the same **dbp** have the same adornment and have the same constants in corresponding bound positions as well as equal variables in correspondence, they yield the same *database formula*, denoted **dbf**. The first two occurrences of the **dbp** *rel* in Fig. 4.1 correspond to the same **dbf**, and the corresponding query in relational algebra is:

$$\sigma_{A=a \wedge C=D} REL$$

During the evaluation of a *Prolog* program, each occurrence of a **dbp** can have several *active instances*. This means that a matching can be performed for different activations of the **dbp** at different levels of recursion. Each active instance of a **dbp** can be mapped to different **dbfs**, because the mapping to constants can be different; even the adornment of different activations of the same **dbp** can be different.

4.1.2 Components of a CPR System

In the architecture of a CPR system, we distinguish four separate subsystems:

a) The *Prolog engine*, which executes *Prolog* programs using the standard *Prolog* inference mechanism.
b) The *Prolog interface*, which is capable of recognizing the *database predicates* and treating them in a special way.
c) One or more *database interfaces*, which provide the communication between the database and *Prolog* engines.
d) The *database engine*, which performs the retrieval or update of data in the mass-memory database.

Though all these modules must be present within a CPR system, their functionalities may vary substantially from system to system. Let us consider each subsystem separately.

Prolog Engine

The *Prolog engine* is a Prolog system capable of executing Prolog goals. Though with the coupling approach we do not expect to change the basic behavior of the *Prolog* engine (depth-first search strategy, unification, and backtracking), *Prolog* engines might have enhanced features in order to adapt to the proposed database environment.

a) Some *Prolog* systems provide *fast access methods in main memory* for searching and accessing a large number of facts with the same format; these techniques include sophisticated indexing and hashing mechanisms.
b) Some *Prolog* systems provide enhanced memory acquisition and management mechanisms, designed for improving the efficiency of the *assertion of facts* within the memory-resident *Prolog* database. This operation is heavily used as a result of the retrieval of facts from mass memory databases, and is currently badly supported by many *Prolog* systems.

c) Finally, another modification applied to several *Prolog engines* concerns returning *the set of all answers* to queries (goals) over database predicates. This can be achieved by using the standard *Prolog* depth-first search strategy and forcing the repetition of the goal execution until failure. Then, the system presents the *set* of all answers collected so far.

Prolog Interface

The *Prolog interface* is capable of recognizing the *database predicates* and treating them in a special way. From the viewpoint of the *Prolog* programmer, the interface can be at various levels of transparency:

a) With *full transparency*, database predicates are recognized by the interface with no user support. In this case, a *program analyzer* accesses database directories in order to recognize database predicates; the analyzer can be activated either during the loading of a *Prolog* program, or at each matching operation.

b) With *intermediate transparency*, the user must provide a declaration of the database predicates; once declared, database predicates appear to the programmer as any other predicates, and the fact that their matching is performed by interfacing the database system is hidden.

Some *Prolog interfaces* guarantee a selective treatment of database predicates, based on additional declarations. For instance, it is possible to declare that database predicates will be dealt with through an *extension table*, thereby asserting all retrieved values for that predicate in main memory; controlling such a directive becomes part of the *Prolog interface*. Similarly, it is possible to indicate that database predicates will always be executed by having some places bound to constant values, and then the *Prolog interface* checks at execution time that the actual goal execution reflects such constraint.

c) With *no transparency*, the *Prolog* programmer is forced to write explicit queries, using the query language of the database engine. For instance, a few systems use explicit queries written in SQL or QUEL.

Database Interface

The *database interface* is specifically designed for interfacing with the database and extracting tuples from it. In a multi-processing environment, there can be several instances of the *database interface* active at the same time, each one active on a specific query. Several alternatives are possible in the design of the *database interface*:

a) A major distinction concerns the *type of queries* which are considered by the *database interface*:
   1) *Single-predicate* (selection) queries are issued for each database predicate separately; each query corresponds to a simple selection.
   2) *Multiple-predicate* (join) queries are issued for multiple database predicates at the same time; these predicates belong to the right side of a rule and have some variables in common, which correspond to join conditions.

3) *Aggregate* queries are issued for special predicates which evaluate aggregate functions over sets or multisets. In *Prolog*, these queries correspond to variations of the *setof* or *bagof* special predicates; in the *database engine*, these queries require the *group-by* operations and the evaluation of *aggregate functions*, which are supported by most relational query languages.
4) *Recursive* queries are issued for recursively defined predicates which involve one or more database predicates in their definitions. In this case, the query passed to the *database engine* contains recursion. This case is not typical of CPR systems, and in fact can be supported only if the *database engine* is itself an *enhanced* system. In the current state of the art, a few database engines support a special subset of recursive queries, those that can be answered by computing the transitive closure of binary relations.

b) Another major feature of the *database interface* concerns *how result tuples are returned*. There are basically two alternatives:
1) Tuples can be returned *one at a time*. Typically, this is implemented through the usage of shared variables by the *Prolog* environment and the *database interface*. At each interaction with the *database interface*, new values are returned to the shared variables; in practice, matching of these variables is under the control of the *database interface*, rather than the *Prolog engine*. This means that the *database interface* must preserve the present retrieval context, and it must be able to return the next values within the shared variables on backtracking.
2) An entire *set of tuples* can be returned; this solution is achieved by *asserting* the tuples in the memory-resident *Prolog* database, and then by resuming execution of the *Prolog engine* on the first tuple of the set of new tuples; during backtracking, the *Prolog engine* will find the next tuple satisfying the same goal already loaded into main memory. Hence, in this case the matching and backtracking over database predicates is under the control of the *Prolog engine*.

Database Engine

The *database engine* is any database software capable of processing queries or updates of the database. The *database engine* can provide a data manipulation language at various levels:

a) With *low-level* (procedural) languages the programmer of the *database engine* must specify the strategy for accessing data. With these languages, it is possible to control the physical access to the database, for instance by selecting the access methods or by controlling the flows of tuples. Hierarchical or network database systems in general provide low-level data manipulation languages.
b) With *high-level* (nonprocedural) languages the programmer of the database must specify the properties of data that need to be retrieved, while the system provides the strategy for executing the retrieval. With these languages, interactions with the database are typically *set-oriented*, though facilities for controlling the flow of tuples are also offered. SQL is the most representative

(and used) language of this kind; in general, relational systems provide high-level languages.
c) Finally, *enhanced* languages provide extensions specifically designed for dealing with special queries; for instance, some systems provide the transitive closure of binary relations. It is possible that enhanced languages will move part of the processing required for recursive goals from the *Prolog* environment to the database environment.

The various subsystems that we have analyzed so far and their features are synthesized in Table 4.1.

Prolog Engine	with/without fast access mechanisms
	with/without special assertion management
	returning one answer/set of answers
Prolog Interface	full/intermediate/no transparency
Database Interface	single/multiple/aggregate/recursive queries
	returning one tuple/set of tuples
Database Engine	low-level/high-level/enhanced query language

Table 4.1. Components of a Coupled Prolog Relational (CPR) system

4.1.3 Architecture of CPR Systems

After analyzing separate components, we now analyze how they can be combined in order to create a CPR system. We recall from Chap. 1 that there are two major alternative approaches, loose coupling and tight coupling.

Loose Coupling

With *loosely coupled systems (LCPR)*, the interaction takes place independently of the actual *Prolog* inference process. Loose coupling is also called *static coupling*, because coupling actions are performed independently of the actual pattern of execution of rules.

Figure 4.2 shows the general schema of a LCPR system. Note that the coupling action (1) takes place at *load time*, when the file containing the *Prolog* program is consulted by the *Prolog engine*; once the relevant database facts have been loaded into the *Prolog* memory, the execution of the *Prolog* program takes place independently of the database.

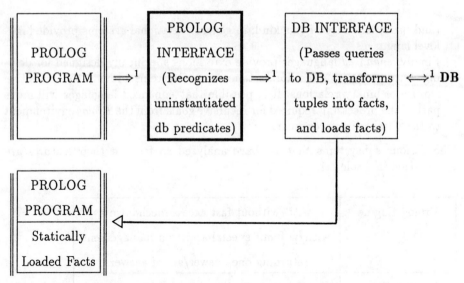

Fig. 4.2. Architecture of a LCPR system.   [1]Coupling at load time

Tight Coupling

With *tightly coupled systems (TCPR)*, the interaction between *Prolog* and the database systems is driven by the inference activity of the *Prolog* system. Tight coupling is also called *dynamic coupling* because coupling actions are performed in the frame of goal execution.

Figure 4.3 shows the general schema of a TCPR system. The coupling action (2) takes place either at *goal presentation time*, when a new goal is proposed by the user to the *Prolog engine*; or at *rule execution time*, when a rule is activated by the *Prolog engine*; or at *database predicate matching time*, when each individual database predicate is matched. These are different alternatives designating situations of decreasing granularity.

Note that there is disagreement in the literature concerning the exact distinction between loose and tight coupling. Various references regard as loosely coupled systems either those which provide no transparency, or those which operate at goal presentation time, or finally those which assert the set of tuples

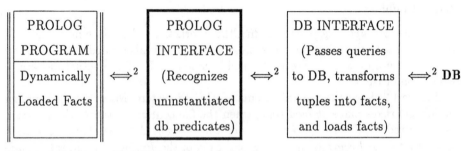

Fig. 4.3. Architecture of a TCPR system.   [2] Coupling at execution time

resulting from a query as main memory facts, rather than communicating them to the *Prolog* program through shared memory. All these different interpretations denote that the terms loose and tight coupling have been quite abused.

4.2 Base Conjunctions

In CPR systems, the most typical interaction between the *Prolog* and the database system takes place for base conjunctions; a *base conjunction* is a sequence of database predicates and arithmetic comparison predicates. An example of base conjunction is:

bc_1 : $db_1(X_1, a, X_3), db_2(X_3, X_4, X_5), (X_4 = b), db_3(X_5, X_6, X_7), (X_6 = c)$

A base conjunction can be part of the body of a *Prolog* rule, but it can also be obtained through the composition of *Prolog* rules; for instance, the above base conjunction bc_1 can be obtained as a result of rule compositions applied to the following rules:

$$go_1(X_1, X_4) : - db_1(X_1, a, X_3), p_1(X_3, X_4).$$

$$p_1(Y_1, Y_4) : - db_2(Y_1, Y_2, Y_3), (Y_2 = b), p_2(Y_3, Y_4).$$

$$p_2(Z_1, Z_3) : - db_3(Z_1, Z_2, Z_3), (Z_2 = c).$$

The use of rule composition to generate large base conjunctions will be addressed in the following section.

A base conjunction bc is *connected* iff it cannot be divided into two parts which have no variable in common. In the following, we assume that base conjunctions are connected; if a base conjunction is disconnected, we simply consider each connected part as a separate base conjunction. Consider the following sequence of database and arithmetic predicates:

$db_1(X_1, a, X_2), db_2(X_2, X_3, X_4), (X_3 = b), db_3(Y_1, Y_2, Y_3), (Y_2 = c)$

The above sequence can be separated into two connected base conjunctions:

bc_2 : $db_1(X_1, a, X_2), db_2(X_2, X_3, X_4), (X_3 = b)$
bc_3 : $db_3(Y_1, Y_2, Y_3), (Y_2 = c)$

A connected base conjunction bc is a *chain* base conjunction if there exists an ordering $db_1, .., db_n$ of **dbps** in the bc such that each db_i in bc shares some variables just with db_{i-1} (when $1 < i \leq n$) and with db_{i+1} (when $1 \leq i < n$); otherwise, the base conjunction is *cyclic*. bc_1 is a chain base conjunction, while the following bc_4 is cyclic:

bc_4 : $db_1(X_1, X_2), db_2(X_2, X_3), db_3(X_1, X_3)$

The *exported variables* of a base conjunction are the variables of the bc which are shared with the remainder of the rule. For instance, in the following rule:

$go_2(X_1) : -db_1(X_1, X_2), db_2(X_2, X_3), db_3(X_1, X_3), p_1(X_3).$

which includes the base conjunction bc_4, the exported variables of bc_2 are X_1 and X_3. The rule go_2 can be rewritten by introducing a separate rule for the base conjunction, having as head the predicate $bc_4(X_1, X_3)$, and as body the sequence of predicates in base conjunction, as follows:

$$go_2(X_1) :- bc_4(X_1, X_3), p_1(X_3).$$

$$bc_4(X_1, X_3) :- db_1(X_1, X_2), db_2(X_2, X_3), db_3(X_1, X_3).$$

Similarly, we can rewrite go_1 as follows:

$$go_1(X_1, X_7) :- bc_1(X_1, X_7).$$

$$bc_1(X_1, X_7) :- db_1(X_1, a, X_3), db_2(X_3, b, X_5), db_3(X_5, c, X_7).$$

Note that we eliminate equality predicates by substituting variables by constants within predicates.

Each base conjunction corresponds to a *join query*, also called a *conjunctive query*, on the database. Join queries are expressed in relational algebra by creating an algebraic expression including: selections over relations corresponding to **dbp**s, when there are **dbp** places bound either to constants, or to variables involved in arithmetic predicates; joins between any two relations corresponding to **dbp**s which share a variable; and a final projection over the attributes corresponding to exported variables of the base conjunction. Chain base conjunctions correspond to chain join queries, and cyclic base conjunctions correspond to cyclic join queries. A formal method for mapping logic programming goals to queries in relational algebra will be shown in Chap. 8, in the framework of the *Datalog* programming language.

For instance, join queries jc_1 and jc_2 correspond to base conjunctions bc_1 and bc_4 in rules go_1 and go_2. We shall use the following notation: database relations R_i correspond to database predicates dbp_i, and attributes A_i correspond to variables X_i.

$jc_1:$ $\quad \Pi_{1,9}((\sigma_{2=a}R_1) \underset{3=1}{\bowtie} (\sigma_{2=b}R_2)) \underset{6=1}{\bowtie} (\sigma_{2=c}R_3)$

$jc_2:$ $\quad \Pi_{1,6}((R_1 \underset{2=1}{\bowtie} R_2) \underset{1=1}{\bowtie} R_3)$

Most CPR systems (including PRO-SQL, EDUCE, ESTEAM, PRIMO, and QUINTUS-PROLOG, see Chap. 5) map base conjunctions to queries formulated in the relational languages SQL and QUEL; for instance, the following queries in SQL correspond to base conjunctions bc_1 and bc_4 (and to algebraic queries jc_1 and jc_2):

SELECT $R_1.A_1$, $R_3.A_7$
FROM R_1, R_2, R_3
WHERE $R_1.A_2 = a$ AND $R_2.A_4 = b$ AND $R_3.A_6 = c$
AND $R_1.A_3 = R_2.A_3$ AND $R_2.A_5 = R_3.A_5$

SELECT $R_1.A_1$, $R_3.A_3$
FROM R_1, R_2, R_3
WHERE $R_1.A_1 = R_3.A_1$ AND $R_1.A_2 = R_2.A_2$
AND $R_2.A_3 = R_3.A_3$

4.2 Base Conjunctions

The optimization of a join query is a classic feature available in all relational DBMSs. DBMS software is responsible for selecting the most efficient access methods in order to compute the resulting set of tuples. Because of the efficiency of DBMSs, the best performance of CPR systems is achieved with large base conjunctions.

In a CPR system, the *Prolog interface* performs the task of determining the largest base conjunctions; the *database interface* creates the corresponding join query and controls its execution. Once result tuples are extracted from the database, the *database interface* returns them to the *Prolog engine*, either one at a time or all together, as discussed in Sect. 4.1.2.

It should be noted that assembling base conjunctions changes the *semantics* of the *Prolog* program, as it causes modified executions with respect to the original program; the proposed program transformations affect the *order* and *replication* of tuples in the result, though the set of tuples in the result remains the same. Thus, a *Prolog engine* that returns the set of answers would mask these differences, while a conventional *Prolog engine* would not.

4.2.1 Determining Base Conjunctions in LCPR Systems

We now examine how base conjunctions are determined by the *Prolog interface*. We consider the *static adornment* of rules, hence presenting transformations that are supported by LCPR systems; we will then discuss how TCPR systems can improve this situation.

Prolog interfaces show great differences in their ability to discover base conjunctions. In the following, we consider cases in which discovering base conjunctions requires dealing with rule composition, predicate exchange, the *cut* special predicate, disjunction, negation, and recursion. Not all *Prolog interfaces*, however, are capable of dealing with all these cases.

Rule Composition

Rule composition is the basic operation used for building base conjunctions. An example of rule composition was shown in the previous section:

$$go_1(X_1, X_4) :- db_1(X_1, a, X_3), p_1(X_3, X_4).$$

$$p_1(Y_1, Y_4) :- db_2(Y_1, Y_2, Y_3), (Y_2 = b), p_2(Y_3, Y_4).$$

$$p_2(Z_1, Z_3) :- db_3(Z_1, Z_2, Z_3), (Z_2 = c).$$

By composing rules for predicates p_1 and p_2, we obtain a single rule:

$$go_1(X_1, X_4) :- db_1(X_1, a, X_2), db_2(X_2, b, X_3), db_3(X_3, c, X_4).$$

Note that variables are renamed after composition; equality predicates $(Y_2 = b)$ and $(Z_2 = c)$ are transformed by including constants b and c within database predicates db_2 and db_3 in place of variables Y_2 and Z_2. Then, the basic conjunction

$bc_1(X_1, X_4)$ is recognized, and the rule go_1 is rewritten as:

$$go_1(X_1, X_4) : - bc_1(X_1, X_4).$$

$$bc_1(X_1, X_4) : - db_1(X_1, a, X_2), db_2(X_2, b, X_3), db_3(X_3, c, X_4).$$

Rule bc_1 is then translated into a database query by the *database interface*.

Predicate Exchange Within Rule Bodies

Exchanging the order of predicates within rules can be useful in order to discover connected base conjunctions. For instance, consider the following go_2 rule:

$$go_2(X_1, Y_1, Y_2) : -db_1(X_1, a, X_2), db_2(Y_1, Y_2), db_3(X_2, X_3, b), db_4(c, Y_2).$$

In the body of the go_2 rule, we observe 4 different **dbps**; db_1 and db_3 share variable X_2; similarly, db_2 and db_4 share variable Y_2; thus, it is possible to recognize two connected base conjunctions, provided that db_2 is exchanged with db_3. This is achieved by the following rewriting of the go_2 rule:

$$go_2(X_1, Y_1, Y_2) : - bc_1(X_1), bc_2(Y_1, Y_2).$$

$$bc_1(X_1) : - db_1(X_1, a, X_2), db_3(X_2, X_3, b).$$

$$bc_2(Y_1, Y_2) : - db_2(Y_1, Y_2), db_4(c, Y_2).$$

Note that bc_1 and bc_2 are passed to the *database interface*, which is responsible for translating them into database queries; these queries can be executed in parallel.

Cut

The special predicate *cut* (!) is used in *Prolog* to control backtracking; as a goal, cut succeeds immediately; its effect is to make the variables occurring to the left of the cut, in the body of the rule executed as current goal, irrevocably instantiated when the cut is executed. Since no backtracking will be operated for those variables, alternative matchings are discarded by the *Prolog engine*.

A *cut* cannot appear within a base conjunction; thus, the effect of a *cut* is to separate a rule into two parts, before and after the cut; base conjunctions are searched within these parts. Consider, for instance, the rule go_3:

$$go_3(X_1, X_4) : -db_1(X_1, a, X_2), db_2(X_2, X_3), !, db_3(X_3, X_4), db_4(c, X_4).$$

Although db_2 and db_3 share variable X_3, they cannot be in the same base conjunction, because they are on opposite sides of a *cut*. Thus, the rule go_3 is rewritten as follows:

$$go_3(X_1, X_4) : - bc_1(X_1, X_3), !, bc_2(X_3, X_4).$$

$$bc_1(X_1, X_3) : - db_1(X_1, a, X_2), db_2(X_2, X_3).$$

$$bc_2(X_3, X_4) : - db_3(X_3, X_4), db_4(c, X_4).$$

In TCPR systems, where the *Prolog interface* is activated in the frame of a specific execution of the go_3 rule, a special strategy can also be used for base conjunction bc_1, consisting in arresting the query execution as soon as the first tuple of the result is produced; in fact, once the cut predicate is executed, the variables X_1 and X_3 get irrevocably bound. Then, the query corresponding to bc_2 is executed with variable X_3 bound.

Since *cuts* affect *Prolog*'s inference mechanism, no predicate exchange can be performed between **dbps** at opposite sides of a *cut*; consider rule go'_2, derived from go_2 by inserting a cut:

$$go'_2(X_1, Y_1, Y_2) : -db_1(X_1, a, X_2), db_2(Y_1, Y_2), !, db_3(X_2, X_3, b), db_4(c, Y_2).$$

In the above example, no predicate exchange is possible.

Disjunction

Disjunctions within rules create obvious problems; the solution used by most *Prolog interfaces* is to build a *disjunctive normal form*, which includes the largest base conjunctions. For instance, consider the following rules:

$$go_4(X_1, X_3) : - p_1(X_1, X_2), p_2(X_2, X_3).$$

$$p_1(Y_1, Y_2) : - db_1(Y_1, a, Y_2); db_2(Y_1, Y_2).$$

$$p_2(Z_1, Z_2) : - db_3(Z_1, b, Z_2); db_4(Z_1, Z_2, c).$$

Two disjunctions appear in predicates p_1 and p_2; the disjunctive normal form is obtained by having as the first rule a disjunction of predicates, each one corresponding to a base conjunction, as follows:

$$go_4(X_1, X_3) : - bc_1(X_1, X_2); bc_2(X_1, X_3); bc_3(X_1, X_3); bc_4(X_1, X_3).$$

$$bc_1(X_1, X_3) : - db_1(X_1, a, X_2), db_3(X_2, b, X_3).$$

$$bc_2(X_1, X_3) : - db_1(X_1, a, X_2), db_4(X_2, X_3, c).$$

$$bc_3(X_1, X_3) : - db_2(X_1, X_2), db_3(X_2, b, X_3).$$

$$bc_4(X_1, X_3) : - db_2(X_1, X_2), db_4(X_2, X_3, c).$$

Predicates bc_1, bc_2, bc_3, and bc_4 are then passed to the *database interface* in order to be translated and executed on the database; the corresponding queries can be executed in parallel.

It is also possible to let the entire goal go_4 be executed under the control of the *database engine*; to this end, it is sufficient to evaluate the relations BC_1, BC_2, BC_3, and BC_4, corresponding to the above base conjunctions, and then to evaluate the algebraic expression:

$$BC_1 \cup BC_2 \cup BC_3 \cup BC_4$$

Resulting tuples, returned by the *database engine*, are the answers to the go_4 goal.

Negation

Negation in *Prolog* rules can refer to database predicates; for instance, in rule go_5, negation applies to the conjunction of predicates db_3 and db_4:

$go_5(X_1) : -db_1(X_1, X_2), db_2(X_2, X_3), not(db_3(X_3, X_4), db_4(X_4))$.

Base conjunctions can be assembled by keeping positive and negated predicates separated, as follows:

$$go_5(X_1) : - bc_1(X_1, X_3), not(bc_2(X_3)).$$

$$bc_1(X_1, X_3) : - db_1(X_1, X_2), db_2(X_2, X_3).$$

$$bc_2(X_3) : - db_3(X_3, X_4), db_4(X_4).$$

In the above example, the treatment of negation is similar to the treatment of cuts: though positive and negated predicates share some variables, they contribute to separate base conjunctions; this schema, however, is not very good.

Note that the base conjunction bc_2 acts as a *filter*, by eliminating some of the tuples of bc_1. This is due to the sharing of variable X_3 between bc_1 and bc_2. On the other hand, this is the natural application of negation within *Prolog* rules; to ensure a correct computation, it is required that all exported variables in the base conjunction of the negation appear to the left of the negation, thus being *bound* when negation is evaluated.

In this situation, the *filtering* action of a negation can be performed by the *database engine*. Let POS and NEG represent the two relations corresponding to the base conjunctions at the *left* of the negation and *within* the negation respectively, and p_i and n_i the position(s) of shared variable(s) X_i in POS and NEG, with exactly one variable p_i for each variable n_i. Then, it is required that all the tuples whose projections on p_i appear in the projections on n_i of NEG be deleted from POS; this is expressed in relational algebra by the following formula, which uses the *semijoin* operation:

$POS - (POS \underset{p_i=n_i}{\ltimes} NEG)$

Recall that the semijoin operation preserves the tuples of POS whose projections on p_i appear in the projections on n_i of NEG; the subtraction of these tuples from POS gives the required filtering action. Note that the above formula is **not** equivalent to:

$POS \underset{p_i \neq n_i}{\ltimes} NEG$

Note that mapping negation into SQL is slightly more natural, due to the possibility of relating two subqueries by the $NOT\ IN$ ($\neq ALL$) block connector; each SQL block represents a base conjunction. Let R_i represent the relation corresponding to database predicate db_i, and let A_i represent attributes corresponding to variables X_i; then a query returning tuples for go_5 is:

$SELECT\ R_1.A_1$
$FROM\ R_1,\ R_2$
$WHERE\ R_1.A_2 = R_2.A_2$

```
        AND R₂.A₃ NOT IN
            SELECT R₃.A₃
            FROM R₃, R₄
            WHERE R₃.A₄ = R₄.A₄
```

We have observed that, in correct *Prolog* programs, the exported variables of negated base conjunctions must appear to the left of negations; this constraint is satisfied by rule go'_5:

$$go'_5(X_1, X_3) :- db_1(X_1, X_2), not(db_2(X_1, a)), db_3(X_2, X_3), not(db_4(X_3, b)).$$

This rule is translated into the following algebraic expression:

$$(R_1 - (R_1 \underset{1=1}{\ltimes} (\sigma_{2=a} R_2))) \underset{2=1}{\bowtie} (R_3 - (R_3 \underset{2=1}{\ltimes} (\sigma_{2=b} R_4)))$$

A predicate exchange between db_2 and db_3 in go'_5 might create a larger positive base conjunction:

$$go'_5(X_1, X_3) :- bc_1(X_1, X_3), not(db_2(X_1, a)), not(db_4(X_3, b)).$$

$$bc_1(X_1, X_3) :- db_1(X_1, X_2), db_3(X_2, X_3).$$

This collection of rules would be naturally translated into the following algebraic expressions:

$$BC_1 - (BC_1 \underset{1=1}{\ltimes} (\sigma_{2=a} R_2)) - (BC_1 \underset{2=1}{\ltimes} (\sigma_{2=b} R_4))$$

where $BC_1 = R_1 \underset{2=1}{\bowtie} R_3$. It is not clear, however, whether this version is more efficient than the previous one.

In general, we observe that negation can be postponed (being pushed to the right) so as to allow the evaluation of largest positive base conjunctions *before* negations; this predicate exchange has the effect of making base conjunctions larger but perhaps more restrictive; the convenience of this transformation should be evaluated on a case-by-case basis.

Recursion

Dealing with recursion is the most difficult problem of CPR systems. Consider the recursive predicates rp_1 and rp_2 (which are syntactic variants of the well-known *ancestor* and *same generation* problems introduced in Chap. 1):

$$rp_1(X_1, X_2) :- db_1(X_1, X_2).$$
$$rp_1(X_1, X_2) :- db_1(X_1, X_3), rp_1(X_3, X_2).$$

$$rp_2(X_1, X_2) :- db_1(X_1, X_2).$$
$$rp_2(X_1, X_2) :- db_2(X_1, X_3), db_3(X_2, X_4), rp_2(X_3, X_4).$$

With loose coupling, it is not possible to reduce database predicates: db_1, db_2, and db_3 are passed to the *database interface* to be retrieved entirely. Clearly, once goals are executed, only a few tuples of db_1, db_2, and db_3 are needed for answering a specific user's goal. Hence, loose coupling deals very badly with recursive predicates. We postpone the treatment of this problem to the next section, where we present the enhancements which are possible in tightly coupled systems with respect to loosely coupled ones.

Program transformations can be applied to programs with recursion in order to improve their evaluation. For instance, consider the following program, apparently with mutual recursion:

$$go_6(X_1, X_2) :- db_1(X_1, a, X_2), rp_3(X_1, X_2).$$

$$rp_3(X_1, X_2) :- db_2(X_1, b, X_2).$$

$$rp_3(X_1, X_2) :- db_3(X_1, X_3), rp_4(X_3, X_2).$$

$$rp_4(X_1, X_2) :- db_4(X_1, X_3), go_6(X_3, X_2).$$

The composition of rp_3 (between the first and second rule) generates the base conjunction bc_1; the compositions of rp_3 (between the first and third rule) and of rp_4 (between the third and fourth rule) generate the base conjunction bc_2. Thus, the program reduces to a simple (linear) recursion over predicate go_6:

$$go_6(X_1, X_2) :- bc_1(X_1, X_2).$$

$$go_6(X_1, X_2) :- bc_2(X_1, X_4), go_6(X_4, X_2).$$

$$bc_1(X_1, X_2) :- db_1(X_1, a, X_2), db_2(X_1, b, X_2).$$

$$bc_2(X_1, X_4) :- db_1(X_1, a, X_2), db_3(X_1, X_3), db_4(X_3, X_4).$$

At this point, with a loosely coupled interface, base conjunctions bc_1 and bc_2 are passed to the *database interface*, and all resulting tuples are inserted in the memory-resident *Prolog* database.

4.2.2 Improving Base Conjunctions in TCPR Systems

With TCPR systems, the determination of base conjunctions is performed in exactly the same way as in LCPR systems; however, with tight coupling we can use additional information either about goal constants, or about the initial bindings of rules; in other words, we use *dynamic adornments* of rules. Thus, the improvement of base conjunctions is obtained through the push of constants from the head to the body of rules. This is quite simple with nonrecursive rules: let X_i be bound to a goal constant qc_i in the head of a nonrecursive rule; then the constant can be pushed to the body of the rule simply by substituting all instances of variable X_i with constant qc_i.

Let us consider all nonrecursive rules examined in the previous section, and let us assume that X_1 is bound to the goal constant qc_1; we indicate the binding by X_1/qc_1. We show that this binding propagates to base conjunctions by substituting qc_1 for X_1 in the body of base conjunctions; the effect of this substitution is to make the base conjunctions more restrictive.

$$go_1(X_1/qc_1, X_4) :- bc_1(X_1/qc_1, X_4).$$
$$bc_1(X_1/qc_1, X_4) :- db_1(qc_1, a, X_2), db_2(X_2, b, X_3), db_3(X_3, c, X_4).$$

$$go_2(X_1/qc_1, Y_1, Y_2) :- bc_1(X_1/qc_1), bc_2(Y_1, Y_2).$$
$$bc_1(X_1/qc_1) :- db_1(qc_1, a, X_2), db_3(X_2, X_3, b).$$
$$bc_2(Y_1, Y_2) :- db_2(Y_1, Y_2), db_4(c, Y_2).$$

$$go_3(X_1/qc_1, X_4) :- bc_1(X_1/qc_1, X_3), !, bc_2(X_3, X_4).$$
$$bc_1(X_1/qc_1, X_3) :- db_1(qc_1, a, X_2), db_2(X_2, X_3).$$
$$bc_2(X_3, X_4) :- db_3(X_3, X_4), db_4(c, X_4).$$

$$go_4(X_1/qc_1, X_3) :- bc_1(X_1/qc_1, X_2); bc_2(X_1/qc_1, X_3);$$
$$bc_3(X_1/qc_1, X_3); bc_4(X_1/qc_1, X_3).$$
$$bc_1(X_1/qc_1, X_3) :- db_1(qc_1, a, X_2), db_3(X_2, b, X_3).$$
$$bc_2(X_1/qc_1, X_3) :- db_1(qc_1, a, X_2), db_4(X_2, X_3, c).$$
$$bc_3(X_1/qc_1, X_3) :- db_2(qc_1, X_2), db_3(X_2, b, X_3).$$
$$bc_4(X_1/qc_1, X_3) :- db_2(qc_1, X_2), db_4(X_2, X_3, c).$$

$$go_5(X_1/qc_1) :- bc_1(X_1/qc_1, X_3), not(bc_2(X_3)).$$
$$bc_1(X_1/qc_1, X_3) :- db_1(qc_1, X_2), db_2(X_2, X_3).$$
$$bc_2(X_3) :- db_3(X_3, X_4), db_4(X_4).$$

The meaning of base conjunctions $bc(X_i/qc_i)$, having just one variable bound to a constant, is a goal satisfaction if a database tuple satisfying the goal $bc(qc_i)$ exists, and a goal failure otherwise.

The above cases are examples of the binding propagation that takes place as an effect of the execution of specific goals when some arguments of these goals are bound to constant values. Binding propagation is performed by the *Prolog interface* at goal execution time.

The previous examples have shown a mechanical propagation of bindings; this, however, is not the general case with recursive rules. In the specific case of rule go_6, a binding on variable X_1 cannot be propagated, while a binding on variable X_2 can be propagated to bc_2 only. The rules for binding propagation within recursive rules (i.e., for *pushing constants within recursive rules*) will be explained in Chap. 10. To convince the reader that the goal constant qc_1 of the goal $go_6(qc_1, X_2)$ cannot be propagated to either bc_1 or bc_2, consider again the goal go_6:

$$go_6(X_1, X_2) : -\ bc_1(X_1, X_2).$$

$$go_6(X_1, X_2) : -\ bc_2(X_1, X_4), go_6(X_4, X_2).$$

Let tuples $bc_2(qc_1, a)$, $bc_2(a, b)$, and $bc_1(b, c)$ be retrieved from the database. The reader should verify that these tuples allow one to deduce that the tuple $go_6(qc_1, c)$ belongs to the result of the goal. But that tuple is not deduced if we apply selections $\sigma_{1=qc_1}$ to either of the base conjunctions bc_1 and bc_2, and use in the deduction process only the tuples resulting from the selections.

Binding propagation on variable X_2 of the first rule of the go_6 predicate, instead, is successful, as shown below; motivations will be given in Chap. 10.

$$go_6(X_1, X_2/qc_2) : -\ bc_1(X_1, X_2/qc_2).$$

$$go_6(X_1, X_2) : -\ bc_2(X_1, X_4), go_6(X_4, X_2).$$

$$bc_1(X_1, X_2/qc_2) : -\ db_1(X_1, a, qc_2), db_2(X_1, b, qc_2).$$

$$bc_2(X_1, X_4) : -\ db_1(X_1, a, X_2), db_3(X_1, X_3), db_4(X_3, X_4).$$

The above example shows that TCPR systems operating at *goal presentation time* can improve only some of the base conjunctions of recursive rules, and at the cost of a rather sophisticated analysis. It is possible, however, to deal with recursive rules at *rule execution time*. For this purpose, consider again predicate rp_1 from the previous section:

$$rp_1(X_1, X_2) : -\ db_1(X_1, X_2).$$

$$rp_1(X_1, X_2) : -\ db_1(X_1, X_3), rp_1(X_3, X_2).$$

Let R_1 be the database relation corresponding to db_1, and let the following tuples belong to R_1:

$$R_1 = \{\ <a, b>, <b, c>, <a, d>, <b, e>, <f, g>, <g, e>,$$

$$<g, h>, <f, h>, <o, p>, <p, q>, <o, q> \}$$

Consider the goal: $?-rp_1(a, X_2)$.

Let us execute the above goal driven by the *Prolog engine*. In order to satisfy this goal, the first rule of predicate rp_1 is activated, generating the query $db_1(a, X_2)$, which is passed to the *database interface*; as a result, facts $db_1(a, b)$ and $db_1(a, d)$ are retrieved from R_1, giving rise to answers $X_2 = b$ and $X_2 = d$.

Then, the subgoal $rp_1(a, X_2)$ is generated by the second rule of predicate rp_1; this generates two new subgoals $rp_1(b, X_2)$ and $rp_1(d, X_2)$, at the second level of recursion. The subgoal $rp_1(b, X_2)$ generates, in turn, the query $db_1(b, X_2)$, which is passed to the *database interface*; as a result, facts $db_1(b, c)$ and $db_1(b, e)$ are retrieved from R_1, giving rise to answers $X_2 = c$ and $X_2 = e$.

This process continues, driven by the *Prolog engine* by generating the queries $db_1(d, X_2)$, $db_1(c, X_2)$, and $db_1(e, X_2)$, which are passed to the *database interface*; but none of these queries produces an answer, and the inference mechanism is finally arrested.

In this way, the interaction with the database consists of five different queries; but only the tuples relevant to the initial *Prolog* goal have been considered by the *Prolog engine*. It should be noted, however, that the proof procedure used by the *Prolog engine* is not minimal; any of the above answers can be generated many times, by following different paths in the search space generated by the initial goal. This means, in particular, that the same queries might be executed on the database several times, giving rise to high execution costs. The techniques discussed in the next section have been developed to solve this problem.

4.3 Optimization of the Prolog/Database Interface

The performance of the *Prolog/Database interface* can be observed from both sides:

a) From the *Prolog side*, good performance is achieved when the *delays required by the interface are minimized*. In other words, a good interface should be capable of masking the fact that some data are in secondary storage, and its performance in retrieving these data and the data from the memory-resident *Prolog* database should be comparable.

b) From the *database side*, good performance is achieved when *execution time and costs of queries generated by the interface are minimized*; the database system sees the CPR system as any one of the various applications requesting database services.

These objectives are not conflicting; they can be restated from the viewpoint of the *Prolog*/database interface by requiring that *queries to the database be maximally restrictive, and be executed only when required*.

We have seen that queries are maximally restricted when the *Prolog interface* produces the largest base conjunctions; with recursive rules, tight coupling is more efficient than loose coupling, producing several instantiated queries rather than a large uninstantiated query. In this section, we are concerned with mechanisms for avoiding repeated execution of queries. To this end, some information can be *cached* into main-memory buffers or into the memory-resident *Prolog* database. Below, we analyze three alternatives, using caching of data, caching of queries, and caching of both data and queries.

4.3.1 Caching of Data

The first caching schema considered consists in asserting in the memory-resident *Prolog* database all tuples retrieved for base conjunctions. With this schema, tuples for base conjunctions are first searched for in main memory, to where they might have been retrieved by a previous query, and then searched for in mass memory, by activating the *database interface*. This behavior must be controlled by the *Prolog interface*. One implementation of the above schema can be described in *Prolog* by the following two meta-rules:

(1) $match(BC) :- BC.$

(2) $match(BC) :- call\_database\_interface(BC).$

The first rule attempts the matching of the current base conjunction by searching through memory-resident tuples; note that the base conjunction can be *instantiated*, as some export variables can be bound to constants. The first rule is satisfied while there are tuples in the *Prolog* memory which unify with the current base conjunction, and then it fails.

The second rule activates the database interface; we assume that the second rule is satisfied when the database query has a nonempty answer, resulting in the loading of new result tuples into the memory-resident *Prolog* database and in the unification of the current base conjunction with the first result tuple. Notice that only tuples not already present in the memory-resident database should be loaded, in order to prevent multiple matching of the same facts; this means that the database relations are considered as sets of tuples (consistently with their definition) and not as a multiset of tuples (as happens in most relational database implementations).

In order to activate meta-rules, it is sufficient to substitute the term

$$match(bc(..))$$

for each occurrence $bc(..)$ of a base conjunction; for instance, predicate rp_1 of Sect. 4.2.2 is rewritten as follows:

$$rp_1(X_1, X_2) :- match(db_1(X_1, X_2)).$$

$$rp_1(X_1, X_2) :- match(db_1(X_1, X_3)), rp_1(X_3, X_2).$$

4.3.2 Caching of Data and Queries

The above schema is improved by caching not only result tuples, but also the identity of queries that have already been executed on the database; the predicate $queried(bc(..))$ indicates that a given instance $bc(..)$ of a base conjunction has already been passed to the *database interface*, hence the query should not be repeated. With this additional information available, the sequence of actions becomes the following:

1) Tuple matching is performed in main memory, until failure.

2) On failure, a test is made to see whether the current instance of base conjunction has already been requested from the database interface. If this were so, a second interaction with the database would not be useful; hence, the matching attempt fails at this point.
3) Finally, the *database interface* is activated. After the completion of the loading of tuples into the *Prolog* memory, the meta-information $queried(bc(..))$ is also asserted in memory, to prevent future executions of the same query.

This schema is implemented by the following three meta-rules, under the control of the *Prolog interface*:

(1) $match(BC) :- BC.$

(2) $match(BC) :- queried(BC), !, fail.$

(3) $match(BC) :- call\_database\_interface(BC), assert(queried(BC)).$

The above rules can be applied, for instance, to the execution of the goal $rp_1(a, X_2)$, discussed in Sect. 4.2.2 (see the definition of rp_1 at the end of Sect. 4.3.1).

The first interaction with the *database engine* brings into memory, as the effect of meta-rule (3), tuples $rp_1(a, b)$ and $rp_1(a, d)$. On completion, the meta-information $queried(rp_1(a, X))$ is asserted. Then, the subgoal $rp_1(b, X)$ has the effect of bringing into memory the tuples $rp_1(b, c)$ and $rp_1(b, e)$, and the meta-information $queried(rp_1(b, X))$ is asserted. Then, interactions with the database produce no additional tuples, but they do assert additional meta-information: $queried(rp_1(c, X))$, $queried(rp_1(d, X))$, and $queried(rp_1(e, X))$. This prevents the database from being accessed several times.

4.3.3 Use of Subsumption

The above schema can be further improved by taking advantage of *subsumption between literals*. Let us first give a precise definition of subsumption between two literals. A literal L *subsumes* a literal M iff there is a substitution θ such that, by applying θ to L, we obtain M: $L\theta = M$. In other words, L is more general than M. Subsumption between generic predicates is formally defined in Chap. 6. For instance:

$pred(X, a, b)$ subsumes $pred(c, a, b)$.
$pred(X, a, Y)$ subsumes $pred(b, a, Z)$.
$pred(X, Y, c)$ subsumes $pred(Z, Y, c)$ and vice-versa.
$pred(X, a, b)$ does not subsume $pred(a, a, Z)$.

Subsumption can be used in the test of rule (2); in order to avoid visiting the database, it is sufficient to find among queried predicates one which subsumes the current one; if this is the case, then the database has already been searched for a more general condition than the current one, hence the tuples answering the current query have already been loaded into memory. The following rules implement this new schema:

(1) $match(BC) :- BC$.

(2') $match(BC) :- queried(Q), subsumes(Q, BC), !, fail$.

(3) $match(BC) :- call\_database\_interface(BC), assert(queried(BC))$.

Note also that testing subsumption between single literals, as required by the above approach, can be reduced to pure matching, which is done by the *Prolog engine*. We are helped by the following property: If L and M are literals, then L subsumes M iff L matches M', where M' is derived from M by substituting a new constant symbol for each variable. Let us explain the above property by means of an example, shown in Fig. 4.4.

QUERIED FACTS
queried(pred2(e,8)).
queried(pred1(d,X,e)).
queried(pred1(c,Y,Z)).
queried(pred2(j,91)).
queried(pred(e,a,Z)).

CURRENT BASE CONJUNCTION
pred1(c,T,e).

Fig. 4.4. Testing subsumption between single literals

In this example we look for a *queried* instance which subsumes $pred1(c, T, e)$. It is easy to see that the tuple $pred1(c, Y, Z)$ satisfies this condition. By applying the above property, we first replace all variables of $pred1(c, T, e)$ with new constants; assume *spec_const* is such a constant, which replaces variable T. We obtain $pred1(c, spec\_const, e)$. Then we match this ground literal against tuples of the *queried* functor; the matching is satisfied by the first tuple, with substitutions $Y/spec\_const$, Z/e. Therefore, the subsumption test is also satisfied.

4.3.4 Caching Queries

The major problem in caching data is the need to assert potentially large data collections in main memory, hence incurring the risk of exhausting the available memory. The problem can be solved by using virtual memory and paging mechanisms, but this solution is quite inefficient. Thus, a different schema for interacting with the database has been designed, consisting in caching into memory only queries; data are retrieved from the database and put into external files, where they are accessed, possibly many times, without interacting with the database.

When a query is made, the predicate $queried(bc(..), filename)$ indicates that a given query $bc(..)$ has been executed, and resulting tuples have been stored in the file whose name appears as second argument of the predicate. A special predicate $unify(Predicate, File)$ is then capable of retrieving tuples one at a time from the external file and unifying them with the current *Predicate*; on subsequent calls, $unify$ succeeds by retrieving subsequent tuples; it fails at the *end-of-file*. The following rules implement the above schema:

(1) $match(BC) :- queried(BC, FILE), unify(BC, FILE)$.

(2) $match(BC) :- call\_database\_interface(BC, FILE),$

$assert(queried(BC, FILE)), unify(BC, FILE)$.

A disadvantage of this schema is the large number of files generated in the frame of execution of a *Prolog* program, thus *polluting* mass memory. These files may be large; note that the same database tuple might contribute to replicated tuples within these files, since queries are not disjoint.

If main memory is exhausted, meta-information about the *queried* predicate can be discarded without compromising the correctness of the interaction: the consequence of discarding one such fact would be the repetition of a query on the database. We assume that recent queries are more likely to be repeated; thus, in the case of main-memory shortage, it is convenient to discard the less recent facts asserted for the *queried* predicate, together with the corresponding files. In this way, the amount of memory space required for storing facts for the *queried* predicate can be controlled.

4.3.5 Parallelism and Pre-fetching in Database Interfaces

We now discuss *parallelism* and *pre-fetching*, two different approaches to increasing the efficiency of the *Prolog/Database* interface. Parallelism is achieved by using multiple *database interfaces* acting at the same time; pre-fetching anticipates the data needs of the *Prolog engine*. They both have the effect of reducing the delays from the *Prolog* side, at the cost of generating extra demands on the database.

With LCPR systems, the parallel execution of queries is clearly possible, since each base conjunction can be executed independently of the other base conjunctions, before activating the *Prolog engine*. In order not to overload the database, it is possible to arrange a fixed number of *database interfaces*, each holding a queue of pending requests.

With TCPR systems, interactions are driven by the *Prolog engine*, which controls rule execution; though several versions of *Prolog* exist which use parallelism, the computation performed by classic *Prolog* systems is inherently sequential. Thus, parallelism is driven by the *Prolog interface* in the following situations:

a) When a rule has several independent base conjunctions (for instance, when the rule has disjunctions or cuts, see Sect. 4.2.1), then each of them can be passed to a different *database interface* for parallel execution.
b) When the first fact is retrieved from a *database interface*, thus allowing the unification of the current base conjunction, then the *Prolog interface* can resume the execution of the *Prolog engine*, while at the same time the *database*

interface proceeds with the loading of the remaining facts. This reduces the delay of the *Prolog* computation, and at the same time generates the potential for parallelism, since the *Prolog engine* can proceed with other goals, which in turn might require the execution of different queries, thus activating other *database interfaces*.

c) Facts can be loaded by the *database interface* organized in *blocks*, or *pages*. In this case, the *database interface* is activated at each *page fault*, namely, when all facts of one page have been matched. The *Prolog interface* can anticipate page faults and pre-fetch the subsequent page of data asynchronously, with classic mechanisms of operating systems.

4.4 Conclusions

CPR systems can be considered a first step in the development of logic-based databases and knowledge bases. We have shown that the development of **CPR** systems on top of an existing database system is quite feasible. We have also shown a number of methods for optimizing the performance of the database interface. In spite of existing methods for optimization, **CPR** systems do not really provide specialized mechanisms for dealing with logic programs, in particular with recursion. Thus, performance of **CPR** systems still need to be assessed; *integrated logic databases*, which will be described in the third part of this book, have the potential for achieving better performance.

4.5 Bibliographic Notes

The distinction between loose and tight coupling is discussed in many papers, such as those by Missikoff and Wiederhold [Miss 84], Bocca et al. [Bocc 86b], and Gardarin and Simon [Gard 87]. Extension tables are defined by Wagner-Dietrich and Warren [Wagn 86]; this technique enables a programmer to define a predicate as an extension table, thereby causing the assertion of all facts deduced for that predicate in the memory-resident *Prolog* database. The definition of base conjunctions is discussed in papers by Cuppens and Demolombe [Cupp 86] (in the framework of loose coupling) and by Denoel, Roelants, and Vauclair [Deno 86] (in the framework of tight coupling).

Bry [Bry 89] has introduced a new algebraic operator, called complement-join, whose result is constituted by all tuples excluded by the result of the corresponding join; this operator can be used to improve the translation of rules with negation, as complement-join may be used instead of the pair of operations semijoin and difference.

The method for caching queries and data and its improvement through subsumption was introduced by Ceri, Gottlob, and Wiederhold in [Ceri 86b] and [Ceri 89]. The method for caching queries is implemented in BERMUDA, described by Ioannidis, Chen, Friedman, and Tsangaris [Ioan 87b]; this paper also discusses the use of concurrent *database engines* and of pre-fetching.

4.6 Exercises

4.1 Determine the largest connected base conjunctions in the following *Prolog* programs:

(1) $p_1(X_1, X_2) :- dbp_1(X_1, X_3), dbp_2(X_2, X_4, \_), p_2(X_3, X_5), p_3(X_4).$

$p_2(X_1, X_2) :- dbp_3(X_1, b, X_2), p_3(X_2).$

$p_3(X_1) : -dbp_4(X_1, b, X_1).$

(2) $p_1(X_1, X_2) :- dbp_1(X_1, X_3), p_2(X_3, X_4), dbp_2(X_4, b, X_2).$

$p_2(X_1, X_2) :- dbp_3(X_1, a, X_2).$

$p_2(X_1, X_2) :- dbp_4(X_1, a, X_3), p_1(X_3, X_2).$

(3) $p_1(X_1, X_2) :- ((p_2(X_1), !, p_2(X_2), dbp_1(X_2, a));$

$(not(p_2(X_1)), dbp_1(X_1, X_2), p_2(X_3))).$

$p_2(X_1) :- dbp_2(X_1, \_, X_1).$

4.2 Translate the base conjunctions of Exercise 4.1 into SQL. Assume the following correspondence:

$$dbp_1 \iff R_1(A_1, A_2)$$

$$dbp_2 \iff R_2(B_1, B_2, B_3)$$

$$dbp_3 \iff R_3(C_1, C_2, C_3)$$

$$dbp_4 \iff R_4(D_1, D_2, D_3)$$

4.3 Consider the goal: $?- p_1(a, X).$ and the programs of Exercise 4.1. Propagate the query constant in nonrecursive cases.

4.4 Consider the following *Prolog* program:

$sg(X_1, X_2) :- (X_1 = X_2).$

$sg(X_1, X_2) :- par(X_1, X_3), sg(X_3, X_4), par(X_2, X_4).$

Let the parent relation be:

$$PAR = \{ <a,b>, <a,c>, <c,e>, <f,e>, <g,e>,$$
$$<d,b>, <e,h>, <m,n>, <o,p>, <p,q>,$$
$$<r,q>, <r,p>, <o,s>, <x,y>, <x,z>, <w,z> \}$$

To get acquainted with the problem, determine all answers to the goal:

$$? - sg(X_1, X_2), (X_1 \neq X_2).$$

(Hint: use a graphic representation or use the Prolog interpreter itself.) Then, consider the following goals:

1) $? - sg(a, X)$. starting from an empty main-memory database.
2) $? - sg(a, f)$. starting from an empty main-memory database.
3) $? - sg(a, X)$. followed by $? - sg(a, e)$.

followed by $? - sg(f, e)$. followed by $? - sg(f, Y)$. executing all goals within the framework of the same session (namely, without erasing the content of the *Prolog* memory between goal executions).

For each goal, do Exercises 1–4 below:

1) Simulate the step-by-step execution of database queries and the loading of facts into memory by using the *caching data* method. Indicate which queries are repeated.
2) Simulate the step-by-step execution of database queries and the loading of facts into memory by using the *caching queries and data* method. Indicate the final number of *queried* facts which are asserted into main memory.
3) Would subsumption between literals improve the execution of the program in the above case 2?
4) Simulate the step-by-step execution of database queries and the content of external files with the *caching queries* method. Indicate the final number of files required.

4.5 Write the meta-interpreter of a *Prolog interface* which caches data and maintains in core memory only the last 100 facts loaded from main memory; use the special predicates *assert* and *retract*. Hint: this is a classical problem of memory management in operating systems. Facts should be associated with the timestamp of the time at which they are asserted into memory.

4.6 Write the meta-interpreter of a *Prolog interface* which caches queries, assuming that the main memory for storing *queried* facts can be exhausted.

Chapter 5
Overview of Systems for Coupling Prolog to Relational Databases

This chapter presents an overview of some of the CPR systems and prototypes which have been developed for coupling *Prolog* to relational databases. We present:

a) PRO-SQL, a system for coupling *Prolog* to the system SQL/DS, developed at the IBM Research Center at Yorktown Heights.
b) EDUCE, a system for coupling *Prolog* to the database system Ingres, developed at the European Computer Industry Research Center in Munich.
c) The ESTEAM interface, developed in the framework of the Esprit Project ESTEAM, for coupling generic *Prolog* and database systems.
d) BERMUDA, a prototype developed at the University of Winsconsin, for coupling *Prolog* to the Britton-Lee Intelligent Database Machine IDM 500.
e) CGW, an architecture for coupling *Prolog* to a database system developed at Stanford University, and PRIMO, a prototype of an interface between ARITY-PROLOG and the database system ORACLE, developed at the University of Modena, Italy.
f) The QUINTUS interface between QUINTUS-PROLOG and the Unify Database System, a product developed by Quintus Computer Systems of Mountain View, California.

Though these systems interface different relational databases (SQL/DS, Ingres, Oracle, Unify, and IDM500), their architecture is in fact system-independent in most cases, and several of these projects state explicitly that the selection of the database system has little impact on their general architecture. The six approaches listed above inspired the previous chapter on CPR systems.

5.1 PRO-SQL

The relevant feature of PRO-SQL is the total absence of transparency. The basic assumption is that PRO-SQL programs should be written by persons who are familiar with both the *Prolog* and SQL languages. Thus, a special predicate SQL is used to include statements which are executed over the SQL/DS database system, as follows:

$$SQL(< SQL - Statement >).$$

The SQL statements supported include data definition, insertion, and retrieval statements, as well as statements for transaction control.

An example of the use of the SQL predicate for a query over the *EMPLOYEE* relation is:

$$SQL('SELECT\ NAME, SAL\ FROM\ EMPLOYEE\ WHERE\ SAL > 50').$$

The effect of this query is to assert, in the form of *Prolog* facts, all tuples which satisfy the query predicate in the memory-resident *Prolog* database. The predicate name used for asserted facts is the name of the *first* relation mentioned in the *FROM* clause. The execution of the SQL query takes place synchronously: the *Prolog* engine is suspended until completion of the loading of selected tuples into main memory.

A second example of use of the SQL predicate is the following:

$$SQL('SELECT\ CHILD, PARENT\ INTO\ X, Y$$
$$FROM\ EMPLOYEE\ WHERE\ SAL > 50').$$

In this case, the variables X and Y are bound, after execution, to values from the *first* tuple extracted from the database; once the first tuple has been retrieved, the control is returned to the *Prolog* engine, while the loading of subsequent tuples of the result is performed asynchronously.

The execution of a *recursive* program takes place by iterating calls to the SQL predicate. These calls are built by using the knowledge of the *adornment* of query predicates (i.e., of the positions that will be bound to query constants). The following program is an example. It computes the *ancestor* relation, and works correctly for evaluating ancestors of a given person, but not descendents:

$$ancestor(X, Y)\ :-\ SQL('SELECT\ CHILD, PARENT\ INTO\ X, Y$$
$$FROM\ FATHER\ WHERE\ CHILD\ =\ X').$$
$$ancestor(X, Y)\ :-\ father(X, Z), ancestor(Z, Y).$$

In this case, the *Prolog* program interacts with the SQL/DS database only when executing the first *ancestor* rule, while the second rule attempts unification with main-memory facts which have been loaded by the first rule. Thus, the second rule is normally executed under the control of the *Prolog* engine.

PRO-SQL supports transaction control. By default, each activation of the SQL predicate is considered a transaction, and is protected by recovery and concurrency control systems from the effects of failures and concurrent execution of other transactions. The programmer can also indicate that several SQL statements should be executed as a single transaction, by means of two SQL/DS statements: *COMMIT WORK* and *AUTOCOMMIT OFF*. The following is an example of a transaction:

$$transaction :- SQL('COMMIT\ WORK'),$$
$$SQL('AUTOCOMMIT\ OFF'),$$
$$SQL('SQL-STATEMENT-1'),$$
$$SQL('SQL-STATEMENT-2'),$$
$$SQL('SQL-STATEMENT-3'),$$
$$SQL('COMMIT\ WORK').$$

In the above example, the first statement ends any prior transaction, the second statement indicates that PRO-SQL controls the commit, the third, fourth, and fifth statement constitute a transaction, and the sixth statement commits the transaction.

Another use of transaction control statements is exemplified by the following query, which operates on the *ancestor* program defined above:

$$?- SQL('COMMIT\ WORK'), SQL('AUTOCOMMIT\ OFF'),$$
$$ancestor(john, X), SQL('COMMIT\ WORK').$$

In this case, all activations of the SQL predicate in the first rule of the *ancestor* program constitute a single transaction.

PRO-SQL runs under the VM/CMS operating system; the *Prolog engine* and SQL/DS run as two independent virtual machines in the same physical machine. Queries submitted to SQL/DS are treated by the "dynamically defined query facility" subsystem, which is typically used for casual queries.

5.2 EDUCE

EDUCE is one of the various projects for integrating logic programming and databases being developed at the European Computer-Industry Research Center (ECRC) of Munich. EDUCE is a CPR system, supporting both loose and tight coupling; another project, called DEDGIN, presents an integrated approach to logic and databases. EDUCE is used as the kernel of PROLOG-KB, a Knowledge Base System supporting a semantic data model. All these systems use the INGRES database system and QUEL query language as target database. This variety of activities makes ECRC one of the most active research environments in the field.

EDUCE supports various user languages and implementation strategies. From a linguistic viewpoint, it provides two different languages: a *loose language*, which is nonprocedural, and a *close language* which is similar in style to *Prolog*. EDUCE also supports two contrasting implementation strategies, a *set-oriented* one and a *tuple-at-a-time* one. These can be intermixed, giving rise to hybrid strategies, though the usual case is to execute queries in the *loose language* through the set-oriented strategy and queries (goals) in the *close language* through the tuple-oriented strategy.

The *close language* is fully transparent; database predicates can be intermixed with other predicates, and recursive goals are allowed. The *loose language* is more intricate, as it includes several built-in predicates with quite peculiar syntax and semantics. For instance, the predicate *query* can have as argument any QUEL query; the predicate succeeds on completion of the query, when result tuples have been loaded into memory. A more complex predicate *retrieve* allows queries with shared variables. With this formalism, transparency is thus lost.

The EDUCE architecture provides two types of coupling, called, respectively, *loose* and *tight*, though these terms have a different interpretation from that of Chap. 4. Loose coupling consists in providing two separate processes (called *Prolog interface* and *database interface* in Sect. 4.2), such that the former passes a query to the latter, and the latter returns result tuples back. Such a mechanism is suited to nonrecursive queries expressed in the *loose language*. Tight coupling consists in including some access methods to the disk database *within* the *Prolog engine*. Thus, specialized methods are used to execute complex (recursive) queries. This approach provides direct translation of the *close language* queries (goals). The EDUCE optimizer is capable of selecting the most appropriate strategy, sometimes by overriding the correspondence mentioned above.

5.3 ESTEAM

The ESTEAM interface has been designed in the framework of the Esprit Project P316 by the Philips Research Laboratory in Brussels and by Onera-Cert in Toulouse. The approach used by the ESTEAM interface is that of recognizing large *base conjunctions*, which should be considered as units of interaction with the database; the algorithms for determining base conjunctions have largely inspired Sect. 4.2, and need not be presented again in this section.

The target database system in ESTEAM is INGRES, and the target query language is QUEL. Consider the following base conjunction:

$bc(X_1, X_4) : -dbp_1(X_1, X_2, X_3), dbp_2(X_3, a, X_4), X_2 > 5.$

Let $R_1(A_1, A_2, A_3)$ and $R_2(B_1, B_2, B_3)$ be the database relations correponding to database predicates dbp_1 and dbp_2 respectively. The translation in QUEL is:

range of w_1 *is* R_1
range of w_2 *is* R_2
retrieve into $bc(w_1.A_1, w_2.B_3)$
where $w_1.A_2 > 5$
and $w_1.A_3 = w_2.B_1$
and $w_2.B_2 = a$

The ESTEAM interface includes an algorithm for building base conjunctions progressively, starting from the *Prolog* goals. With loose coupling, all interactions with the database take place prior to the activation of the *Prolog engine*. With tight coupling, the building of base conjunction queries is interleaved with the computation of the *Prolog* program. The ESTEAM interface supports disjunction, negation, recursion, and the cut special predicate as discussed in Sect. 4.3; further, it supports aggregate queries.

5.4 BERMUDA

The BERMUDA prototype has been developed at the University of Wisconsin; BERMUDA is a sophisticated acronym which stands for: **B**rain **E**mployed for **R**ules, **M**uscles **U**sed for **D**ata **A**ccess. The focus of the project is on architectural issues, in particular those rising from employing *multiple database interfaces* at the same time.

Parallelism is achieved by means of two basic mechanisms:

a) The *Prolog engine* is activated as soon as the first result tuple is retrieved from the database; the *database* and *Prolog interfaces* continue to load data asynchronously. This enables the *Prolog engine* to request other queries before the current one has been completed.

b) Each *database interface* does not retrieve the entire query result from the database, but rather a few blocks (pages) of data, which are stored in main memory; before the *Prolog engine* reaches the end of the data page, the *database interface* is activated again, to get the subsequent page or pages of data. This *pre-fetching* policy is designed and tuned so as to reduce the amount of data loaded in memory and at the same time to reduce waiting times of the *Prolog engine*; pre-fetching also has the potential for increasing parallelism, since many *database interfaces* could be activated at the same time.

BERMUDA recognizes *base conjunctions* in a few of the cases described in Sect. 4.3, and passes each base conjunction as a single query to the *database interface*. The execution of the extra-logical predicate *cut* determines a control message that suspends loading actions pending for predicates to the left of the cut, since the corresponding variables get irrevocably instantiated.

BERMUDA caches queries as described in Sect. 4.3.4: tuples of the result relation are stored in an external file, and the system keeps track of the queries that have been already made, in order not to repeat their execution, but rather to retrieve result tuples directly from the files. No use is made of subsumption between predicates, hence only identical queries are recognized; but subsumption could in fact be used by BERMUDA, as indicated in Sect. 4.3.4. As was observed in Chap. 4, caching queries has the obvious disadvantage of polluting mass memory with a large number of result files.

Architecturally, there is one single *Prolog interface*, called the BERMUDA AGENT, which receives query requests from the *Prolog engine* and passes them to *database interfaces*, called LOADERS; there are a fixed number of such LOAD-

ERS, created at system invocation time and remaining alive until the session is ended. When pending queries exceed the number of available LOADERS, queries are queued and then served with a First-In-First-Out (FIFO) policy. The BERMUDA agent knows about queries which have been issued and files which have been used for storing the query results since the system's activation; hence, it has enough information to enable queries to be cached. Several *Prolog engines* might be active at the same time, each executing its own *Prolog* program and communicating with the BERMUDA AGENT.

5.5 CGW and PRIMO

The CGW approach was developed at Stanford University in the framework of the *KBMS* project. The rationale of the CGW approach is to reduce accesses to the database by cacheing queries and data, as discussed in Sect. 4.3.2. The proposed approach uses a low-level interface to interact with the database, and control access to pages of data; *tracers* are specifically designed *Prolog* data structures which indicate how many pages of data have been currently accessed for a given access method. A *meta-interpreter* plays the role of *Prolog interface*, controlling the matching of individual database predicates. A simplified version of the meta-rules used by the interpreter is as follows:

(1) $match(DBP) :- DBP.$

(2) $match(DBP) :- queried(DBQ), subsumes(DBQ, DBP), !, fail.$

(3) $match(DBP) :- tracer(DBQ, PAGE), subsumes(DBP, DBQ),$
$call\_database\_interface(DBQ).$

(4) $match(DBP) :- call\_database\_interface(DBP).$

The rationale behind rules (1) and (2) was explained in Sect. 4.3.2, and has not been changed here; matching is first attempted on memory-resident facts, then a test is made to see whether a more general query has already been performed and the *queried* predicate has been asserted for it. Rules (3) and (4) use tracers. First, rule (3) attempts to use an access path which has already been partially followed for a predicate DBQ subsumed by (less general than) the current query. If rule (3) succeeds, then the *database interface* is called for the predicate DBQ; if rule (3) fails, then rule (4) calls the *database interface* for the current DBP. The rationale behind rule (3) is to retrieve facts that are potentially useful to another database predicate, and to eliminate current tracers (thus reducing the number of queries which are partially executed).

The *database interface* retrieves one data page from the database and asserts it into the memory-resident *Prolog* database. If the query is completed, it also asserts the predicate $queried(dbp)$; otherwise, it asserts the predicate $tracer(dbp, page)$, whose second argument indicates the identifier of the page currently loaded from disk.

The CGW approach also shows that subsumption can be reduced to pure matching in the two cases shown above, and describes a policy for tracer management which is quite efficient, exploiting, as it does, formal properties of tracers. Finally, it considers how memory management should be performed if main memory is exhausted by database facts and meta-information ("*queried*" and "*tracer*" predicates).

The main limitation of the approach lies in the separate matching of each database predicate, instead of *base conjunctions*. As a consequence of this choice, joins are executed in main memory rather than being executed by the database system; in practice, the database treats each database predicate separately from the others. This approach causes potential savings when the same database predicate is joined to various other database predicates in the same program; but it also causes potential inefficiency, since join selectivity is lost, and therefore large amounts of data need to be stored.

The CGW approach has been followed in the design of PRIMO, a **P**rolog - **R**elational **I**nterface developed at the University of **MO**dena. The main design goals of PRIMO include portability, modularity, and transparency. As a result of portability, the PRIMO interface can be established between any two *Prolog* and relational systems, provided that the former supports a call to the operating system and the latter supports SQL. In the current implementation, PRIMO uses *Arity-PROLOG* and *Oracle*. The PRIMO system is fully transparent, as it supports pure *Prolog*. This is accomplished by providing a program analyzer which transforms the initial *Prolog* program into a modified *Prolog* program that interacts with PRIMO, without requiring any user support. In particular, the analyzer accesses the database catalog to recognize *database predicates*, i.e., predicates whose ground rules (facts) are stored in the database.

The PRIMO system uses both loose and tight coupling. Loose coupling is used whenever the database predicates in the user program are highly selective, and therefore few facts need to be retrieved for a particular predicate; in this case, a small amount of core memory is sufficient for storing the facts required by the user program. Tight coupling is used for all remaining predicates.

The general architecture of the CGW approach has been modified and/or improved in the following directions:

a) The *database interface* operates at high level, using the SQL language; thus, direct control over page access mechanisms is lost. PRIMO does not use tracers, and caches queries and data using the meta-interpreter discussed in Sect. 4.3.2, with no modifications. This ensures portability of PRIMO.

b) PRIMO considers base conjunctions of database predicates as a single *join* query. Thus, the *database interface* uses join selectivity to reduce the number of tuples to be loaded into main memory. On the other hand, facts are asserted into memory using the same format as the original database predicates, and the *Prolog engine* repeats the execution of the join in main memory. This is done in order to achieve maximum transparency: the *Prolog engine* executes the original user program, rather than a program rewritten with explicit rules for base conjunctions.

c) PRIMO supports several variations of the *Prolog interface*, which are implemented using standard *Prolog* (in particular, standard *assert* predicates for loading facts into memory). The various alternatives are compared in terms of efficiency of access and amount of memory used.

5.6 QUINTUS-PROLOG

QUINTUS-PROLOG, developed by Quintus Computer Systems Inc., provides an interface to the Unify database system, running on Sun workstations and on Vax-Unix machines. The interface operates at two levels: the *relation level*, providing a tight, one-tuple-at-a-time retrieval of tuples from underlying relations, and the *view level*, in which an entire *Prolog* rule is translated into a single query to the database, including joins as well as aggregate operations.

The connection with the database is provided by a special predicate $db\_connect$; prior to calling that predicate, several other facts in the *Prolog* database have to be prepared, declaring relations and views. This information is used by $db\_connect$ for setting up the link; after connection, relation and view predicates can be used just as any other predicate, with full transparency.

In the following, we exemplify the connection to the database emp_db by means of the two well-known relations *emp* and *dept*. The predicate db_name is used to indicate the existence of the database and the physical file containing the database within UNIFY:

$db\_name(emp\_db, unify, 'unify/emp\_db/file.db')$.

The schema of each database relation is defined through a *db* fact, as follows:

$$db(emp, emp\_db, emp('Empnum' : integer$$
$$'Name' : string$$
$$'Salary' : integer$$
$$'Deptnum' : integer)).$$
$$db(dept, emp\_db, dept('Deptnum' : integer$$
$$'Production' : string$$
$$'Location' : string)).$$

In the above definitions, attribute names should be exactly as in the database catalogs; some of the attributes of relations could be omitted. Several *db* facts can be defined for the same relation, hence providing various projections.

These definitions become effective after executing the connection, through the predicate $db\_connect$:

$$db\_connect(emp\_db).$$

The *view interface* can be used for the execution of rules whose body includes only database predicates and aggregate predicates (defined below). In this case, the rule is translated into a complex database query, which is then executed taking advantage of the query processing ability of the database system.

Each view definition is complemented by a *mode specification*, consisting of a term whose main functor is composed using the name of the predicate at the head of the view rule. The fields of this term give information on the *dynamic adornment* of the view predicate. They are either '+' or '-'; a '+' indicates that the corresponding position will be bound to a constant value when the view predicate is executed. This information enables an optimized view compilation to be made; if the view predicate is called with a dynamic adornment different from the one described by the mode specification, then an error is indicated and the view predicate fails.

In the following, we show the definition of a simple join view between the two predicates *emp* and *dept*. The view gives the location and production of the department of a given employee. Note that we expect to execute *view*1 with the first argument bound to a constant.

$$:- db\_modeview1(+,-,-).$$

$$view1(Empnum, Loc, Prod) :- emp(Empnum, \_, \_, Dept),$$

$$dept(Dept, Prod, Loc).$$

The view interface also supports aggregate predicates. A ternary predicate *aggregate* is used; the first argument indicates the aggregate function(s) to be evaluated, selected from the usual aggregate functions *min, max, average, sum,* and *count*; the second argument indicates how aggregation should take place; the third argument indicates the name(s) of variables in the head of the view rule that should be set equal to the result of aggregate function evaluation. For instance, consider a view giving, for all employees having a given manager, the average salary aggregated by departments. This view is defined as:

$$:- db\_modeview2(+,-,-).$$

$$view2(Mgr, Deptnum, Sumsal) :- dept(Deptnum, Mgr, \_),$$

$$aggregate(sum(Sal),$$

$$(E, N), emp(E, N, Sal, Deptnum),$$

$$Sumsal).$$

View declarations should be added to relation declarations, so that they may be processed when the *db_connect* predicate is executed; a special *compile_view* predicate allows more view definitions to be added.

Notice that the interface does not cache either data or queries; at each call to a database relation or view, the communication takes place through shared variables, which get bound. The *database interface* controls the result of queries and provides unification to the next tuple on backtracking. However, the system performs query compilation and binding at *connection time*, thus maintaining efficient access methods for relations and views throughout the session.

The interface supports insertion and deletion of facts from the disk-resident database, using the predicates *db_ assert(DBP)*, *db_ retract_first(DBP)*, and *db_ _retract_ all(DBP)*. The effect of the second predicate is to delete exactly one matching tuple (randomly selected), while the effect of the third predicate is to delete all matching tuples.

5.7 Bibliographic Notes

PRO-SQL is described in a paper by Chang and Walker [Chan 84]; EDUCE is described in papers by Bocca [Bocc 86a] and Bocca et al. [Bocc 86b]; the Esprit project ESTEAM is described in papers by Cuppens and Demolombe [Cupp 86] and Denoel, Roelants, and Vauclar [Deno 86]; BERMUDA is described in a paper by Ioannidis, Chen, Friedman, and Tsangaris [Ioan 87b]; the CGW approach is described in papers by Ceri, Gottlob, and Wiederhold [Ceri 86b], [Ceri 89]; the PRIMO system is described in a paper by Ceri, Gozzi, and Lugli [Ceri 88c]; the QUINTUS-PROLOG interface is described in a reference manual [Quin 87].

Part II
Foundations of Datalog

Part II of this book is dedicated to the formal study of some fundamental aspects of the *Datalog* language. *Datalog* is a recently developed logical programming language which allows complex queries to a database to be expressed.

We first explain the syntax of *Datalog* and introduce several useful concepts from the fields of Logic Programming and Automated Theorem Proving. Then, we carefully define the model-theoretic semantics of *Datalog*. We present a sound and complete proof theory of *Datalog* which directly leads to a bottom-up evaluation procedure for *Datalog* programs. We also explain the fixpoint theory of *Datalog* and show how *Datalog* programs can be evaluated backwards in a top-down fashion.

In Part II we are mainly interested in the logical basis of *Datalog* and in some fundamental principles and algorithms of fact inference. We disregard all aspects which are strictly related to data retrieval from an external database and to efficient database access. All methods and algorithms presented in this part are formulated independently of any database structure. We assume that all the data which are needed during the processing of a *Datalog* program are directly available.

Later, in Part III of this book, we show how these principles can be combined with relational database technology in order to efficiently evaluate *Datalog* queries against a secondary storage database. In particular, it will be shown how we can profit from set-oriented techniques (relational algebra) and how programs and algorithms can be optimized in order to minimize database access.

Chapter 6
Syntax and Semantics of Datalog

In this chapter we give an exact syntactic and semantic definition of the *Datalog* query language.

First, in Sect. 6.1, we formally define several basic concepts and introduce *Datalog* programs as sets of *Horn clauses*. We also explain important notions, such as *Herbrand base, extensional database, substitution*, and *unification*.

In Sect. 6.2, we then define the purely logical semantics of *Datalog* programs by using concepts of *model theory*. Model theory allows us to define the concepts of *logical truth* and *logical consequence* in a very elegant and intuitively comprehensible way. Using this theory, we are able to give a precise characterization of *what* a *Datalog* program computes. However, we do not explain *how* the output of a *Datalog* program can be computed, deferring this subject to Chap. 7.

Note that *Datalog* is with many respects a simplified version of *Logic Programming*. During our discussion, we point out some of the major differences between these two formalisms. Several of our definitions and results are simplifications of corresponding issues in the context of Logic Programming. References to literature on Logic Programming are given at the end of this chapter.

6.1 Basic Definitions and Assumptions

6.1.1 Alphabets, Terms, and Clauses

We first define three infinite alphabets, *Var, Const*, and *Pred* for denoting variables, constants, and predicates:

- *Var* consists of all finite alphanumeric character strings beginning with an upper-case letter. Some elements of *Var* are, for instance, X, $X1$, *Dog*, *CATCH33*.
- *Const* consists of all finite alphanumeric character strings which are either numeric or consist of one lower-case letter followed by zero or more lower-case letters or digits. Some elements of *Const* are: 10 , *a* , *xylophone* , *zanzibar* , 115 , *a1b2c3*.
- *Pred* consists of all finite alphanumeric character strings beginning with a nonnumeric lower-case character. *Pred* and *Const* have a nonempty intersection; however, in clauses or in a *Datalog* program, it is always clear from the context whether a symbol stands for a predicate or a constant. Each predicate symbol

represents a predicate, i.e., a mapping from a domain D^n to the set of truth values $\{true, false\}$, where n is an integer greater than zero.

In Logic Programming another important category of symbols exists: the *function symbols*. The absence of function symbols is one of the most important characteristics of *Datalog*.

We assume that a mapping *arity* from *Pred* to the set $\mathcal{N}^+$ of all strictly positive integers exists. If $arity(p) = n$, then p is a symbol for denoting n-ary predicates, i.e., mappings from a domain D^n to $\{true, false\}$. The *arity* function is session- and system-dependent, since predicates are defined by users or database administrators. Note that in Logic Programming 0-ary predicates are also admitted. This is not the case in *Datalog*.

A *term* is either a constant or a variable. The set *Term* of all terms is thus $Term = Const \cup Var$. A term t is *ground* iff no variable occurs in t. In the context of *Datalog*, this means that a term is ground iff it is a constant. The set *Const* of all ground terms is also called the *Herbrand Universe*. Note that in the context of general Logic Programming a term can be a complex structure built from function symbols, variables and constants; in that context, ground terms are not necessarily constants and the Herbrand Universe does not coincide with the set of all constants.

An *atom* $p(t_1, \ldots, t_n)$ consists of an n-ary predicate symbol p and a list of arguments $(t_1, \ldots, t_n)$, such that each t_i is a term. Examples of atoms are: *fatherof(john,henry)* , *ancestorof(Everybody,eve)* , *tired(susan)* , $q(a, X, 12, Y)$. A *ground atom* is an atom which contains only constants as arguments.

A *literal* is an atom $p(t_1, \ldots, t_n)$ or a negated atom $\neg p(t_1, \ldots, t_n)$. Correspondingly, a *ground literal* is a ground atom or a negated ground atom. Examples of literals are: $fatherof(john, henry)$, $\neg fatherof(joe, mary)$, $eats(joe, Food)$, $\neg eats(mary, X)$; the first two literals are ground literals, while the latter two are not ground, since they contain variables. Literals which are atoms are also called *positive literals*, while literals consisting of negated atoms are called *negative literals*.

A *clause* is a finite list of literals. Clauses are often defined as *sets* and represented in set notation, for instance $\{\neg p(X, a), p(Y, b)\}$. We also use set notation for representing clauses. Nevertheless, we do not want to exclude the possibility of duplicate literals in a clause; furthermore, when we are manipulating *Datalog* clauses through algorithms, we are interested in the order of literals appearing in a clause (while the ordering of literals within a clause is irrelevant to its semantics). Hence we always conceive of a clause as a list, or, equivalently, as an ordered set with the possibility of duplicates and not merely as a set. Horn clauses (which are defined below) can also be represented in Prolog notation.

Up to now, our definitions have been purely syntactical. It is important, at this point, that the reader get some feeling about the meaning of the notion of clause. For this purpose, we temporarily use the formalism of first-order predicate logic, with which we assume the reader to be familiar. According to this formalism, each clause C corresponds to a formula $\underline{C}$ of first-order logic. $\underline{C}$ is of the form $Q(M)$, where Q is a prefix of quantifiers and M is a quantifier-free part (also called the "matrix"). Q contains a universal quantifier for each variable occurring in C and M consists of the disjunction of all literals of C. Consider, for instance, the clause $C = \{\neg p(X,a), p(Y,b)\}$. This clause corresponds to the first-order formula $\underline{C}: \forall X\ \forall Y\ (\neg p(X,a) \lor p(Y,b))$.

Let us proceed now with the definition of the syntax of *Datalog*. A *ground clause* is a clause which does not contain any variables. In other words, a ground clause is a finite set of ground literals. An example of a ground clause is $D = \{p(a,b,c), q(r1), \neg p(i,j,k)\}$. The corresponding first-order formula is $\underline{D}: p(a,b,c) \lor q(r1) \lor \neg p(i,j,k)$. Clauses consisting of one single literal are called *unit clauses*. Clauses containing only negative literals are called *negative clauses*, while clauses containing only positive literals are called *positive clauses*. The clause $\{p(a,X,c)\}$, for instance, is a positive unit clause, while $\{\neg q(a,b)\}$ is a negative ground unit clause.

A *Horn clause* is a clause containing at most one positive literal. There exist three types of Horn clauses:

a) *Facts.* Facts are positive unit clauses. They express unconditional knowledge, for example: $\{father(john, harry)\}$, $\{eats(joe, X)\}$. We can also represent facts in Prolog notation, i.e., without set brackets, but with a period at the end, as follows:

$$father(john, harry).$$

$$eats(joe, X).$$

b) *Rules.* A rule is a clause with exactly one positive literal and with at least one negative literal. A rule represents conditional knowledge of the type "if p is true then q is true". An example of a rule is:

$$R: \quad \{\neg grandparent(X), grandfather(X), \neg male(X)\}.$$

Since this form of representing rules is not very intuitive, we often write rules in Prolog notation. The above rule is written as:

$$grandfather(X)\ :-\ grandparent(X), male(X).$$

This is read as:

"$grandfather(X)$ if $grandparent(X)$ and $male(X)$".

Indeed, it is not hard to see that rule R corresponds to such an implication. Consider the logical formula $\underline{R}$ corresponding to R:

$$\underline{R}: \quad \forall X\ (\neg grandparent(X) \lor grandfather(X) \lor \neg male(X)).$$

By commuting the literals in $\underline{R}$ and by factorizing the negation according to De Morgan's law, we get the following first-order formula, which is equivalent to $\underline{R}$:

$$\underline{R'}: \quad \forall X\ (\neg(grandparent(X) \land male(X)) \lor grandfather(X)).$$

Now, since $\neg A \lor B$ is the same as $A \Rightarrow B$, we can rewrite this formula equivalently as an implication:

$$R'' : \quad \forall X \, ((grandparent(X) \land male(X)) \Rightarrow grandfather(X)).$$

The Prolog notation is just another way of expressing this implication. After this example, it should be clear why every rule of the form

$$\{p(\ldots), \neg p_1(\ldots), \ldots, \neg p_k(\ldots)\}$$

can be rewritten in Prolog notation as:

$$p(\ldots) :- p_1(\ldots), \ldots, p_k(\ldots).$$

The literal on the left-hand side of " $:-$ " is called the *head* of the rule, while the set of all literals on the right-hand side of " $:-$ " is called the *body* of the rule.

c) *Goal clauses.* A goal clause (or *goal*) is a negative clause, i.e., a clause consisting only of negative literals. Unit goals are goal clauses with only one literal. An example of a unit goal is: $\{\neg fatherof(john, X)\}$. This clause can be represented in Prolog notation as follows:

$$?- fatherof(john, X).$$

The name "goal clause" stems from resolution-oriented theorem proving, where, in order to prove that a certain formula holds, one assumes the negation of this formula and shows that this assumption leads to some contradiction. For instance, if we want to prove the formula $\exists X \, (fatherof(john, X))$, we assume the negation of this formula, i.e, $\forall X \, (\neg fatherof(john, X))$. This latter formula is represented by the goal clause "$?- fatherof(john, X)$.". For the moment, this should be sufficient to explain the name "goal clause". We will come back to the resolution method in Sect. 7.3.

It is very important to understand that clauses, and in particular Horn clauses, represent *closed formulas*, i.e., formulas which do not contain any free variables. Indeed, as we have seen, all variables appearing in a clause are bound by a universal quantifier in the corresponding first-order formula. It follows that the name of a variable which appears in a clause C makes sense only within that clause, but its meaning does not extend to whatever may exist outside the clause. Consider, for instance, two clauses C and D:

$$C : \quad grandfatherof(X, Y) :- fatherof(X, Z), \, fatherof(Z, Y).$$

$$D : \quad grandfatherof(X, Y) :- motherof(X, Z), \, fatherof(Z, Y).$$

Here the variables X, Y, Z appearing in clause C have absolutely nothing to do with the variables X, Y, Z of clause D. In order to emphasize this, we can rewrite clause D as an equivalent clause D', such that C and D' have no variables in common:

D' : $grandfatherof(U,V)$:− $motherof(U,W)$, $fatherof(W,V)$.

It is clear that $\underline{D}$ is logically equivalent to $\underline{D'}$, and since D and D' differ only in the names of their variables, D and D' are called *variants*. In general, we say that two clauses B and B' are *equivalent* iff their corresponding formulas $\underline{B}$ and $\underline{B'}$ are logically equivalent. We say that B' is a *variant* of B iff B' differs only by the name of its variables from B and different variables of B correspond to different variables of B' and viceversa. If B' is a variant of B then B is a variant of B'; thus we may simply say that B and B' are variants. If B and B' are variants, then B and B' are equivalent. Note that the converse does not hold: Let C be the clause "$p(a) : -p(X), p(Y)$." and let D be the clause "$p(a) : -p(Z)$"; then C and D are equivalent, but neither is a variant of the other.

If $S = \{C_1, \ldots C_n\}$ is a set of clauses, then S represents the following first-order formula: $\underline{S} : \underline{C_1} \wedge \ldots \wedge \underline{C_n}$.

6.1.2 Extensional Databases and Datalog Programs

We assume that the alphabet *Pred* is partitioned into two sets $EPred$ and $IPred$, such that $EPred \cup IPred = Pred$ and $EPred \cap IPred = \emptyset$. The elements of $EPred$ denote extensionally defined predicates, i.e., predicates that correspond to relations which are effectively stored in the database, while the elements of $IPred$ denote intensionally defined predicates, i.e., predicates corresponding to relations which are defined by *Datalog* programs. The actual subdivision of *Pred* into $EPred$ and $IPred$ may vary from system to system.

If S is a set of positive unit clauses, then $E(S)$ denotes the *extensional part of* S, i.e., the set of all unit clauses in S whose predicates are elements of $EPred$. On the other hand, $I(S) = S - E(S)$ denotes the *intensional part of* S, i.e., the set of all unit clauses in S whose predicate belongs to $IPred$.

The Herbrand base HB is the set of all positive ground unit clauses that can be formed using predicate symbols in *Pred* and constants in *Const*. We denote the extensional part $E(HB)$ of the Herbrand base by EHB and the intensional part $I(HB)$ by IHB.

An *extensional database* (EDB) is a finite subset of EHB. In other words, an extensional database is a finite set of positive ground facts.

Example 6.1. The following set E_1 is an EDB:

{$par(john,mary)$, $par(john,reuben)$, $par(terry,john)$, $par(terry,lisa)$,
$par(lisa,david)$, $par(lisa,joe)$, $par(joe,jeff)$, $par(joe,susan)$,
$lives(john,wash)$, $lives(mary,ny)$, $lives(terry,paris)$, $lives(reuben,ny)$,
$lives(lisa,ny)$, $lives(david,athens)$, $lives(joe,chicago)$, $lives(jeff,ny)$}

Here we suppose that the two predicate symbols "*par*" and "*lives* " are both elements of $EPred$. □

An extensional database represents the data that are effectively stored in the relations of a relational database. An EDB is also called an *instance* of a relational database. If E is an EDB, then all clauses of E having the same predicate symbol are stored in the same relation. The correspondence between relations and predicates has been considered in Part I and will be considered in Part III, but for the purposes of Chaps. 6 and 7, the notion of EDB as a set of ground facts is sufficient.

A *Datalog program* P is a finite set of Horn clauses such that for all $C \in P$, either $C \in EDB$ or C is a rule which satisfies the following conditions:

a) The predicate occurring in the head of C belongs to $IPred$.
b) All variables which occur in the head of C also occur in the body of C.

Condition a) ensures that the extensional database is not altered through the rules of a *Datalog* program. Condition b) is called a *safety condition*. This condition, together with our requirement that each fact belonging to a *Datalog* program be a ground fact, ensures, as we will see later, that only a finite number of facts can be deduced from a *Datalog* program. Each clause which is either a ground fact or a rule satisfying conditions a) and b) is called a *Datalog clause*.

When we write a *Datalog* program, we omit the set brackets and usually place each single clause of the program on a single line. Since each *Datalog* program P is a particular set of clauses, the logical formula $\underline{P}$ corresponding to P consists of the conjunction of all formulas $\underline{C}$, such that C is a clause of P.

Example 6.2. An example of a *Datalog* program is:

P_1: anc(X,Y) :- par(X,Y).
   anc(X,Y) :- par(X,Z) , anc(Z,Y).
   anc(X,adam) :- person(X).
   person(X) :- lives(X,Y).

Here, *person* and *anc* are predicates belonging to $IPred$, while *par* and *lives* belong to $EPred$. □

A *Datalog goal* for a given program consists of a unit goal. Note that we do not require that the predicate of this goal occur within the *Datalog* program. However, in the more interesting cases, the goal predicate is a predicate occurring in the head of one of the clauses. A goal is an additional specification that we may add to a *Datalog* program. For instance, the goal "?-anc(terry,X)" specifies that we are only interested in the ancestors of *terry*. The result of program P_1 activated with this goal and applied to the EDB E_1 from Example 6.1 is: { anc(terry,john), anc(terry,lisa), anc(terry,mary), anc(terry,reuben), anc(terry,david), anc(terry,joe), anc(terry,jeff), anc(terry,susan), anc(terry,adam)}.

Often we consider *Datalog* programs without particular goals. In these cases we are interested in *all* the ground clauses from IHB that the program computes on the base of a given EDB.

A *Datalog* program, with or without goal clause, takes as input an EDB and computes as output a subset of IHB, i.e, a set of positive ground unit clauses, whose predicate symbols are in $IPred$. A precise definition of the semantics of

6.1.3 Substitutions, Subsumption, and Unification

A *substitution* θ is a finite set of the form $\{X_1/t_1, \ldots, X_n/t_n\}$, where each X_i is a distinct variable and each t_i is a term, such that $X_i \neq t_i$.

Each element X_i/t_i of a substitution is called a *binding*. The set of variables $\{X_1, \ldots X_n\}$ is called the *domain* of θ, while the set of terms $\{t_1, \ldots t_n\}$ is called the *co-domain* of θ.

Recall that in our model terms are either constants or variables (note that this is not the case in general logic programming, where terms can be complicated nested structures built from variables, constants, and function symbols). If all terms $t_1, \ldots, t_n$ are constants, then θ is called a *ground substitution*.

If θ is a substitution and t is a term, then $t\theta$ denotes the term which is defined as follows:

$$t\theta = \begin{cases} t_i, & \text{if } t/t_i \in \theta; \\ t, & \text{otherwise.} \end{cases}$$

If L is a literal then $L\theta$ denotes the literal which is obtained from L by simultaneously replacing each variable X_i that occurs in L by the corresponding term t_i, iff X_i/t_i is an element of θ. For instance, let L be the literal $\neg p(a, X, Y, b)$ and let $\theta = \{X/c, Y/X\}$; then $L\theta = \neg p(a, c, X, b)$.

If $C = \{L_1, \ldots, L_n\}$ is a clause, then $C\theta = \{L_1\theta, \ldots, L_n\theta\}$. If C and D are clauses and if a substitution θ with $C\theta = D$ exists, then D is called an *instance* of C. The unit clause "$P(X, a)$.", for example, is an instance of the unit clause "$P(X, Y)$.".

Let $\theta = \{X_1/t_1, \ldots, X_n/t_n\}$ and $\sigma = \{Y_1/u_1, \ldots, Y_m/u_m\}$ be two substitutions. The *composition* $\theta\sigma$ of θ and σ is obtained from the set

$$\{X_1/t_1\sigma, \ldots, X_n/t_n\sigma, \ Y_1/u_1, \ldots, Y_m/u_m\}$$

by eliminating each binding of the form ξ/ξ and by eliminating each binding Y_i/u_i, where $Y_i = X_j$, for some j.

If C is a clause, then applying $\theta\sigma$ to C has the same effect as first applying θ to C, yielding $C\theta$, and then applying σ to $C\theta$. We thus have: $C(\theta\sigma) = (C\theta)\sigma$. It is also easy to see that the composition of substitutions is associative, i.e., if θ, γ, and σ are substitutions, then $(\theta\gamma)\sigma = \theta(\gamma\sigma)$. However, note that in general it is not true that $\theta\gamma = \gamma\theta$, i.e., the composition of substitutions is not commutative.

Example 6.3. Consider the clause

$$C : anc(X, Y) :- par(X, Z), anc(Z, Y).$$

and the substitutions

$$\theta = \{X/Y, Y/U, Z/V\} \quad \text{and} \quad \sigma = \{Y/john, U/T, V/T, X/reuben\}.$$

We have:

$\theta\sigma = \{X/john, Y/T, Z/T, U/T, V/T\}$
$C\theta = anc(Y, U) :- par(Y, V), anc(V, U).$
$(C\theta)\sigma = C(\theta\sigma) = anc(john, T) :- par(john, T), anc(T, T).$nz $(C\sigma)\theta = C(\sigma\theta) = anc(reuben, john) :- par(reuben, V), anc(V, john).$ □

Note that in the above example $C\theta$ is a variant of C. Using the concept of substitution, we can give an elegant characterization of variants: C is a variant of D iff substitutions θ and σ exist, such that $C\theta = D$ and $D\sigma = C$.

A clause C *subsumes* a clause D, denoted by $C \triangleright D$ iff a substitution θ exists, such that $C\theta \subseteq D$.

Example 6.4. Consider the following clauses:

$$C : p(X, Z) :- p(X, Y), p(Y, Z).$$

$$D : p(a, a) :- p(a, a), p(b, b).$$

and the substitution $\theta = \{X/a, Y/a, Z/a\}$. Clearly, $C\theta \subseteq D$, hence $C \triangleright D$. □

If for a pair of literals L and M a substitution θ exists, such that $L\theta = M\theta$, then we say that L and M are *unifiable* and the substitution θ is called a *unifier*. Of course, for a pair of literals, several (possibly infinitely many) different unifiers may exist.

Example 6.5. Consider the literals $L = p(X, a, Z)$ and $M = p(V, W, b)$. The following substitutions are examples of unifiers for L and M:

$\theta_1 = \{X/a, Z/b, V/a, W/a\},$
$\theta_2 = \{X/b, Z/b, V/b, W/a\},$
$\theta_3 = \{X/V, Z/b, W/a\},$
$\theta_4 = \{Z/b, V/X, W/a\},$
$\theta_5 = \{X/c, Z/b, V/c, W/a\},$
$\theta_6 = \{X/U, Z/b, V/U, W/a, T/g, U/k\},$
$\theta_\alpha = \{X/\alpha, Z/b, V/\alpha, W/a\},$ for any constant or variable α.

On the other hand, it is easy to see that the literals $N : q(X, a, X, b)$ and $M : q(Y, a, a, Y)$ are not unifiable. □

Let θ and γ be substitutions. We say that θ is *more general* than γ iff a substitution λ such that $\theta\lambda = \gamma$ exists. Let L and M be two literals; a *most general unifier* (**mgu**) of L and M is a unifier which is more general than any other unifier.

Note that the property of being more general is not antisymmetric, i.e., unifiers θ_1 and θ_2 exist, such that $\theta_1 \neq \theta_2$ and θ_1 is more general than θ_2 and θ_2 is more general than θ_1. Hence, the relationship "is more general than" does not constitute an ordering in the traditional sense on the set of all unifiers of L and M. This is the reason why a **mgu** of two literals L and M, if it exists, is not unique.

6.1 Basic Definitions and Assumptions

Example 6.6. The unifier θ_3 of Example 6.5 is more general than the unifier θ_1, since a substitution $\gamma = \{V/a\}$ exists, such that $\theta_1 = \theta_3 \gamma$. Furthermore, it can be seen that if α is a variable, then θ_α is a **mgu** for L and M. Hence, L and M have infinitely many **mgus**. □

The concept of **mgu** has been introduced in much more general contexts, where terms may contain function symbols. In these contexts, the unification problem is a bit less trivial than in *Datalog*. In the context of *Datalog*, there is an extremely simple algorithm which generates a **mgu** for each pair of literals L and M if they are unifiable, and otherwise outputs a dummy symbol ∇. The rationale of this algorithm is to compare first the predicate symbols and the arities of L and M and then, successively, all arguments of L and M. If the predicate symbols or arities are not equal, or if some arguments in corresponding positions do not match, then the algorithm stops with output ∇. Otherwise a **mgu** for L and M is constructed by dynamically composing the partial substitutions which are generated step by step during the unification of the terms in the corresponding argument positions. Our algorithm is formulated in pseudo-code as a function procedure:

FUNCTION MGU(L, M)

INPUT: literals $L = \psi(t_1, \ldots, t_n)$, and $M = \psi'(t'_1, \ldots, t'_m)$
OUTPUT: a **mgu** θ for L and M if they are unifiable,
otherwise a dummy symbol ∇.

BEGIN
IF $\psi \neq \psi'$ OR $n \neq m$ THEN RETURN ∇
ELSE BEGIN
$\quad \theta := \{\}$; /* initialize θ with empty substitution */
$\quad i := 1$;
$\quad unifies := true$;
$\quad$ REPEAT
$\quad\quad$ IF $t_i \theta \neq t'_i \theta$
$\quad\quad$ THEN IF $t'_i \theta$ is a variable
$\quad\quad\quad$ THEN $\theta := \theta \{t'_i \theta / t_i \theta\}$
$\quad\quad\quad$ ELSE IF $t_i \theta$ is a variable
$\quad\quad\quad\quad$ THEN $\theta := \theta \{t_i \theta / t'_i \theta\}$
$\quad\quad\quad\quad$ ELSE $unifies := false$;
$\quad\quad i := i + 1$;
$\quad$ UNTIL $i > n$ OR NOT $unifies$;
$\quad$ IF $unifies$ THEN RETURN θ ELSE RETURN ∇
$\quad$ END
END.

Example 6.7. Consider the two literals $L: p(X, Z, a, U)$ and $M: p(Y, Y, V, W)$ and the function call MGU(L, M). The successive values θ_i of the program variable θ immediately before the i-th execution of the while loop are:

$$\theta_1 = \{\},$$
$$\theta_2 = \{Y/X\},$$
$$\theta_3 = \{Y/X\}\{X/Z\} = \{Y/Z, X/Z\},$$
$$\theta_4 = \{Y/Z, X/Z\}\{V/a\} = \{Y/Z, X/Z, V/a\}$$
$$\theta_5 = \{Y/Z, X/Z, V/a\}\{W/U\} = \{Y/Z, X/Z, V/a, W/U\}$$

Hence, the **mgu** for L and M computed by the MGU algorithm is

$$\{Y/Z, X/Z, V/a, W/U\}. \qquad \square$$

6.2 The Model Theory of Datalog

In this section we give a precise answer to the question: "What does a *Datalog* program compute ?" We conceive of a *Datalog* program as a *query* which acts on extensional databases and produces a *result*. The result of a *Datalog* program P applied to an EDB E consists of all ground facts which are logical consequences of the clause set $P \cup E$, and whose predicate symbol belongs to $IPred$. Hence the meaning (or semantics) of a *Datalog* program can be specified as a mapping which associates a subset of IHB to each possible EDB, i.e., to each subset of EHB.

A *Datalog* program is indeed similar to what is commonly called a query: it associates a result to each instance of a database (recall that we have pointed out the equivalence between an EDB and an instance of a relational database). The only substantial difference is that the result of a *Datalog* program may contain facts for *several* intensional predicates, while the result of a query usually consists of one single relation. However, if we specify a *goal* in addition to a *Datalog* program, then the result consists of ground facts for one predicate only.

In the remainder of this section we will make the above ideas more precise. First of all, we have to define formally what we mean when we say that a ground fact is a logical consequence of a set of ground clauses and *Datalog* rules. For this purpose, we first introduce the concepts of Herbrand interpretation and Herbrand model. Later we show that, among all the Herbrand models of a *Datalog* program, there is one particular Herbrand model which is smaller than all others, called the "Least Herbrand Model". We show that the concept of least Herbrand model is tightly related to the semantics of *Datalog* programs.

6.2.1 Possible Worlds, Truth, and Herbrand Interpretations

In Mathematical Logic two approaches to the definition of truth exist: the model-theoretic approach and the proof-theoretic approach. Roughly speaking, in Model Theory a logical sentence is (tautologically) true if and only if it is true in all possible situations, or in all *possible worlds*. On the other hand, in Proof Theory a sentence is true if and only if it can be derived by the application of some given rules from a given set of axioms.

From a formal point of view, the two approaches are equivalent: exactly the same sentences are recognized as true by both methods. However, model theory is, in general, considered as the more intuitive approach to defining truth. Proof theory, on the other hand, often provides more efficient computational methods for establishing whether a fact is true or not in a given formalism.

In this section we are interested in the model theory of *Datalog*. First we must define the notion of *interpretation* or *possible world*.

As we have seen, the *Datalog* language is made up of different types of symbols, in particular, variables, predicate symbols, and constants. Variables are always interpreted as "placeholders" for constants. However, particular constants and predicates may be interpreted in different ways by different programmers.

Consider, for instance the *Datalog* clause

$$C: \quad l(t,X) : -g(X,f,t).$$

For a programmer A, constant symbols and variables may denote animals or persons. In particular, the constant and predicate symbols occurring in this clause C have the following meaning:

$$\text{Interpretation A}: \begin{cases} t: & \text{the cat} \\ f: & \text{food} \\ l(\alpha,\beta): & \alpha \text{ loves } \beta \\ g(\alpha,\beta,\gamma): & \alpha \text{ gives } \beta \text{ to } \gamma. \end{cases}$$

For another programmer, B, the same symbols may have a completely different meaning. Constants and variables may range over integers; in particular, the symbols occurring in clause C have the following signification:

$$\text{Interpretation B}: \begin{cases} t: & \text{the integer 10} \\ f: & \text{the integer 5} \\ l(\alpha,\beta): & \alpha \text{ is less than } \beta \\ g(\alpha,\beta,\gamma): & \alpha+\beta \text{ is greater than } \gamma. \end{cases}$$

Thus, in the world of A, the clause C is interpreted as follows: "the cat loves everybody who supplies him with food". Knowing the cat we may assume that, in the world of A, this sentence is true. In other words, *interpretation* A *satisfies clause* C. On the other hand, for programmer B, the clause C has the following meaning: $\forall X \in \mathcal{Z} : (X+5 > 10 \Rightarrow 10 < X)$, where $\mathcal{Z}$ denotes the set of all integers. Obviously, this is a false proposition (just try $X = 6$), hence the interpretation B does not satisfy C. So, the clause C is true in one interpretation, but false in another interpretation. Hence we cannot say that C is generally true.

As a definition of general (tautological) truth we may adopt the following: a clause is generally true iff it is true in every possible interpretation.

An example of a generally true *Datalog* clause is $p(a) : -p(a)$. Another example of a tautology is $q(X) : -p(a), q(X), r(b)$. Note that facts such as $\{p(a)\}$ are never tautologies, since for each such fact, one can always find a falsifying interpretation.

We can generalize our concept of truth from single clauses to sets of clauses: a set S of clauses is true in a particular interpretation $\Im$, (or satisfied by $\Im$) iff each clause of S is true in $\Im$.

If a clause C is satisfied by an interpretation $\Im$, we also say that $\Im$ is a *model* for C. In the same way, if a set of clauses is satisfied by an interpretation $\Im$, then we say that $\Im$ is a *model* for S. Thus, a clause C is tautologically true iff every interpretation is a model for C. A set S of clauses is tautologically true iff every interpretation is a model for S.

In order to define the semantics of *Datalog* programs, we are not just interested in the absolute (tautological) truth values of clauses or clause sets, but we need to establish under which circumstances a fact F follows from a set S of *Datalog* clauses. Let us first consider the following simple definition:

A ground fact F is a *consequence* of a set S (denoted by $S \models F$) of *Datalog* clauses iff each interpretation satisfying each clause of S also satisfies F. In other words, F is a consequence of S iff every model for S is also a model for F.

As an example, consider a set S consisting of the clauses $C1 : \{p(a,b)\}$ and $C2 : p(X,Y) : -p(Y,X)$ and consider a fact $F : \{p(b,a)\}$. Clearly, for each possible interpretation of our constant and predicate symbols, whenever $C1$ and $C2$ are satisfied then F is also satisfied, hence, $S \models F$.

Up to this point, we have given some rather informal definitions of the concepts "truth" and "consequence". Though intuitively sound, these definitions do not allow us to perform any concrete computation. It is inconceivable that, in order to test whether a given *Datalog* fact is a consequence of a given *Datalog* program, one has to check an infinity of interpretations dealing with cats, humming birds, integers, and myriads of ants. To solve this problem, we first give an exact formal definition of the terms "interpretation" and "model" leading to a rigorous definition of the semantics of *Datalog* programs. In a second step (in the next section), we address the problem of computability. Our notion of "interpretation" can be formalized through mathematical set theory, capturing formally the relevant aspects of all possible worlds. However, such a general concept of interpretation is not necessary.

It was independently shown by the famous logicians J.Herbrand and T. Skolem that, without loss of generality, we can limit ourselves to considering a particular type of interpretation, which is much more amenable than general interpretations. Such special interpretations are now referred to as *Herbrand interpretations*. All Herbrand interpretations assign the same meaning to the constant symbols: each constant symbol (i.e., each element of *Const*) is interpreted as "itself", i.e., as the lexicographic string which makes up the constant symbol. Each unary predicate symbol of *Pred* is interpreted as mapping from *Const* to the set of truth value $\{true, false\}$; and, more generally, each n-ary predicate symbol of *Pred* is interpreted as a mapping from $Const^n$ to the set $\{true, false\}$. Thus, different Herbrand interpretations always interpret constant symbols in the same way; they differ only in the interpretations of predicate symbols. For instance, in one Herbrand interpretation $\Im$ the literal $p(a,b)$ may evaluate to *true*, while in another Herbrand interpretation $\mathfrak{I}'$, the same literal $p(a,b)$ may evaluate to false. This means that Herbrand interpretations differ from one another only in the truth values of positive ground literals (or, equivalently, in the truth values of positive ground unit clauses, i.e., ground facts). If we assign a truth value to each ground fact, then we have determined a particular Herbrand interpretation.

This observation gives rise to a very elegant characterization of Herbrand interpretations: each Herbrand interpretation is characterized by the set of all ground facts whose truth value is *true*. Note that the set of all ground unit clauses is the Herbrand base HB. So, each different Herbrand interpretation corresponds to a different subset of HB. Hence we may *identify* Herbrand interpretations with subsets of HB, yielding the following alternative definition:

A *Herbrand interpretation* is a subset of the Herbrand base HB.

Given a particular Herbrand interpretation $\mathfrak{I} \subseteq HB$, we define the truth value of *Datalog* facts or rules under $\mathfrak{I}$ as follows:

- If G is a ground fact then the truth value of G under $\mathfrak{I}$ is *true* if $G \in \mathfrak{I}$; otherwise, the truth value of G under $\mathfrak{I}$ is *false*.
- Let R be a *Datalog* rule of the form $L_0 : -L_1, \ldots, L_n$. The truth value of R under $\mathfrak{I}$ is *true* iff for each ground substitution θ for R, whenever $\{L_1\theta\} \in \mathfrak{I} \wedge \ldots \wedge \{L_n\theta\} \in \mathfrak{I}$ then it is also true that $\{L_0\theta\} \in \mathfrak{I}$. Otherwise, the truth value of R under $\mathfrak{I}$ is *false*. (Note that this definition is based upon the assumption that a *Datalog* rule is considered an implication and that variables are considered to be universally quantified.)

If a clause C has truth value *true* under a Herbrand interpretation $\mathfrak{I}$, we say that $\mathfrak{I}$ *satisfies* C. Accordingly, if $\mathfrak{I}$ satisfies all clauses of a clause-set S, then we say that $\mathfrak{I}$ satisfies S. A Herbrand interpretation which satisfies a clause C or a set of clauses S is called a *Herbrand model* for C or, respectively, for S. It follows from the results of Skolem and Herbrand (which apply to a much more general context than *Datalog*) that we can concentrate on Herbrand models and disregard any other types of model.

Example 6.8. Consider the following Herbrand interpretations $\mathfrak{I}_1$ and $\mathfrak{I}_2$:

$\mathfrak{I}_1 = \{\{loves(bill, mary)\}, \{loves(mary, bill)\}\}$.
$\mathfrak{I}_2 = \{\{loves(bill, mary)\}, \{loves(mary, tom)\}, \{loves(tom, mary\}\}$.

It is easy to see that the *Datalog* rule $R: loves(X, Y) : -loves(Y, X)$ is true in interpretation $\mathfrak{I}_1$; thus $\mathfrak{I}_1$ is a Herbrand model for R. The same rule is false in interpretation $\mathfrak{I}_2$. Consider the ground substitution $\theta = \{X/mary, Y/bill\}$ we then have: $\{loves(Y, X)\theta\} \in \mathfrak{I}_2$ but $\{loves(X, Y)\theta\} \notin \mathfrak{I}_2$. Thus the truth value of R under $\mathfrak{I}_2$ is *false*, and hence $\mathfrak{I}_2$ is not a Herbrand model for R. □

In the light of these results, the notion of logical consequence can be reformulated as follows: A ground fact F is a *consequence* of a set S of *Datalog* clauses (denoted by $S \models F$) iff each Herbrand interpretation satisfying each clause of S also satisfies F. In other terms, F is a consequence of S iff every Herbrand model for S is also a Herbrand model for F.

Although we did not introduce a rigorous formalization of general models, our definition of Herbrand model is mathematically accurate. Thus we are finally aware of a precise characterization of the "$\models$" concept. This concept allows us to give an exact definition of the semantics of *Datalog* programs.

As noted earlier, a *Datalog* program P takes as input an extensional database EDB and produces as output a set of intensional ground facts. The output consists of exactly those intensional ground facts which are logical consequences of the ground facts of $P \cup EDB$. Hence, we represent the semantics of a *Datalog* Program P as a function $\mathcal{M}_P$ from the power set of EHB to the power set of IHB such that:

$$\forall V \subseteq EHB : \mathcal{M}_P(V) = \{G \mid G \in IHB \wedge (P \cup V) \models G\}.$$

When a goal "$? - H$" is specified together with the *Datalog* program the output is limited to all *instances* of this goal which are consequences of $P \cup EDB$. In other words, the output then consists of all ground facts which are consequences of $P \cup EDB$ and which are *subsumed* by H. More formally, if a *Datalog* goal "$? - H$" is specified together with a *Datalog* program P, then the semantics of the pair $< P, H >$ is a function $\mathcal{M}_{P,H}$ from the power set of EHB to the power set of HB, such that

$$\forall V \subseteq EHB : \mathcal{M}_{P,H}(V) = \{C \mid C \in HB \wedge (P \cup V) \models C \wedge H \triangleright C\}.$$

Note that when a goal H is specified, the solution space $\mathcal{M}_{P,H}(V)$ consists either only of intensional ground facts or only of extensional ground facts, depending on whether the predicate symbol of H is intensional or extensional. We accept extensional predicate symbols in goals, because we do not want to exclude the possibility of issuing simple queries directly to the EDB.

We are now aware of an exact definition of the semantics of *Datalog* programs. Though our definition is precise, it unfortunately still does not hint at *how* the answer to a *Datalog* program should be computed. Even worse, the above definitions do not even provide a reasonable method for testing whether a given ground fact F is a logical consequence of a set of *Datalog* clauses S. Indeed, when trying to enumerate all Herbrand interpretations in order to check whether each interpretation satisfying S also satisfies F, we are faced with three different types of infinity: (1) we would have to check an infinite number of different Herbrand interpretations, (2) many of the Herbrand interpretations have an infinite number of elements, and (3) if R is a rule of S containing variables then we must consider an infinite number of ground substitutions for R.

In order to overcome these difficulties, let us first show that among the many Herbrand models of a set S of *Datalog* clauses, one distinguished Herbrand model exists, the *least Herbrand model*, which is a subset of each other Herbrand model of S.

6.2.2 The Least Herbrand Model

If S is a set of *Datalog* clauses (i.e., ground facts and *Datalog* rules), then let $cons(S)$ denote the set of all ground facts which are consequences of S, i.e.,

$$cons(S) = \{F \mid F \in HB \land S \models F\}.$$

If we know how to compute $cons(S)$ for finite sets S, then we know how to compute the result $\mathcal{M}_P(EDB)$ of the application of a *Datalog* program P to an extensional database EDB. Indeed, both P and EDB are finite sets of clauses; hence their union $P \cup EDB$ is also finite and the result $\mathcal{M}_P(EDB)$ of applying P to EDB is

$$\mathcal{M}_P(EDB) = I(cons(P \cup EDB)) = cons(P \cup EDB) \cap IHB.$$

Let us present a new, interesting characterization of $cons(S)$.

Theorem 6.1. $cons(S) = \bigcap \{\Im \mid \Im \text{ is a Herbrand model of } S\}$. In other words, $cons(S)$ is equal to the intersection of all Herbrand models of S.

PROOF. Let F be an element of $cons(S)$. Since F is a consequence of S, F must be satisfied by each Herbrand model of S. Since F is a ground fact, this means that F is an *element* of each Herbrand interpretation of S. Hence F belongs to the intersection of all Herbrand interpretations of PS. On the other hand, assume that F is an element of each Herbrand interpretation of S. Then F is a ground fact and each model of S is also a model of F, and hence $S \models F$, i.e., $F \in cons(S)$. □

The set $cons(S)$ is a set of ground facts. Thus it is a subset of HB and hence it is itself a Herbrand interpretation. The next theorem shows that $cons(S)$ is also a Herbrand *model* of S. Note that, by definition, $cons(S)$ is a subset of each Herbrand model for S. This is the reason why $cons(S)$ is also called *the least Herbrand model of S*.

Theorem 6.2. $cons(S)$ is a Herbrand model of S.

PROOF. We have to show that each clause of S is satisfied by $cons(S)$. S may contain two types of clause: ground facts and *Datalog* rules. Each ground fact of S belongs to each model of S, and hence also to the intersection $cons(S)$ of all models of S. Now assume that S contains a *Datalog* rule R of the form $L_0 : -L_1, \ldots, L_n$. Let θ be a ground substitution for R. Assume that $\forall 1 \leq i \leq n : \{L_i\theta\} \in cons(S)$. Then, for each Herbrand model $\Im$ of S, $\forall 1 \leq i \leq n : \{L_i\theta\} \in \Im$. Hence, for each Herbrand model $\Im$ of R, $\{L_0\theta\} \in \Im$. Since each Herbrand model of S is also a Herbrand model of R, $\{L_0\theta\}$ is an element of each Herbrand model for S. Hence $\{L_0\theta\}$ is an element of the intersection $cons(S)$ of each Herbrand model for S. Thus $cons(S)$ is a Herbrand model of R. Thus $cons(S)$ is a Herbrand model for each fact and each rule of S. We conclude that $cons(S)$ is a Herbrand model for S. □

One can slightly generalize Theorem 6.1 and show that the intersection of an arbitrary number of Herbrand models for a set of *Datalog* clauses S is itself a

Herbrand model for S. This "model intersection property" is not only valid in the context of *Datalog*, but, more generally, in the context of logic programming. This property is a particular characteristic of Horn clauses. For sets of general clauses a similar property does not exist.

6.3 Conclusions

In this chapter we have presented a formal definition of *Datalog*. Our main goal was to give a precise characterization of the semantics of this language. For this purpose, we introduced some basic concepts of Model Theory and we characterized the output of a *Datalog* program by use of Herbrand models. In particular, we have shown that the set $cons(S)$ of all ground facts which logically follow from a set S of *Datalog* clauses is identical to the least Herbrand model of S.

The main results presented in this chapter are purely descriptive. They do not immediately lead to constructive methods for evaluating *Datalog* programs. Such methods will be presented in the next chapter.

6.4 Bibliographic Notes

Many of the definitions of this chapter have their origins in the theory of resolution-based mechanical theorem proving. [Chan 73] and [Love 78] are two standard textbooks which both present an excellent overview of this field.

Horn clauses were first investigated in a purely logical context. Much later they were used as the basic constituents of logic programs. The fundamental theory of the semantics of logic programs is developed in [VanE 76] and [Apt 82]. The standard textbook on the theory of logic programming is [Lloy 84]. This book offers a very good presentations of all the relevant issues connected to logic programming. In particular, several results presented in Chaps. 6 and 7 can be found in a more general form in [Lloy 84]. Clause subsumption is discussed in [Gott 85], [Gott 85a], and [Gott 87].

An important event bridging the gap between logic programming and databases was the 1977 Symposium on Logic and Databases in Toulouse (France). The proceedings of this conference have been published in book form [Gall 78]. Function-free Horn clauses for expressing recursive database queries have been considered independently by several authors since then. The term "*Datalog*" was invented by a group of researchers from Oregon Graduate Center, MCC, and Stanford University. Early publications on *Datalog* are [Ullm 85a], [Banc 86a], and [Banc 86b], [Maie 88].

6.4 Exercises

6.1 What is the logical formula $\underline{P_1}$ corresponding to the *Datalog* program P_1 presented in Example 6.2 ?

6.2 Prove that the composition of substitutions is associative.

6.3 Find a pair of clauses C, D such that $C \neq D$ and $C \triangleright D$ and $D \triangleright C$. Find a pair of rules R and S, such that the head of R subsumes the head of S, the body of R subsumes the body of S, but R does not subsume S.

6.4 a) Consider the literals L and M and the unifiers $\theta_1 \ldots \theta_6, \theta_\alpha$ defined according to Example 6.5. List all pairs (θ_i, θ_j) of unifier, such that θ_i is more general than θ_j. Justify your result.   b) Show that if L_1 is an arbitrary literal and L_2 is a ground literal, then at most one **mgu** of L_1 and L_2 exists.

6.5 Show that there is a Herbrand interpretation $\Im$, such that $\Im$ is a Herbrand model for every possible *Datalog* clause. Show that this interpretation is unique.

Chapter 7
Proof Theory and Evaluation Paradigms of Datalog

The aim of this chapter is to show *how* the answer to a *Datalog* program can be computed. In order to achieve a computational approach to *Datalog*, in Sect. 7.1 we develop the *proof theory* of this language. We show how new facts can be *inferred* from an extensional database by use of a *Datalog* program. We then demonstrate that our proof-theoretic method is *sound* and *complete*, i.e., that it computes exactly what it should according to the model-theoretic semantics of *Datalog*. The proof theory of *Datalog* leads to a first algorithm "INFER" for processing *Datalog* programs. This algorithm iteratively computes new facts from facts that have already been established by using a *forward chaining* technique.

After briefly introducing some basic elements of fixpoint theory, in Sect. 7.2, we show that the semantics of each *Datalog* program P can be described as the least fixpoint of a transformation T_P. We make clear that the INFER algorithm described in Sect. 7.1 can be conceived as a *least fixpoint iteration* which precisely computes the least fixpoint of T_P. We show, however, that it is possible to define a simpler transformation T'_P which leads to a slightly more efficient algorithm. Further optimizations of (the algebraic version of) this algorithm will be discussed in Sect. 9.1.

In Sect. 7.3 we discuss an alternative paradigm for processing *Datalog* programs: the *backward chaining* approach, also called the *top-down approach*. This approach is particularly useful in cases where a goal G is specified together with a *Datalog* program, and we are only interested in those facts which are instances of G. After describing a rudimentary general backward chaining method, we introduce the *resolution method*. Resolution is an important technique, which was originally introduced in the context of automated theorem proving with general clauses. A particular resolution strategy, *SLD-resolution*, is used in the context of logic programming. We define SLD-resolution in the context of *Datalog* and show how it can be used to answer *Datalog* queries.

7.1 The Proof Theory of Datalog

In this section we show how *Datalog* rules can be used to produce new facts from given facts. We define the notion of "fact inference" and introduce a proof-theoretic framework which allows one to infer all ground facts which are consequences of a finite set of *Datalog* clauses. In this way we obtain an algorithm for computing the output of any *Datalog* program.

7.1.1 Fact Inference

Consider a *Datalog* rule R of the form $L_0: -L_1, \ldots, L_n$ and a list of ground facts $F_1, \ldots, F_n$. If a substitution θ exists such that, for each $1 \leq i \leq n$, $L_i\theta = F_i$, then, from rule R and from the facts $F_1 \ldots F_n$, we can *infer* in one step the fact $L_0\theta$. The inferred fact may be either a new fact or it may be already known.

What we have just described is a general inference rule, which produces new *Datalog* facts from given *Datalog* rules and facts. We refer to this rule as *Elementary Production (EP)*. In some sense, EP can be considered a *meta-rule*, since it is independent of any particular *Datalog* rules, and it treats them just as syntactic entities.

Example 7.1. Consider the *Datalog* rule $R: p(X, Z): -p(X, Y), p(Y, Z)$ and the ground facts $\{p(a, b)\}$ and $\{p(b, c)\}$. Then, by EP, we can infer in one step the fact $\{p(a, c)\}$ using the substitution $\theta = \{X/a, Y/b, Z/c\}$. This is a new fact. Now consider the *Datalog* rule $R': p(X, Y): -p(Y, X)$ and the fact $\{p(a, a)\}$. Obviously, by applying EP, we cannot infer anything new, but only $\{p(a, a)\}$, i.e., the fact itself. □

The following algorithm tests whether the EP inference rule applies to a *Datalog* rule and a list of ground facts. If EP applies, then the algorithm outputs the ground fact produced by EP; otherwise it outputs the dummy symbol ∇. This algorithm makes use of the **MGU** function, as defined in Sect. 6.1.3. The **MGU** function is used for finding a unifier between a ground fact and a literal belonging to a rule body. (Note that if such a unifier exists, it is unique; see Exercise 6.4.b.)

```
FUNCTION PRODUCE(R, F₁ ... Fₙ)
INPUT: a Datalog rule R of the form L₀: -L₁, ..., Lₙ and
       a list of ground facts F₁ ... Fₙ.
OUTPUT: the ground fact resulting from the application of EP,
        if EP applies; the dummy symbol ∇ otherwise.
BEGIN
FOR i:=0 TO n DO Kᵢ := Lᵢ;    /* make a copy of rule R */
FOR i:= 1 TO n DO
   BEGIN
   λ := MGU(Kᵢ, Fᵢ);
   IF λ = ∇
   THEN RETURN ∇;
   ELSE  FOR j:= 0 TO n DO Kⱼ := Kⱼλ
   END
RETURN K₀
END.
```

The correctness of the PRODUCE algorithm is easy to see. Let λ_i denote the substitution λ produced by the statement "$\lambda := MGU(K_i, F_i)$" for each distinct value of i during the execution of PRODUCE. If the algorithm terminates positively (i.e., with a result different from ∇), then the substitution $\lambda_1 \lambda_2 \ldots \lambda_n$ accomplishes exactly the task of the substitution θ of the EP inference rule. On the other hand, it is not hard to see that whenever PRODUCE terminates negatively (i.e., with output ∇), then EP does not apply.

Example 7.2. Let us apply the PRODUCE algorithm to the *Datalog* rule R : $p(X, Z): -p(X, Y), p(Y, Z)$ and to the ground facts $\{p(a, b)\}$ and $\{p(b, c)\}$. Here we have $n = 2$ and $L_0 = p(X, Z)$, $L_1 = p(X, Y)$, $L_2 = p(Y, Z)$, $F_1 = p(a, b)$, and $F_2 = p(b, c)$. First the literals of R are copied to $K_0 : p(X, Z)$, $K_1 : p(X, Y)$, and $K_2 : p(Y, Z)$. Now we enter the first iteration of the second for-loop with i=1. We evaluate $\lambda := MGU(K_1, F_1)$, getting $\lambda = \{X/a, Y/b\}$. By applying the substitution λ to the K-literals, we get $K_0 := p(a, Z)$, $K_1 := p(a, b)$, $K_2 := p(b, Z)$. Now we enter the second iteration of the for-loop with i=2. We evaluate $\lambda := MGU(K_2, F_2)$, getting $\lambda = \{Z/c\}$. By applying the substitution λ to the K-literals, we get $K_0 := p(a, c)$, $K_1 := p(a, b)$, $K_2 := p(b, c)$. The algorithm then stops returning the ground fact $K_0 : p(a, c)$. □

Note that the rule EP as well as the algorithm PRODUCE are sensitive to the order in which the input ground facts are submitted.

Example 7.3. Let us apply the PRODUCE algorithm to the *Datalog* rule R : $p(X, Z): -p(X, Y), p(Y, Z)$ and to the ground facts $\{p(b, c)\}$ and $\{p(a, b)\}$ in this order. The first matching produces the substitution $\lambda = \{X/b, Y/c\}$. However no further matching can be obtained. The algorithm stops with output ∇. □

Given a set S of *Datalog* clauses, let us denote by *infer1(S)* the set of all ground facts which can be inferred in one step from S by applying EP. For finite S, this set can be computed in the following way:

```
FUNCTION INFER1(S)
INPUT : a finite set S of Datalog clauses
OUTPUT: the set of all facts which can be inferred in one
        step from S by applying EP.
BEGIN
result := ∅
FOR EACH rule R : L_0 ... L_n OF S DO
    FOR EACH n-tuple < F_1, ... F_n > of ground facts of S DO
      BEGIN
        new:=PRODUCE(R, F_1 ... F_n);
        IF new ≠ ∇ THEN result := result ∪ {new}
      END;
RETURN result
END.
```

It is obvious that the above algorithm effectively computes all facts which can be inferred in (exactly) one step from S by application of the inference rule EP, since for each *Datalog* rule of S it considers each possible combination of facts and tests whether these facts match (unify) with the literals on the right-hand side of the rule.

Example 7.4. Consider a set S of *Datalog* clauses consisting of the rules $R_1 : p(X,Z) \colon -p(X,Y), p(Y,Z)$ and $R_2 : p(X,Y) \colon -p(Y,X)$ and of the ground facts: $\{p(a,b)\}, \{p(b,c)\}, \{p(c,d)\}, \{p(d,e)\}$. Applying the INFER1 algorithm to S, we get the following set of ground facts:

$$infer1(S) = \{\{p(a,c)\}, \{p(b,d)\}, \{p(c,e)\}, \{p(b,a)\}, \{p(c,b)\},$$
$$\{p(d,c)\}, \{p(e,d)\}\}.$$

□

Let us finally define the concept of *inferred ground fact*. Let S be a set of *Datalog* clauses. Informally, a ground fact G can be *inferred from* S, denoted by $S \vdash G$ iff either $G \in S$ or G can be obtained by applying the inference rule EP a finite number of times. The relationship "$\vdash$" is more precisely defined by the following recursive rules:

- $S \vdash G$ if $G \in S$.
- $S \vdash G$ if a rule $R \in S$ and ground facts $F_1 \ldots F_n$ exist such that, $\forall\, 1 \leq i \leq n$, $S \vdash F_i$ and G can be inferred in one step by the application of EP to R and $F_1 \ldots F_n$.
- in no other cases does $S \vdash G$ hold.

Example 7.5. Reconsider the set S of Example 7.4. Since $\{p(a,b)\} \in S$, it holds that $S \vdash \{p(a,b)\}$. As shown, we can infer in one step $\{p(b,a)\}$, hence $S \vdash p(b,a)$. Applying EP, we infer the ground fact $\{p(a,a)\}$ from R_1 and from the two ground facts $\{p(a,b)\}, \{p(b,a)\}$. Hence $S \vdash \{p(a,a)\}$.

□

The sequence of applications of EP which is used to infer a ground fact G from S is called a *proof* of G. Any proof can be represented as a *proof tree* with different levels. The nodes of the proof tree are of two different types: some of them are labeled with clauses of S and others are labeled with ground facts inferred from S. The bottom level consists of clauses of S; the next level consists of clauses of S or clauses which can be inferred in one step from S, and so on. Each level contains only clauses which appear in S or on some lower level or which can be inferred in one step from clauses appearing on the next lower level.

Each inference step is represented as a set of edges connecting the inferred ground clause to its antecedents on the next lower level. Each node of a Proof tree is connected to at most one node of the next higher level. If a clause is used in several inferences then several copies of this clause appear in the Proof tree. A level may be connected to the next higher level by several inferences. The depth of a Proof tree is the number of levels minus 1. Any ground fact appearing in S has a Proof tree of depth 0.

Example 7.6. The Proof tree for the inference of $\{p(a,a)\}$ according to Example 7.5 is depicted in Fig. 7.1. This tree has depth 2. □

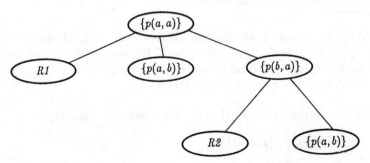

Fig. 7.1. Proof tree for p(a,a)

7.1.2 Soundness and Completeness of the Inference Rule EP

Now that we have a Proof-theoretic framework which allows us to infer new ground facts from an original set of *Datalog* clauses S, we want to compare this approach to the model-theoretic approach presented in the last section. It turns out that both approaches are equivalent, since, for each S and G, $S \models G$ iff $S \vdash G$. We show this important equivalence in two steps: first we show that $S \vdash G \Rightarrow S \models G$; in other words, we prove the *soundness* of the EP rule. Then we show that $S \models G \Rightarrow S \vdash G$; in other words, we prove the *completeness* of the EP rule.

Theorem 7.1 (Soundness of EP). *If S is a set of Datalog clauses and G is a ground fact, then $S \vdash G \Rightarrow S \models G$.*

Proof. We show by induction on the depth of Proof trees that whenever a ground fact G can be derived from S using a Proof tree, then $S \models G$.

Induction Base: If G can be derived by a Proof tree of depth 0 then $G \in S$. Thus G belongs to each Herbrand model of S and hence $S \models G$.

Induction Step: Assume G has a Proof tree of depth $i+1$. Then a *Datalog* rule $R: L_0 :- L_1 \ldots L_n$ at level i and ground facts $F_1 \ldots F_n$ at level i exist such that G can be inferred in one step by EP from R and $F_1 \ldots F_n$. This means that a substitution θ exists such that for $1 \leq k \leq n$ $L_k\theta = F_k$ and $G = L_0\theta$. Since the facts $F_1 \ldots F_n$ appear on level i of the Proof tree, a Proof tree of length $\leq i$ exists for each of these facts. Hence, by the induction hypothesis, for $1 \leq k \leq n$ we have $S \models F_k$. Thus, for each Herbrand model $\Im$ of S, $F_1, \ldots, F_n \in \Im$. In other words, for each Herbrand model $\Im$ of S, $L_1\theta \in \Im$ and $L_2\theta \in \Im$ ... and $L_n\theta \in \Im$. Since $R \in S$, it follows (by the definition of Herbrand model) that for each Herbrand model $\Im$, $G = L_0\theta \in \Im$. Hence $S \models G$. □

Theorem 7.2 (Completeness of EP). If S is a set of *Datalog* clauses and G is a ground fact, then $S \models G \Rightarrow S \vdash G$.

Proof. Consider the set $infer(S) = \{G \mid G \text{ ground fact and } S \vdash G\}$, i.e., the set of all ground facts which can be inferred from S. We first show that $infer(S)$ is a model for S. By the definition of "$\vdash$", each ground fact $F \in S$ is also an element of infer(S). Now consider any rule $R : L_0 :- L_1 \ldots L_n$ of S. Assume that for some substitution θ, $L_1\theta, \ldots L_n\theta \in infer(S)$, i.e., $S \vdash L_1\theta$ and $S \vdash L_2\theta$ and $\ldots S \vdash L_n\theta$. By the second rule in the definition of "$\vdash$" it clearly follows that $S \vdash L_0\theta$, and hence $L_0\theta \in infer(S)$. Hence $infer(S)$ is a Herbrand model of S. In order to prove our theorem, assume that $S \models G$. This means that G is an element of each Herbrand model of S. In particular, G is an element of $infer(S)$. Hence, $S \vdash G$. □

Theorem 7.3. If S is a set of *Datalog* clauses and G is a ground fact then $S \models G$ iff $S \vdash G$.

Proof. The theorem follows immediately from Theorems 7.1 and 7.2. □

Recall from Chap. 6 that $cons(S)$ denotes the set of all ground facts which follow from a set S of *Datalog* clauses. The next corollary provides a nice characterization for $cons(S)$. The corollary follows immediately from Theorem 7.3.

Corollary If S is a set of *Datalog* clauses and G is a ground fact then $cons(S) = \{G \mid G \text{ ground fact and } S \vdash G\}$.

For finite sets S, $cons(S)$ can be computed by the following simple algorithm INFER:

```
FUNCTION INFER(S)
INPUT: a finite set S of Datalog clauses
OUTPUT: cons(S)
BEGIN
old:= ∅;
new:= S;
WHILE new ≠ old DO
  BEGIN
  old := new;
  new := new ∪ INFER1(new)
  END;
result:= all facts of new;
RETURN result
END.
```

Theorem 7.4. For finite S the INFER algorithm eventually halts with output $cons(S)$.

Proof. During each execution of the while loop (except the last), new ground facts are added to the set variable "new". These facts are built from a finite vocabulary of predicate symbols and constant symbols: only the predicate symbols and constant symbols which occur in S may occur in each of the generated facts. Thus only a finite number of new facts can be inferred from S. All inferred facts are accumulated in "new". Hence, after a finite number of executions of the while loop, $new \cup infer1(new) = new$ and the algorithm stops. From the structure of the INFER algorithm, it is obvious that the output of INFER(S) consists of the set of all ground facts which can be inferred from S. By the corollary of Theorem 7.3, this set is equal to $cons(S)$. □

Corollary. If S is finite then $cons(S)$ is finite.

Example 7.7. Consider again the set of *Datalog* clauses S used in Example 7.4. We show how INFER works by displaying the value that the variable "new" takes during the computation of INFER(S) after each execution of the while loop. Let new_i denote the value of new after the i-th execution of the while loop. We have:

$new_1 = S \cup infer1(S) = \{ \{p(a,b)\}, \{p(b,c)\}, \{p(c,d)\}, \{p(d,e)\},$
$\{p(a,c)\}, \{p(b,d)\}, \{p(c,e)\}, \{p(b,a)\},$
$\{p(c,b)\}, \{p(d,c)\}, \{p(e,d)\} \};$
$new_2 = new_1 \cup \{ \{p(a,d)\}, \{p(a,a)\}, \{p(b,e)\}, \{p(b,b)\}, \{p(c,c)\},$
$\{p(d,d)\}, \{p(c,a)\}, \{p(d,b)\}, \{p(a,e)\}, \{p(e,e)\}, \{p(e,c)\} \};$
$new_3 = new_2 \cup \{ \{p(e,a)\}, \{p(d,a)\}, \{p(e,b)\} \};$
$new_4 = new_3.$

The algorithm stops after four executions of the while-loop. □

The computation performed by the INFER algorithm is called a *bottom-up computation*, because the algorithm starts by considering the ground facts of S, which are at the bottom of the Proof trees, and works "upwards" by producing first all facts which can be inferred in one step from the bottom clauses, then all facts which can be computed in one step from the bottom clauses or from facts which were inferred in one step, and so on. After the i-th execution of the while-loop, all ground facts for which a Proof tree with i or less levels exists are accumulated.

The algorithm INFER1 is also classified as *forward chaining* because, if we write *Datalog* rules as implications (see Sect. 6.1.1), then they are processed "forward" in the sense of the implication sign.

Now that we have an algorithm for computing $cons(S)$ for finite sets S of *Datalog* clauses, we are able to build a rudimentary processor for *Datalog* programs. Assume P is a *Datalog* program which has to be applied to an extensional database EDB. In order to compute the output, we first apply INFER to $P \cup EDB$ and then take all clauses of the result whose predicate symbols correspond to intensional predicates. In other words, the result of applying P to EDB is equal to $I(infer(P \cup EDB))$. The INFER algorithm, being the core of such an interpreter, is also called the underlying *inference engine* of the computation.

7.2 Least Fixpoint Iteration

Fixpoint Theory was first developed in the field of recursion theory – a subfield of mathematical logic – as a tool for explaining recursive functions. Later, fixpoint theory was used in several other mathematical disciplines such as algebra and functional analysis. In computer science, this theory can be profitably applied wherever a formal description of the semantics of recursive programs or systems is required. In particular, the method of *Denotational Semantics* is based on fixpoint theory.

One of the main characteristics of logic programming languages is their ability to express recursion. Hence fixpoint theory is well suited for describing their semantics. This kind of semantics – usually called *Fixpoint Semantics* – has the advantage of being nonprocedural, though at the same time it directly leads to a precise computation method.

In this section we first present some elementary results of fixpoint theory in a very general setting. We then show how these results relate to *Datalog*. In particular, we show that the output of each *Datalog* program can be described as the least fixpoint of a transformation and that the INFER algorithm precisely corresponds to computing this fixpoint through a computation method called "least fixpoint iteration".

7.2.1 Basic Results of Fixpoint Theory

Let V be a set. A *partial order* "$\leq$" on V is a binary relation on V which satisfies the following conditions:

- Reflexivity: $\forall x \in V : (x \leq x)$
- Antisymmetry: $\forall x, y \in V : ((x \leq y \land y \leq x) \Rightarrow x = y)$
- Transitivity: $\forall x, y, z \in V : ((x \leq y \land y \leq z) \Rightarrow x \leq z)$.

If V is a set and "$\leq$" is a partial order on V, then V is called a *partially ordered set* or briefly *poset*. In order to indicate a poset, we specify a pair consisting of the set itself and the partial order $(V, \leq)$.

Example 7.8. Let H be a set and V a set which contains as elements some subsets of H. The set-theoretic inclusion "$\subseteq$" is a partial order on V.   □

Let $(V, \leq)$ be a poset and let $R \subseteq V$ be a subset of V. An element a of V is an *upper bound* of R iff for each element $x \in R$, $x \leq a$. If a is an upper bound of R and if, for each other upper bound b of R, $a \leq b$, then a is called the *least upper bound* of R. Clearly, if the least upper bound of R exists, it is unique. We denote the least upper bound of R (if it exists) by $sup(R)$. An element a of V is a *lower bound* of R iff for each element $x \in R$, $a \leq x$. If a is a lower bound of R and if for each other lower bound b of R it is true that $b \leq a$, then a is called the *greatest lower bound* of R. Clearly, if the greatest lower bound of R exists, it is unique. We denote the greatest lower bound of R (if it exists) by $inf(R)$.

Example 7.9. Consider the set $\mathcal{N}^+$ of all strictly positive natural numbers. The relation "$|$", such that $x|y$ iff x divides y is clearly a partial order on $\mathcal{N}$. It is not hard to see that each finite or infinite subset V of $\mathcal{N}^+$ has a greatest lower bound with respect to "$|$": $inf(V)$ is the greatest common divisor of the integers appearing in V. On the other hand, only finite subsets of $\mathcal{N}^+$ have a least upper bound. □

A partially ordered set $(V, \leq)$ is a *lattice* iff for each pair of elements $a, b \in V$, both $inf(\{a,b\})$ and $sup(\{a,b\})$ exist.

Example 7.10. $(\mathcal{N}^+, |)$ is a lattice. □

A partially ordered set $(V, \leq)$ is a *complete lattice* iff for each subset R of V, both $inf(R)$ and $sup(R)$ exist. Each complete lattice $(V, \leq)$ has a *bottom element* $\bot_V = inf(V)$ and a *top element* $\top_V = sup(V)$.

Example 7.11 Let us first note that $(\mathcal{N}^+, |)$ is not a complete lattice, because infinite subsets of $\mathcal{N}^+$ don't have a least upper bound. Now consider the set V_{12} of all divisors of 12, i.e., $V_{12} = \{1, 2, 3, 4, 6, 12\}$. It is easy to see that $(V_{12}, |)$ forms a complete lattice. We have $\bot_{V_{12}} = 1$ and $\top_{V_{12}} = 12$. □

The above example suggests that problems with completeness may arise only in the case of infinite posets. This conjecture is confirmed by the following theorem.

Theorem 7.5. Any finite lattice is complete.

Proof. Let $(V, \leq)$ be a finite lattice. Let W be a subset of V. Since V is finite, W is of the form $W = \{a_1, \ldots, a_n\}$. Let $s_1 = sup(\{a_1, a_2\})$ and, for $2 \leq i \leq n-1$, $s_i = sup(\{s_{i-1}, a_{i+1}\})$. It follows by the definition of sup and by a simple induction argument that $s_{n-1} = sup(W)$. In a similar way, by setting $t_1 = inf(\{a_1, a_2\})$ and, for $2 \leq i \leq n-1$, $t_i = inf(\{t_{i-1}, a_{i+1}\})$, it can be shown that $t_{n-1} = inf(W)$. □

Note that the converse of Theorem 7.5 does not hold. As we see in the following Theorem 7.6, there are infinite lattices which are complete.

If U is a set, then the *subset-lattice of* U consists of the power set $\mathcal{P}(U)$ of U ordered by the set inclusion "$\subseteq$", i.e., it consists of the poset $(\mathcal{P}(U), \subseteq)$.

Theorem 7.6. Any subset-lattice is complete.

Proof. Consider the subset-lattice $(\mathcal{P}(U), \subseteq)$ of a set U. Any subset R of $\mathcal{P}(U)$ consists of a collection of subsets of U: $R = \{U_i \subseteq U \mid i \in I\}$. It is easy to verify that for any such R, $sup(R) = \bigcup R = \bigcup_{i \in I} R_i$ and $inf(R) = \bigcap R = \bigcap_{i \in I} R_i$. □

If $(V, \leq)$ is a lattice, then a *transformation* T on V is a mapping from V to V. A transformation T on V is called *monotonic* iff for all $a, b \in V$, $a \leq b$ implies $T(a) \leq T(b)$.

Example 7.12. Consider the following transformation T_1 defined on the lattice $(\mathcal{N}^+, |)$: for each $n \in \mathcal{N}^+$, $T_1(n)$ is the greatest common divisor of 17 and n itself. For instance, $T_1(5) = 1$, $T_1(20) = 1$, $T_1(34) = 17$, $T_1(35) = 1$. It is not hard to see that T_1 is a monotonic transformation. Assume that $n|m$. Then all divisors of n are also divisors of m. Thus $T_1(n)$, i.e., the greatest common divisor of n and 17, is also a common divisor of m and 17. However, in general, a greater common divisor of m and n might exist. Hence $T_1(n) \leq T_1(m)$. On the other hand, consider the transformation T_2 on $(\mathcal{N}^+, |)$ defined as follows: for each $n \in \mathcal{N}^+$, $T_2(n) = n + 1$. It is easy to see that transformation T_2 is not monotonic: we have $5|25$, but not $T_2(5)|T_2(25)$, since 6 is not a divisor of 26. □

A *fixpoint* of a transformation T defined on a poset $(V, \leq)$ is an element $a \in V$ such that $T(a) = a$. If T has a fixpoint b, such that, for all other fixpoints x of T, $b \leq x$, then b is called the *least fixpoint* of T. If T has a least fixpoint then this least fixpoint is unique, denoted by $lfp(T)$.

Example 7.13. Consider the transformation T_1 and T_2 of the previous example. T_1 has two fixpoints: 1 and 17. Since $1|17$, 1 is the least fixpoint of T_1. T_2 has no fixpoint. □

Theorem 7.7 (Fixpoint Theorem). Let T be a monotonic transformation on a complete lattice $(V, \leq)$. Then T has a nonempty set of fixpoints. In particular, T has a least fixpoint $lfp(T)$ such that $lfp(T) = inf(\{x \in V \mid T(x) \leq x\})$.

Proof. Let $U = \{x \in V \mid T(x) \leq x\}$. For each $x \in U$, $inf(U) \leq x$. By the monotonicity of T it follows that, for each $x \in U$, $T(inf(U)) \leq T(x)$. Hence, $T(inf(U))$ is a lower bound of U and thus $T(inf(U)) \leq inf(U)$. Observe that $inf(U)$ satisfies the condition for being a member of U, hence $inf(U) \in U$. Furthermore, since $T(inf(U)) \leq inf(U)$, by the monotonicity of T it follows that $T(T(inf(U))) \leq T(inf(U))$, hence $T(inf(U)) \in U$. By definition of "inf", we have $inf(U) \leq T(inf(U))$ and thus $inf(U) = T(inf(U))$. Hence $inf(U)$ is a fixpoint of T. Since all fixpoints of T are elements of U, and $inf(U)$ is smaller than each other element of U, $inf(U)$ is smaller than each other fixpoint of T, i.e., $inf(U)$ is the least fixpoint of T. □

Let us turn to the problem of *computing* least fixpoints. Note that it is not always possible to compute the least fixpoint of a given transformation: for instance, if the least fixpoint consists of an infinite set, it is impossible to enumerate the elements of this set in a finite number of steps. However, we present an algorithm LFP with the following property: LFP receives as input a transformation T; if LFP terminates, then its output consists of the fixpoint of T. The underlying principle of the LFP algorithm is commonly known as *least fixpoint iteration*, or *LFP-iteration* for short.

FUNCTION LFP(T)
INPUT : a monotonic transformation T on a complete
    lattice $(V, \leq)$.
OUTPUT: $lfp(T)$ [if the algorithm stops].
BEGIN
  old:=$\perp_V$;
  new:=$T(\perp_V)$;
  WHILE new $\neq$ old DO

        BEGIN
            old := new;
            new := T(new);
        END;
        RETURN new
    END

Theorem 7.8. Whenever LFP terminates, its output is $lfp(T)$.

Proof. Assume the algorithm halts after n executions of the while-loop. Let $T^0(\bot_V) = \bot_V$ and for each integer $i > 0$, let $T^i(\bot_V) = T(T^{i-1}(\bot_V))$. After n executions of the while-loop, the algorithm then outputs $T^n(\bot_V) = T^{n+1}(\bot_V)$. Of course, $T^n(\bot_V)$ is a fixpoint of T, since

$$T(T^n(\bot_V)) = T^{n+1}(\bot_V) = T^n(\bot_V).$$

It now remains to show that for each fixpoint a of T, we have $T^n(\bot_V) \leq a$. We show by induction that for each integer $i \geq 0$, $T^i(\bot_V) \leq a$.

Induction basis: for $T^0(\bot_V) = \bot_V$ the proposition is obvious, since $\bot_V$ is the least element of V.

Induction step: Assume $T^j(\bot_V) \leq a$. By the monotonicity of T, $T(T^j(\bot_V)) \leq T(a)$. Since a is a fixpoint of T, $T(a) = a$ and hence $T^{j+1}(\bot_V) = T(T^j(\bot_V)) \leq a$. Thus for each i, $T^i(\bot_V) \leq a$ and hence, in particular, $T^n(\bot_V) \leq a$. This proves that $T^n(\bot_V)$ is the least fixpoint of T. □

We have now introduced all concepts and results of fixpoint theory which are necessary in order to characterize the semantics of a *Datalog* program in terms of least fixpoints. Note that fixpoint theory can be used to define the semantics of almost all languages which allow for recursion. In particular, in Chap. 9 we will apply fixpoint theory to solving recursive equations in relational algebra.

7.2.2 Least Fixpoints and Datalog Programs

In order to establish the fixpoint semantics of *Datalog* programs, consider the lattice $V_{DAT} = (\mathcal{P}(HB), \subseteq)$, where $\mathcal{P}(HB)$ stands for the power set of the Herbrand base HB and "$\subseteq$" denotes the set-inclusion predicate. Of course, V_{DAT} is a subset-lattice.

We show that for each set S of *Datalog* clauses there is a monotonic transformation T_S on V_{DAT}, such that $cons(S) = lfp(T_S)$.

If S is a set of *Datalog* clauses, then let $FACTS(S)$ denote the set of all (ground) facts of S and let $RULES(S)$ denote the set of all *Datalog* rules of S. Obviously, we have $S = FACTS(S) \cup RULES(S)$. The *transformation* T_S *associated to* S is a mapping from $\mathcal{P}(HB)$ to $\mathcal{P}(HB)$ defined as follows:

$$\forall W \subseteq HB : T_S(W) = W \cup FACTS(S) \cup infer1(RULES(S) \cup W).$$

Theorem 7.9. For each set S of *Datalog* clauses, T_S is a monotonic transformation on V_{DAT}.

Proof. First let us note that $infer1$ is monotonic on V_{DAT}; this follows easily from the definition of $infer1$. Now assume that $W_1 \subseteq W_2 \subseteq HB$. We have to show that $T_S(W_1) \subseteq T_S(W_2)$. Clearly, $RULES(S) \cup W_1 \subseteq RULES(S) \cup W_2$ and hence, by the monotonicity of $infer1$,

$$infer1(RULES(S) \cup W_1) \subseteq infer1(RULES(S) \cup W_2).$$

Thus,

$$T_S(W_1) = W_1 \cup FACTS(S) \cup infer1(RULES(S) \cup W_1)$$
$$\subseteq W_2 \cup FACTS(S) \cup infer1(RULES(S) \cup W_2) = T_S(W_2).$$
□

In the next theorem, we show an interesting relationship between the Herbrand models of S and the fixpoints of T_S.

Theorem 7.10. Let S be a set of *Datalog* clauses and let $I \subseteq HB$. I is a Herbrand model of S iff I is a fixpoint of T_S.

Proof. Let us first state a very simple reformulation of the concept of Herbrand model: I is a Herbrand model of S iff

$$FACTS(S) \subseteq I \quad \text{and} \quad infer1(RULES(S) \cup I) \subseteq I.$$

This follows trivially from the definition of "Herbrand model" and from the definitions of "$infer1$", "$FACTS(S)$" and "$RULES(S)$". Now let us prove both directions of our theorem:

1) **"if"** Assume I is a fixpoint of T_S, i.e., $T_S(I) = I$. Then $I = I \cup FACTS(S) \cup infer1(RULES(S) \cup I)$. Thus $FACTS(S) \subseteq I$ and $infer1(RULES(S) \cup I) \subseteq I$. Hence, I is a Herbrand model of S.

2) **"only if"** Assume I is a Herbrand model of S. Then $FACTS(S) \subseteq I$ and $infer1(RULES(S) \cup I) \subseteq I$.
Hence, $T_S(I) = I \cup FACTS(S) \cup infer1(RULES(S) \cup I) = I$. Thus I is a fixpoint of T_S.
□

Among the fixpoints of T_S, the least fixpoint is of particular importance. As we show in the next theorem, this fixpoint consists of exactly the set $cons(S)$ of all ground facts which are logical consequences of S.

Theorem 7.11. For each set S of *Datalog* clauses, $cons(S) = lfp(T_S)$.

Proof. Recall from Sect. 6.2.2 that, among the Herbrand models of S, $cons(S)$ is the smallest one w.r.t. the set-inclusion ordering "$\subseteq$". By Theorem 7.10, the Herbrand models of S exactly coincide with the fixpoints of T_S. Hence $cons(S)$ is the smallest fixpoint of T_S w.r.t. the partial order "$\subseteq$", i.e., $cons(S)$ is the least fixpoint of T_S in the lattice V_{DAT}.
□

For any set S of *Datalog* clauses, $cons(S)$ can thus be obtained by LFP-iteration by computing $LFP(T_S)$.

If P is a *Datalog* program and if EDB is an extensional database, then $P \cup EDB$ is a (finite) set of *Datalog* clauses; hence $T_{P \cup EDB}$ is a monotonic transformation on V_{DAT}. The following corollary of Theorem 7.11 gives a fixpoint characterization of the semantics $\mathcal{M}_P$ of P.

Corollary. For each finite $EDB \subseteq E(\text{HB})$:

$$\mathcal{M}_P(EDB) = lfp(T_{P \cup EDB}) \cap IHB.$$

Proof. The corollary follows immediately from Theorem 7.11 and from the fact that $\mathcal{M}_P(EDB) = cons(P \cup EDB) \cap IHB$. □

It is not very surprising that the resulting LFP algorithm is quite similar to the INFER algorithm. Indeed, the variables *old* and *new* of both algorithms at each step contain the same *facts*. Both algorithms, in each step, apply $INFER1$ in order to generate new facts and add these new facts to the ones which have been generated in previous steps.

More surprisingly, we may observe that the transformation T_S can be simplified, i.e., one can find a transformation T'_S on V_{DAT} which is less complicated than T_S and such that $lfp(T_S) = lfp(T'_S)$. The computation LFP(T'_S) is slightly more efficient than the activation LFP(T_S) of the LFP algorithm.

To a given set S of *Datalog* clauses, let us define T'_S as follows:

$$\forall W \subseteq HB : T'_S(W) = FACTS(S) \cup infer1(RULES(S) \cup W).$$

Thus T'_S differs from T_S by the omission of the term W in the union.

Theorem 7.12. For each set S of *Datalog* clauses, T'_S is a monotonic transformation on V_{DAT} and $lfp(T'_S) = lfp(T_S) = cons(S)$.

Proof. The monotonicity of T'_S can be shown in a similar way to the monotonicity of T_S (see Theorem 7.9). Let us now show that both transformations have the same least fixpoint. By the Fixpoint Theorem (Theorem 7.7) we have:

$$lfp(T_S) = inf\{W \subseteq HB | W \cup FACTS(S) \cup infer1(RULES(S) \cup (W)) \subseteq W\}.$$

Due to basic set-theoretic laws, $W \cup FACTS(S) \cup infer1(RULES(S) \cup (W)) \subseteq W$ is equivalent to the simpler condition $FACTS(S) \cup infer1(RULES(S) \cup (W)) \subseteq W$. Hence, $lfp(T_S) = inf\{W \subseteq HB | FACTS(S) \cup infer1(RULES(S) \cup (W)) \subseteq W\} = lfp(T'_S)$. □

It follows that $cons(S) = lfp(T'_S)$. Thus, if S is finite, we can compute $cons(S)$ by a LFP-iteration of the transformation T'_S. The corresponding algorithm LFP(T'_S) is given below:

FUNCTION LFP(T'_S);
INPUT : T'_S defined through a finite set S of *Datalog* clauses
OUTPUT: $cons(S)$
BEGIN
   old:=∅;
   new:=$FACTS(S)$;

```
WHILE new ≠ old DO
  BEGIN
    old := new;
    new := FACTS(S) ∪ infer1(RULES(S)∪old);
  END;
  RETURN new
END.
```

It can be seen that the $LFP(T'_S)$ algorithm computes at each step the same set of facts as the $LFP(T_S)$ algorithm and the INFER algorithm. However, unlike the other two algorithms, the $LFP(T'_S)$ algorithm, at each step, after generating the new facts, performs the union of the set of new facts with the set $FACTS(S)$ and not with the larger set *old*. This may lead to more efficient implementations.

7.3 Backward Chaining and Resolution

In this section we present an important alternative to the bottom-up computation (or forward chaining) of *Datalog* programs. We outline how *Datalog* rules can be processed backwards, by starting from a goal clause and building a Proof tree in a top-down fashion. This method is also referred to as *backward chaining*. After presenting a general method for backward chaining, we show how the well-known *resolution method* can be used to answer *Datalog* queries.

7.3.1 The Principle of Backward Chaining

Imagine that you have a *Datalog* program P and an extensional database EDB and a goal $G : ? - p(t_1, \ldots, t_n)$, where p denotes an intensional predicate. You want to know which ground instances of G are satisfied by $P \cup EDB$.

One possibility for answering this question would be first to compute the set $\mathcal{M}_P(EDB)$ of *all* intensional ground facts which are consequences of $P \cup EDB$ by a bottom-up computation, and then throw away all those facts which are not subsumed by G. This method, however, is quite inefficient, since in general we compute many more facts than requested. First, $\mathcal{M}_P(EDB)$ may contain a large number of ground facts whose predicates are different from p. Second, constants appearing in our goal G are not exploited in order to reduce the number of facts generated, i.e., if the goal G has the form $? - p(a, Y, b)$, first all answers to the more general goal $? - p(X, Y, Z)$ are computed and then those answers which are not subsumed by G are deleted.

Very interesting techniques can be used for reducing the number of irrelevant facts generated during a bottom-up computation. These methods ("Magic Sets", "Counting" and many others) are presented in Part III of this book, since they are considered to be optimization techniques. In this section we discuss a radically different computing paradigm, called *top-down computation* or *backward chaining*.

A top-down computation constructs Proof trees by starting from the top level with the goal clause G and ending with the bottom level containing nodes labeled by clauses of $P \cup EDB$. We start by considering the top goal G. If some instances of G already appear as ground facts of $P \cup EDB$, then, for each of these ground facts we construct a separate Proof tree consisting of one single node labeled by the ground fact. Each of these Proof trees is of depth 0 and produces one answer to our goal query.

Once we have produced all Proof trees of depth 0, in order to produce more solutions, we look for Proof trees of depth 1 by expanding G as follows. Choose a *Datalog* rule $R: L_0 :- L_1, \ldots, L_k$ from P such that G and L_0 are unifiable. Let $\theta = mgu(G, L_0)$. Attach k child-nodes labeled $L_1\theta, \ldots, L_k\theta$ to the node labeled G; furthermore, attach a child-node labeled R to the node labeled G. In doing so, we obtain a tree of depth 1. The root node labeled G and its children labeled $L_1\theta, \ldots, L_k\theta$ are called *goal-nodes*; the child of G labeled R is called a *rule-identifier*. For different choices of R, we get different trees of depth 1. These trees are called *search trees* and are, in general, different from Proof trees.

In order to transform search trees into Proof trees, we look for substitutions λ such that for each goal-leaf of the tree, labeled by some clause C, $C\lambda \in P \cup EDB$. Then, by applying λ to all labels of the goal-nodes of the entire search tree we get a Proof tree of depth 1. For different substitutions λ, in general, we get different Proof trees. In particular, for each suitable substitution λ, $G\lambda$ is a solution to the query "? – G".

Assume we are aware of all search trees and corresponding Proof trees of depth i. In order to generate Proof trees of depth $i + 1$, we expand the search trees of depth i in the same way by adding suitable child-nodes to some or all leaves and by determining all substitutions λ such that for each goal-leaf of the new search tree, labeled by some clause C, $C\lambda \in P \cup EDB$.

Whenever we choose a rule R to expand some leaf of a search tree T, we should take care that the variables occurring in R are disjoint from those occurring in T. This can always be obtained by *renaming* the variables of R which also occur in T.

In any search tree, the label G of the root is called the *top goal*. All other labels of goal-nodes are called *subgoals*. Note that this is a slight misuse of the concept of "goal", since in Sect. 6.1.1 we have defined a goal as a negative clause. Note also that the goals and subgoals occurring as labels within a search tree should not be logically interpreted as universally quantified positive sentences according to the interpretation of positive unit clauses (see Sect. 6.1.1). Instead, they should merely be regarded as uninterpreted data structures.

Example 7.14. Consider a *Datalog* program P consisting of the rules $R_1 : p(X, Z):- p(X, Y), p(Y, Z)$ and $R_2 : p(X, Y):- p(Y, X)$ and an extensional database EDB consisting of the ground facts:

$$\{p(a,b)\}, \{p(b,c)\}, \{p(c,d)\}, \{p(d,e)\}.$$

(Note that the set of clauses $S = P \cup EDB$ has already been used in some examples of Sect. 7.1 to illustrate the principle of forward chaining.)

7.3 Backward Chaining and Resolution

Consider the goal "$? - p(a, X)$". The only Proof tree of depth 0 for this goal is the one consisting of a single node labeled $p(a, b)$.

There are two possibilities of expanding the node labeled $\{p(a, X)\}$ to a search tree of depth 1. First, we may unify $p(a, X)$ with the left-hand side of rule R_1. However, before doing so, we rewrite R_1 as $R_1 : p(X', Z)\!:\!-p(X', Y), p(Y, Z)$ in order to avoid conflicts with the variable names. The resulting search tree $T_{1,1}$ is depicted in Fig. 7.2.

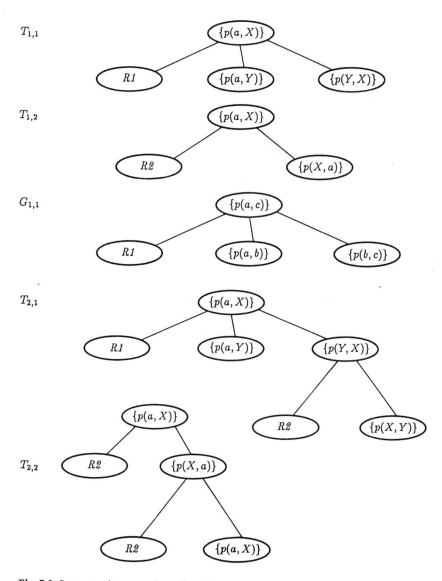

Fig. 7.2. Some search trees and one Proof tree

A second possibility is to unify $p(a,X)$ with the left-hand side of R_2. Again, before doing so, we rewrite R_2 as $R_2 : p(X',Y) \colon -p(Y,X')$, since the variable name X already occurs in $p(a,X)$. The resulting search tree $T_{1,2}$ is shown in Fig. 7.2.

The search tree $T_{1,1}$ can be transformed in a unique way into a Proof tree: the only substitution λ for the variables X and Y such that $p(a,Y)\lambda \in P \cup EDB$ and $p(Y,X)\lambda \in P \cup EDB$ is $\lambda = \{X/c, Y/b\}$. The resulting Proof tree $G_{1,1}$ is depicted in Fig. 7.2. On the other hand, there is no way of transforming $T_{1,2}$ into a Proof tree, because no ground clause of EDB matches with $\{p(X,a)\}$. Fig. 7.2 also displays two search trees of depth 2: $T_{2,1}$ and $T_{2,2}$. $T_{2,1}$ is an expansion of $T_{1,1}$ and can be transformed in a unique way into a Proof tree for $\{p(a,a)\}$; this Proof tree is depicted in Fig. 7.1 of Sect. 7.1.1. $T_{2,2}$ is an expansion of $T_{1,2}$ and can be transformed in a unique way into a Proof tree for $\{p(a,b)\}$. Note, however, that $\{p(a,b)\}$ has already been proven using a Proof tree of depth 0. $\square$

Note that not all possibly existing Proof trees can be generated by backward chaining, but only those whose leaves are all labeled with ground facts belonging to $P \cup EDB$; such Proof trees are called *full Proof trees*. However, it is not hard to see that whenever a Proof tree for a ground fact G exists, then there is also a full Proof tree for G.

It is intuitively clear that each ground fact which can be derived from $P \cup EDB$ in a bottom-up fashion by constructing a full Proof tree G, can also be derived in a top-down fashion (via backward chaining) by first constructing a search tree T and then transforming T into G. Since, for each ground fact which is a consequence of $P \cup EDB$, a full Proof tree exists, we conclude that each such fact can be derived through backward chaining. This result, also known as the "completeness of backward chaining", can be formally proven by induction on the depth of G. Moreover, it can be shown that each ground fact of the form $p(a_1, \ldots, a_n)$ which logically follows from $P \cup EDB$ can be derived via backward chaining by constructing a search tree starting with the top goal $p(a_1, \ldots, a_n)$ or starting with any top goal which is *more general* than $p(a_1, \ldots, a_n)$, and, in particular, with the most general top goal $p(X_1, \ldots, X_n)$.

On the other hand, it is trivial to see that whenever a ground fact G is deduced via backward chaining from $P \cup EDB$, then $G \in cons(P \cup EDB)$, since, in this case, a Proof tree for G exists. This simple result is also known as the "soundness of backward chaining". Hence, being complete and sound, backward chaining is an alternative to forward chaining for deriving ground facts from a *Datalog* program and an EDB. A formal completeness and soundness Proof for backward chaining is conceptually simple, but involves some rather lengthy technical arguments which we want to omit here. However, we formulate our observations above as a theorem.

Let us denote by $BACK_{P,EDB}(G)$ the set of all ground facts which are obtained by backward chaining from a given *Datalog* program P and a given extensional database EDB by starting with the top goal G. Note that if G is a ground fact then $BACK_{P,EDB}(G)$ is either empty or consists of the singleton $\{G\}$.

Theorem 7.13. Let P be a *Datalog* program and let EDB denote an extensional database.

a) For any ground fact G, $(P \cup EDB) \models G$ iff $BACK_{P,EDB}(G) = \{G\}$.
b) If a goal G is specified together with P, then

$$\mathcal{M}_{P,G}(EDB) = BACK_{P,EDB}(G).$$

c) Let IP denote the set of all intensional predicates occurring in P. It holds that:

$$\mathcal{M}_P(EDB) = \bigcup_{p \in IP} BACK_{P,EDB}(p(X_1, \ldots, X_{arity(p)})).$$

Part a) of Theorem 7.13 states that backward chaining is complete and sound. Part b) treats the case in which a particular goal G is specified together with the program P and we are interested in instances of G only. Part (c) states that each ground fact which logically follows from $P \cup EDB$ can be obtained by a top-down derivation starting with a top goal whose arguments are all variables.

Theorem 7.13 suggests the following method for computing $\mathcal{M}_P(EDB)$: let $RESULT$ be a set-variable used for accumulating step by step the ground facts of $\mathcal{M}_P(EDB)$. $RESULT$ is initially empty. Start by considering the top goals $p(X_1, \ldots, X_{arity(p)})$ for all intensional predicates appearing in P. Construct all Proof trees of depth 0 from these goals and add the labels of their nodes to $RESULT$. Expand the top goals to all possible search trees of depth 1. Transform these search trees in all possible ways into Proof trees of depth 1. Add the labels of all Proof trees obtained to $RESULT$. Expand the search trees of depth 1 in all possible ways to search trees of depth 2, transform these search trees in all possible ways into Proof trees, and so on.

Unfortunately, Theorem 7.13 doesn't tell us when to stop this process. Note also that the process in general doesn't stop by itself: search trees exist which can be expanded to any arbitrary level (consider, for instance, Tree $T_{2,2}$ of Fig. 7.2). Of course, at a certain point, $RESULT$ contains all ground facts of $\mathcal{M}_P(EDB)$. However, how do we know when?

Let us show that for each *Datalog* program P and for each extensional database EDB an upper bound which limits the depth of all search trees (and, accordingly, of all Proof trees) that we have to consider exists. Let $nconst(P, EDB)$ denote the number of distinct constant symbols which occur in $P \cup EDB$, let $npreds(P)$ denote the number of distinct predicate symbols occurring in P and let $maxargs(P)$ denote the maximum arity of all predicates in P. Let $maxdepth(P, EDB) = npreds(P) \times nconst(P, EDB)^{maxargs(P)}$.

Theorem 7.14. Backward chaining remains complete if the depth of the search trees considered is limited to $maxdepth(P, EDB)$.

Proof Sketch. Let us first note that $maxdepth(P, EDB)$ constitutes an upper bound for the number of distinct ground facts which may appear as labels in a Proof tree built from P and EDB. Indeed, it is easy to see that the ground facts labeling the nodes of any such Proof tree are all made of predicate symbols which occur in P and of constants which occur in $P \cup EDB$. It is obvious that no more than $maxdepth(P, EDB)$ distinct ground facts of this type can exist. (Note, however, that several nodes can be labeled with the same ground fact.)

Observe, furthermore, that we may limit our attention to full Proof trees containing at each level at least one "new fact", i.e., a node which is labeled by a ground fact which does not occur as a label in any lower level: if a Proof tree contains a level with no new facts, then this level is useless and an equivalent full Proof tree with fewer levels exists.

Since there are at most $maxdepth(P, EDB)$ distinct ground facts which may appear as labels, any full Proof tree which contains at least one new fact at each level cannot have more than $maxdepth(P, EDB)$ levels. This shows that we may limit our attention to trees whose depth is less than or equal to $maxdepth(P, EDB)$. □

The generation of all search trees and Proof trees of depth $\leq maxdepth(P \cup EDB)$ can be performed according to different control strategies. One possibility is to first generate all trees of depth 0, then all trees of depth 1, then all trees of depth 2 and so on, until all trees of depth $\leq maxdepth(P \cup EDB)$ are produced. This strategy is called *level saturation strategy* and corresponds to a searching technique which is well known as *breadth-first search*. Another possibility would be to generate all trees in a *depth-first* order by using recursion and backtracking.

Several methods for evaluating *Datalog* programs and *Datalog* goals in a top-down fashion through backward chaining have been presented in the literature. In principle, these methods are all similar to the one that we have just outlined. However, the methods may differ from one another (and from the method put forward here) in the following characteristics:

a) *Derivation Trees.* Not all methods can be adequately described using the separate concepts of search tree and Proof tree. Several methods construct slightly different types of tree. For instance the resolution method, which we discuss in the next section, constructs a unique type of tree which combines the features of search and Proof trees. In most methods, the information about the rules used during a derivation (such as the rule-identifier nodes in our search trees) is not maintained.

b) *Data Structures.* Search trees, Proof trees, and other types of trees are relevant, as they describe the computation from a conceptual viewpoint. However, they often do not appear as explicit data structures. In most systems, the subgoals (i.e., the labels that would appear on the leaves of a search tree), the answers to *Datalog* queries, and all the intermediate results are handled either as sets or as lists.

c) *Tree Traversal.* Some methods use breadth-first search, other methods use depth-first search. Among the latter, most use left-to-right backtracking, but a few use more complicated methods based on heuristics.

d) *Halting Condition.* The halting condition presented here, i.e., the bound $maxdepth(P, EDB)$ which limits the depth of the trees generated, can be classified as a (rather weak) *static halting condition*, because this bound is established before the evaluation of a *Datalog* program takes place and is independent of the computation itself. Many top-down strategies use more sophisticated *dynamic halting conditions*, which are sensitive to the computation behavior and strongly depend on the values that some program variables take during execution. We have already used such halting conditions in our forward chaining algorithms (for instance, the condition "*new $\neq$ old*" of the INFER algorithm). In the case of backward chaining, the halting condition is usually formulated in terms of both the set of subgoals that have been visited and the set of results that have been generated in the course of the previous computation.

e) *Subtree Factoring.* A subgoal may occur in various places within a search tree or Proof tree giving rise to identical subtrees. It is possible to avoid the repeated recomputation of such identical subtrees by keeping track, at each instant, of all subgoals that have been processed previously. Similarly – and more generally – one can avoid computing subtrees corresponding to subgoals which are *subsumed* by other subgoals that have already occurred earlier in the computation. This strategy of avoiding redundant computations is referred to as "subtree factoring". A well known efficient strategy which uses subtree factoring is the *Query-Subquery* method (QSQ) by Vieille. For this method an iterative and a recursive variant exist; the recursive variant will be presented in Chap. 9.

In the next section, we will discuss a particularly interesting class of top-down evaluation strategies: the *resolution method*.

7.3.2 Resolution

Resolution is a well-known technique which was originally used for automated theorem proving in the context of general clauses (and not just Horn clauses). The theoretical basis for resolution, i.e., the soundness and completeness of the resolution principle for general clauses was established in 1965 in a seminal paper by J.A. Robinson (see the bibliographic notes at the end of this chapter). Since then, resolution has conquered the world of symbolic computation like no other technique, and several refinements of resolution have been studied and implemented. In particular, the entire discipline of logic programming is based on the paradigm of resolution. The specific resolution technique used in logic programming is called SLD-resolution, which is a shorthand for "Linear Resolution with Selection Function for Definite Clauses". SLD-resolution does not stand for a particular algorithm, but for an entire class of algorithms. Among the different resolution methods, this class is also the most appropriate one for answering *Datalog* queries.

In this section, we briefly describe how *Datalog* queries can be answered using SLD-resolution. Readers who are interested in a more rigorous treatment of the theoretical aspects of SLD-resolution are referred to standard sources, referenced in the bibliographic notes; in particular, the book by Lloyd contains the Proofs of most of the theorems presented below.

Before giving a formal definition of SLD-resolution, let us outline the main characteristics of this method.

1) *Logical Goals Interpretation.* In the context of the resolution method, goals are negative clauses which have to be *disproved* by the system. For instance, if we have a *Datalog* program P, an extensional database EDB and a goal "$?-p(a,X)$", then this goal represents the assertion $\forall X : \neg p(a,X)$. Via the resolution method, one tries to *disprove* this assertion by generating *counterexamples*, i.e., by finding constants α, such that $(P \cup EDB) \models p(a,\alpha)$. Of course, the existence of each such α contradicts the assertion $\forall X : \neg p(a,X)$. We then say that $p(a,\alpha)$ is a counterexample to our goal. The set $\{p(a,\alpha)\}$ of all such counterexamples to the goal "$?-p(a,X)$" is the desired solution to our query. Each different way of disproving the goal "$?-p(a,X)$" leads to a single element of the solution. If "$?-p(a,X)$" cannot be disproved by resolution, then the solution is the empty set. Each Proof of the falsehood of a goal G, i.e., each way of disproving G, is called a *refutation* of G out of $P \cup EDB$.

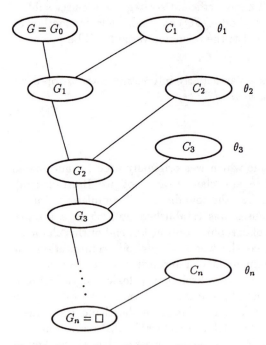

Fig. 7.3. General shape of a linear refutation tree

2) *Linear Refutation Tree.* In order to find a refutation of a goal G out of $P \cup EDB$, one builds a linear refutation tree, as depicted in Fig. 7.3. Note that such trees, unlike most other types of tree, are graphically represented in a very "natural" way with the root at the bottom. However, linear refutation trees are built in a top-down fashion starting with a top node labeled with the

original goal $G = G_0$ (top goal), and finishing with the root node labeled with a symbol □, denoting a logical contradiction (indeed, it is our aim to *disprove* the goal G, i.e., to show that from this goal one can derive a contradiction). All clauses G_i consist of negated literals only and represent *derived goals* which are obtained through the application of the resolution rule. The C_i are clauses from $P \cup EDB$ or variants thereof. Linear refutation trees, or parts of such trees, combine the functions of both search trees and Proof trees. Each such tree corresponds to the refutation of a particular goal via SLD-resolution. A precise definition is given below.

3) *Tree Traversal and Halting Condition.* The type of tree traversal used for finding all refutation trees to a goal is implementation-dependent and is not part of the resolution principle itself. Most systems use depth-first search. In a similar way, the halting condition is system-dependent. Some systems, such as Prolog interpreters, do not use any particular halting condition. In such systems, there is no a priori guarantee for completeness; rather, completeness issues and termination problems are deferred to the responsibility of the programmer.

Let us now describe more formally the basic concepts of the resolution method.

Recall that a goal is a clause consisting of one or more negated literals. A *selection function* is a mapping, which associates to each goal one of its literals. Intuitively, the literal selected is the one to be considered "next" in the course of a refutation procedure.

Example 7.15. The most widely used selection function is the one which associates to each goal its leftmost (i.e., its first) literal. This selection function is used by Prolog interpreters. Another selection function can be defined by choosing the most instantiated literal from each goal, i.e., the literal containing the maximum number of constants (if several such literals exist, then the leftmost one). □

We are now going to define the concept of resolvent. In *Datalog*, a goal H (possibly a compound goal) can produce a resolvent either with a *Datalog* rule or with a *Datalog* fact. The resolvent itself is a (possibly compound) goal or the empty clause.

Let $E = \{\neg K_1, \ldots, \neg K_i, \ldots, \neg K_k\}$ be a (possibly compound) goal and let R be a *Datalog* rule of the form $L_0\colon -L_1, \ldots L_n$. Let s be a selection function such that $s(E) = K_i$. Assume furthermore that a **mgu** θ of K_i and L_0 exists (the concept of **mgu** was introduced in Sect. 6.1.3). The *resolvent of the goal E and the rule R* according to selection function s is the goal
$H : \{\neg K_1\theta, \ldots, \neg K_{i-1}\theta, \neg L_1\theta, \ldots, \neg L_n\theta, \neg K_{i+1}\theta, \ldots, \neg K_k\theta\}$.

If $F = \{L\}$ is a fact and a **mgu** θ of K_i and L exists, then the *resolvent of the goal E and the fact F* according to selection function s consists of the clause $J : \{\neg K_1\theta, \ldots, \neg K_{i-1}\theta, \neg K_{i+1}\theta, \ldots, \neg K_k\theta\}$, i.e., the goal obtained from E by dropping K_i and by applying θ to all literals.

Since more than one **mgus** of a couple of literals may exist, there is, in general, more than one resolvent of E and R (or of E and F). However, as long as we use the same selection function, these resolvents are all variants of one another, i.e., they are all equal up to the renaming of variables. Also, if we use a particular algorithm for finding **mgus** (for instance the **mgu** algorithm of Sect. 6.1.3), then the resolvent of two clauses according to a selection function s is uniquely determined.

Recall that the variables occurring in different clauses are completely independent from one another. In order to avoid name clashes during the resolution process, it is sufficient to make sure that the clauses which are candidates for producing a resolvent do not have any variable in common. This can easily be obtained by renaming variables.

Example 7.16. Consider the following goal E and clauses R and F:

$$E: \{\neg p(a, X), \neg p(b, X)\}.$$
$$R: p(X, Z) :- p(X, Y), p(Y, Z).$$
$$F: \{p(b, c)\}.$$

Let s_l denote the selection functions which always selects the leftmost literal of a goal and let s_r denote the selection function which always selects the rightmost literal of a goal.

First of all, we note that the variable X of R also occurs in E. For this reason, we rename X in R by $X1$, getting a variant R' of R:

$$R': p(X1, Z) :- p(X1, Y), p(Y, Z).$$

Applying our **mgu** algorithm, it is easy to see that the first literal of E can be unified with the head of R' via the **mgu** $\{X1/a, Z/X\}$. Therefore, a resolvent H of E and R' exists, according to s_l:

$$H: \{\neg p(a, Y), \neg p(Y, X), \neg p(b, X)\}.$$

From the same clauses E and R', by using selection function s_r, we get a resolvent H' defined as follows:

$$H': \{\neg p(a, X), \neg p(b, Y), \neg p(Y, X)\}.$$

Now consider the pair of clauses E and F. Clearly, the unique literal of fact F does not unify with the first literal of E. Hence E and F do not produce any resolvent according to s_l. On the other hand, E and F do produce a resolvent J when the selection function s_r is used:

$$J: \{\neg p(a, c)\}. \qquad \square$$

Let S be a set of *Datalog* clauses, let G be a *Datalog* goal and let s be a selection function. An *SLD-refutation* of $S \cup \{G\}$ via s consists of a finite sequence of goals $G = G_0, \ldots, G_n = \square$, a sequence $C_1, \ldots, C_n$ of clauses, and a sequence of substitutions $\theta_1, \ldots, \theta_n$ such that the following conditions are satisfied:

- Each clause C_i, for $1 \leq i \leq n$ is either a clause of S or a variant of a clause of S.
- Each goal G_i, for $1 \leq i \leq n$ is a resolvent of the clauses G_{i-1} and C_i according to selection function s.
- For $1 \leq i \leq n$, θ_i is the **mgu** used in the resolution of G_{i-1} and C_i.

If the underlying set S of *Datalog* clauses is understood, then we may speak about an SLD-refutation of G instead of an SLD-refutation of $S \cup \{G\}$.

Each SLD-refutation can be represented by a linear refutation tree as depicted in Fig. 7.3.

Example 7.17. Consider again the *Datalog* program P consisting of the rules R_1: $p(X, Z) :- p(X, Y), p(Y, Z)$ and R_2: $p(X, Y) :- p(Y, X)$ and an extensional database EDB consisting of the ground facts:

$$\{p(a, b)\}, \{p(b, c)\}, \{p(c, d)\}, \{p(d, e)\}.$$

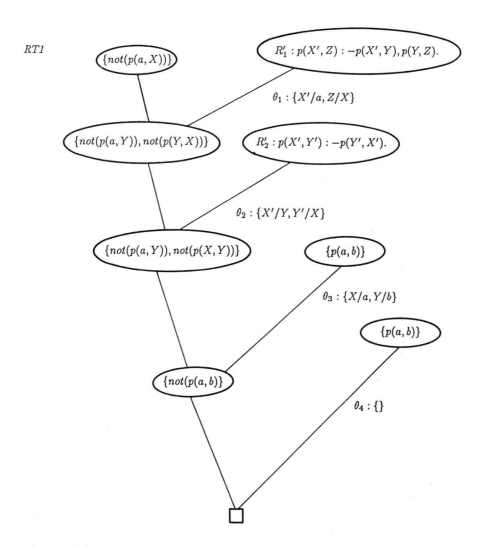

Fig. 7.4a. Refutation tree RT1

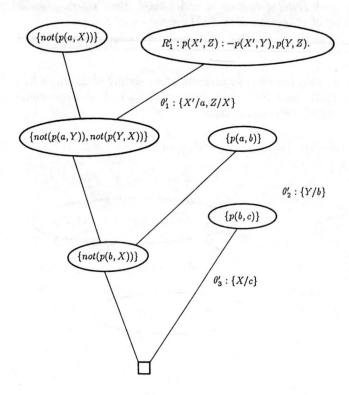

Fig. 7.4b. Refutation tree RT2

Figs. 7.4a and 7.4b show two different linear refutation trees RT_1 and RT_2 for the goal $G: ?-p(a, X)$. RT_1 is a refutation tree according to the selection function s_r which always selects the rightmost literal from a clause. RT_2, on the other hand, represents a refutation according to the selection function s_l, which from each goal selects the leftmost literal. Intuitively, one can see that RT_1 corresponds to a refutation of of the goal $? - p(a, X)$, i.e., of the clause $\{\neg p(a, X)\}$ by proving that the fact $\{p(a, a)\}$ can be derived from $P \cup EDB$. Of course, the validity of $\{p(a, a)\}$ is a counterexample to the validity of our goal G, which states that for all X, $\neg p(a, X)$ holds. On the other hand, RT_2 represents a refutation of the same goal by proving that $\{p(a, c)\}$ follows from $P \cup EDB$. The fact $\{p(a, c)\}$ is just another counterexample to this goal. □

Refuting a *Datalog* goal $G = \{\neg p(t_1, \ldots, t_k)\}$ by SLD-resolution is equivalent to proving that a ground instance of $\{p(t_1, \ldots, t_k)\}$ can be derived from the underlying set S of *Datalog* clauses. Such a ground instance is called a *counterexample* to the goal G. Indeed, the existence of such an instance contradicts the validity of G. As we have seen in Example 7.17, different refutations of the same goal G may lead to different counterexamples. Each refutation generates exactly one counterexample. This counterexample consists of the ground instance of $\{p(t_1,\ldots,t_k)\}$, which is obtained by successively applying to $\{p(t_1,\ldots,t_k)\}$ all the unifiers $\theta_1, \theta_2, \ldots \theta_n$ which appear in the derivation. In other words, the counterexample is equal to $\{p(t_1,\ldots,t_k)\}\theta$, where $\theta = \theta_1\theta_2\cdots\theta_n$.

Example 7.18. Consider the two refutations depicted in Figs. 7.4a and 7.4b. For the first one (corresponding to the tree RT_1), we have $\theta = \theta_1\theta_2\theta_3\theta_4 = \{X'/b, Z/a, Y'/a, X/a, Y/b\}$, and hence the counterexample generated is $p(a, X)\theta = p(a, a)$. For the second derivation (see tree RT_2), we have $\theta' = \theta'_1\theta'_2\theta'_3 = \{X/c, X'/a, Y/b, Z/c\}$ and thus the counterexample generated is $p(a, c)$. □

Let us call the substitution $\theta = \theta_1\theta_2\cdots\theta_n$ associated to a given SLD-refutation the *answer substitution* generated by this refutation. What we have said before about refutations can be stated as a theorem, which is generally known as the *soundness theorem for SLD-resolution*. Proofs of Theorem 7.15 and of the following Theorem 7.16 are omitted, and can be found in the references (see the bibliographic notes).

Theorem 7.15 (Soundness of SLD-resolution). Let S be a set of *Datalog* clauses and G a goal of the form $\{\neg p(t_1,\ldots,t_k)\}$. For any SLD-refutation of $S \cup \{G\}$ with the answer substitution θ, it holds that: $S \models \{p(t_1,\ldots,t_k)\}\theta_1\theta_2\cdots\theta_n$.

Let us now draw our attention to the completeness of SLD-resolution. Completeness, in this context, means that whenever a counterexample to a *Datalog* goal G exists on the base of a set S of *Datalog* clauses, then an SLD-refutation of $S \cup \{G\}$ which generates precisely this counterexample also exists. In particular, one can show that SLD-resolution remains complete, even if a specific selection function is enforced:

Theorem 7.16 (Completeness of SLD-resolution). Let S be a set of *Datalog* clauses, let G be a *Datalog* goal of the form $\{\neg p(t_1,\ldots,t_k)\}$ and let s be a selection function. If there is a ground instance C of $\{p(t_1,\ldots,t_k)\}$ such that $S \models C$ (i.e., if there is a counterexample C to G), then an SLD-refutation of $S \cup \{G\}$ exists, according to s with an answer substitution θ, such that $C = \{p(t_1,\ldots,t_k)\theta\}$.

SLD-resolution is particularly useful for computing answers to a *Datalog* program P which operates on an extensional database EDB and to which a *Datalog* goal G has been specified.

Let us call the *success set* and denote by $success(S, G, s)$ the set of all counterexamples to G that can be generated by all possible refutations of $S \cup \{G\}$ according to a selection function s. Recall from Chap. 6 that we denote the output of a *Datalog* program P with goal G operating on an extensional database EDB by $\mathcal{M}_{P,G}(EDB)$. The following corollary can easily be derived from Theorem 7.15 and Theorem 7.16:

Corollary. (to Theorems 7.15 and 7.16) Let P be a *Datalog* program, let G be a *Datalog* goal, and let EDB denote an extensional database. For any selection function s it holds that: $\mathcal{M}_{P,G}(EDB) = success(P \cup EDB, G, s)$.

The obvious method for computing $\mathcal{M}_{P,G}(EDB)$ by SLD-resolution is to choose a selection function s and to generate all possible refutations of G according to s by iterative or recursive techniques. Of course, in doing so, we are faced with the same "halting problem" as in our general backward chaining method. This problem, in the context of resolution, can be solved in a similar way by showing that for each *Datalog* program P and for each extensional database EDB, a constant $maxdepth'(P, EDB)$ exists such that it suffices to consider only those linear refutation trees whose depth is less than $maxdepth'(P, EDB)$. Note, however, that most resolution-based systems do not use such a bound. They use either more sophisticated halting conditions based on the dynamic behavior of the program (see for instance the QSQ method described in Part III of this book) or they do not use any halting condition at all, renouncing completeness.

Note that the resolution method is very similar to the general top-down method described in the previous section. It is not hard to see that in a linear refutation tree, the set of all resolution steps in which a goal or subgoal is resolved against a *rule* corresponds to a search tree. Indeed, in these steps the goal or subgoal is expanded. On the other hand, whenever a subgoal is resolved against a *ground fact* (and shrinks in size), this corresponds to transforming a piece of the search tree into a piece of a Proof tree. The "search steps" and the "Proof steps" of an SLD-refutation are, in general, interleaved. Note also that, due to the linear structure of an SLD-refutation, we expect that the depth of a linear refutation tree is exponential in the depth of a corresponding Proof tree in the worst case.

7.4 Conclusions

In this chapter we have presented different paradigms for evaluating *Datalog* programs including the INFER algorithm, which represents the simplest bottom-up evaluation strategy. Then we presented some basic results of Fixpoint Theory and showed that the computation performed by the INFER algorithm corresponds to least fixpoint evaluation. Finally we considered the class of top-down methods based on the principle of backward chaining.

The principles and algorithms considered in this chapter are formulated independently of all issues related to data retrieval and database access. It is important, however, to understand how these methods can be used when the facts of the EDB are stored in mass memory. In this case, particular optimization methods are needed for minimizing the number of accesses to secondary storage. Chaps. 8 through 10 of this book will deal with these problems.

7.5 Bibliographic Notes

A general introduction to the evaluation of *Datalog* programs can be found in a survey paper by Gardarin and Simon [Gard 87]. The Proof theory of *Datalog*, as developed in Sect. 7.1, has not appeared as yet in the literature.

The Fixpoint Theorem (Theorem 7.7) is due to Tarski [Tars 55], where it appears in a more general form. A historical overview of the development of fixpoint theory is given in [Lass 82]. The fixpoint semantics of logic programs was first described by Van Emden and Kowalski in [VanE 76] and is also well explained in [Lloy 87].

An interesting paper on backward chaining is the one by Smith, Genesereth, and Ginsberg [Smit 86]. This paper deals in particular with the problem of how far search trees must be expanded in order to produce all answers to a given query. A complete backward chaining algorithm is presented which uses quite sophisticated halting conditions. Another important approach to backward chaining is the Query-Subquery (QSQ) method by Vieille [Viei 86a, 87] which we will discuss in detail in Chap. 9.

The resolution principle for general clauses was established by Robinson [Robi 65] in the context of automated theorem proving. A good overview of several resolution techniques in that context can be found in [Chan 73] and [Love 78]. SLD-resolution is introduced in [VanE 76] and extensively described in [Lloy 87]. Completeness and soundness Proofs for SLD-resolution can be found in these references.

7.6 Exercises

7.1 Let S be defined as in Example 7.4. Find two different Proof trees for $S \vdash \{p(a,e)\}$.

7.2 Write a processor for *Datalog* programs based on the INFER algorithm. Use Prolog or Lisp as your programming language. You may assume that both the *Datalog* program P and the extensional database EDB are available as lists of clauses in main memory.

7.3 Let G be a directed graph with the set of vertices $v(G)$ and let "$\longrightarrow$" be a binary relation on $v(G)$ defined as follows: for all x and y in $v(G)$, $x \longrightarrow y$ iff $x = y$ or a directed path from x to y exists. Under which condition is this relation a partial order on $v(G)$?

7.4 Find a finite set V and a partial order on V, such that not all subsets of V have a least upper bound and not all subsets of V have a greatest lower bound.

7.5 As a generalization of the transformation T_1 defined in Example 7.12 consider a generic poset $(V, \leq)$ and show that for each $a \in V$ the transformations T_a defined by $\forall x \in V : T_a(x) = inf(a,x)$ is monotonic. What about the transformations T^a defined by $\forall x \in V : T^a(x) = sup(a,x)$? *Hint: use induction on i and exploit the monotonicity of T.*

7.6 Show that the values $\bot = T^0(\bot), T^1(\bot), \ldots, T^i(\bot) \ldots$ computed successively by the LFP algorithm form a chain with respect to the ordering "$\leq$" of the underlying lattice $(V, \leq)$, i.e., $\bot = T^0(\bot) \leq T^1(\bot) \leq \ldots \leq T^{i-1}(\bot) \leq T^i(\bot) \leq T^{i+1}(\bot) \leq \ldots$.

7.7 We have shown in Theorem 7.10 that for each set S of *Datalog* clauses, the fixpoints of T_S are precisely the Herbrand models of S. Show that a similar statement does not hold for T'_S. In order to establish a characterization of the Herbrand models of S in terms of T'_S, show that $I \subseteq HB$ is a Herbrand model of S iff $T'_S(I) \subseteq I$.

7.8 Modify the *Datalog* processor built according to Exercise 7.2 by replacing the main while loop with the LFP algorithm. Test both algorithms using sample data and compare their runtime behavior. (Note: we expect just a slight improvement.)

7.9 Fig. 7.2 does not display all search trees of depth 2 according to Example 7.14. Which trees of depth 2 are not depicted in the figure?

7.10 Each of the search trees mentioned in Example 7.14 can be transformed in no more than one way into a Proof tree. Maintaining the same extensional database EDB and the same *Datalog* program P as used in the example, show that search trees which can be transformed into different Proof trees exist. *Hint: try another top goal).*

7.11 Find a Proof tree which is not full and indicate an equivalent full Proof tree.

7.12 Write a metainterpreter, on the top of a Prolog interpreter, which processes Prolog clauses from right to left, i.e., according to the selection function s_r which always chooses the rightmost literal from each goal.

7.13 Let P and EDB be as in Example 7.17. Indicate two different refutation trees for the goal $p(a, e)$ which both use the same selection function.

7.14 Write an interpreter for *Datalog* programs in Prolog which exploits the resolution mechanism of Prolog and which, for a given goal, builds all linear refutation trees up to a specified depth. A call to this interpreter should have the form $Datalog(Goal, Depth)$. You may assume that all clauses of the *Datalog* program and of the extensional database have been loaded just like regular Prolog clauses into the Prolog clause database.

Part III
Optimization Methods for Datalog

In Part II of this book we have properly defined the meaning of computing a program (or query) in *Datalog*, and we have overviewed various alternative models and interpretations. Some of them (the proof-theoretic model, the fixpoint interpretation, backward chaining) also provide methods for computing the result of a query. Our emphasis, however, has been on providing an *operational semantics*, and hence on stressing the understanding of meaning, rather than efficiency of computation.

In Part III of this book, efficiency rather than meaning becomes our major concern. We consider *Datalog* programs operating on potentially large databases, and we propose methods for their efficient computation. The methods are classified according to various criteria: we highlight the difference between program transformations and program evaluations, between bottom-up and top-down strategies, between algebraic and logic methods, and so on.

The underlying motivation for all these methods is to generate efficient execution strategies, where efficiency is evaluated in terms of the interaction with the database system. Thus, a common denominator of all methods is to exploit the additional selectivity that becomes available when a user specifies a query over an existing set of *Datalog* rules. The proposed methods take full advantage of the query optimization techniques designed for traditional database systems, but the problems with *Datalog* are much harder due to the presence of recursion.

After an extensive discussion of the different computation methods and optimization strategies, several extensions and enhancements of pure *Datalog* are introduced. In particular, we will draw our attention to the use of built-in predicates, negation, and complex objects in *Datalog* programs. Finally, the last chapter of this book is dedicated to a survey of research prototypes for the integration of relational databases and logic programming.

Chapter 8
Classification of Optimization Methods for Datalog

In this chapter, we provide a classification of optimization methods for evaluating *Datalog* goals. We use orthogonal classification criteria in order to identify various classes of methods; then, each class will be dealt with in a separate section. A systematic overview of methods is required, since optimization can be achieved using a variety of techniques, and understanding their relationships is not obvious.

This chapter is organized as follows: first we describe several criteria that can be used to classify optimization methods, then we group methods into homogeneous classes, and we present an overall view of the optimization of *Datalog* goals. Then, we show a systematic method for translating logic rules and goals into equations of relational algebra. Finally, we present an optimization-oriented classification of the rules and program structures that can occur in *Datalog*. This classification enables us to identify optimization methods specific for each class.

8.1 Criteria for the Classification of Optimization Methods

We classify optimization methods according to the *formalism*, the *search strategy*, the *objective of the optimization*, and the *type of information considered*.

8.1.1 Formalism

As we already know, the body of a *Datalog* rule can be translated into an algebraic expression; more generally, each set of *Datalog* rules which define the same intensional predicate (which we henceforth also call *Intensional Database (IDB) predicate*) corresponds to an *equation* of relational algebra, and a *Datalog* program, defining several IDB predicates, corresponds to a *system of equations* of relational algebra. This means that optimizing *Datalog* goals is equivalent to optimizing the search for the solution of a system of algebraic equations. Thus, we consider in this book two alternative formalisms, that we regard respectively as *logical* and *algebraic*, and we discuss optimization methods that belong to both worlds. The *translation* from the logical formalism to the algebraic formalism is described in the third section of this chapter.

8.1.2 Search Strategy

Let us restrict our attention to logic approaches. We recall from Chap. 7 that the evaluation of a *Datalog* goal requires building a proof tree. This tree can be constructed in two different ways: *bottom-up*, starting from the existing facts and inferring new facts, thus proceeding towards the conclusions, or rather *top-down*, trying to verify the premises which are needed in order for the conclusion to hold.

This distinction was introduced in Chap. 7: the INFER algorithm, which builds the answer to a *Datalog* program through forward chaining is an example of a bottom-up method, while refutation procedures using resolution are examples of top-down evaluation. Here we discuss the features of these two approaches from the perspective of *optimization*.

In fact, these two evaluation strategies represent different interpretations of a rule. *Bottom-up* evaluations consider rules as *productions*, which apply the initial program to the extensional database and produce all the possible consequences of the program, until no new fact can be deduced. The answer is the subset of the consequence set formed by the tuples that satisfy the goal. Bottom-up methods can naturally be applied in a set-oriented fashion, i.e., taking as input the entire relations of the extensional databases. This is a desirable feature in the *Datalog* context, where large quantities of data must be retrieved from mass memory. On the other hand, bottom-up methods do not take immediate advantage of the existence of bound arguments in the goal predicate.

In top-down evaluation, instead, rules are seen as *problem generators*. Each goal is considered a problem that must be solved. The initial goal is unified with the left-hand side of some rule, and *generates* other problems corresponding to the right-hand side literals of that rule; this process is continued until the proof tree is completed. In this case, if the goal contains some bound argument, then only facts that are somehow related to the goal constants are involved in the computation. Thus, this evaluation mode performs a relevant optimization, because the computation automatically disregards many of the facts which are not useful for producing the result. On the other hand, in top-down methods it is more natural to produce the answer *one-tuple-at-a-time*, and this is an undesirable feature in *Datalog*.

If we restrict our attention to the top-down approach, we can further distinguish two search methods: *breadth-first* and *depth-first*.

With depth-first search, the generation of subproblems due to the effect of the right-hand side of a rule occurs according to some ordering of the right-hand side subgoals, and produces search trees which grow *in depth*. The disadvantage of this approach is that the chosen order of literals in rule bodies strongly affects the performance of methods. This happens in *Prolog*, where not only efficiency, but even *termination* of programs, is affected by the left-to-right order of subgoals in the rule bodies.

With breadth-first search, the generation of all the subproblems from the right hand side of a rule is done at the same time, thus producing a balanced growth of the search tree. *Datalog* goals seem to be more naturally executed through breadth-first techniques, so that the result of the computation is neither affected by the order of predicates within the right-hand sides (RHS) of rules, nor by the order of rules within the program.

8.1.3 Objectives of Optimization Methods

Our third classification criterion is based on the different *objectives* of optimization methods. Some methods perform *program transformation*, namely, they transform a program into another program which is written in the same formalism, but which yields a more efficient computation when one applies an evaluation method to it; we refer to these as *rewriting methods*. These methods contrast with the *pure evaluation methods*, which propose effective evaluation strategies, where the optimization is performed during the evaluation itself.

Since the input program and the output program produced by a rewriting method are written in the same formalism, it is possible to compare them, thus evaluating the *efficiency* of the transformation; in particular, some methods work poorly on certain input programs, producing output programs which are either less efficient or equally efficient but more complex than the input programs. In these cases, a "blind" acceptance of the results is not advised.

8.1.4 Type of Information Considered

Optimization methods differ in the *type of information used in the optimization*.

Syntactic optimization is the most widely used; it deals with those transformations to a program which descend from the program's syntactic features. In particular, we distinguish two kinds of structural properties. One is the analysis of the *program structure*, and in particular the type of rules which constitute the program. For example, some methods exploit the *linearity* (see Sect. 8.4) of the rules to produce optimized forms of evaluation. The second one is the *structure of the goal*, and in particular the selectivity that comes from goal constants. These two approaches are not mutually exclusive: it is possible to build syntactic methods which combine both cases.

Semantic optimization, instead, concerns the use of additional semantic knowledge about the database in order to produce an efficient answer to a query; the combination of the query with additional semantic information is performed automatically. Semantic methods are often based on integrity constraints, which express properties of valid databases. For instance, a constraint might state that "*all intercontinental flights directed to Milan land in Malpensa airport*". This constraint can be used to produce the answer of a goal asking for "*the arrival airport of the intercontinental flight AZ747 from New York to Milan"* without accessing the EDB. Though we think that semantic optimization has the potential for significant improvements of query processing strategies, we do not further consider semantic optimization methods in this book; a review of the proposals presented in the literature can be found in the bibliographic notes.

8.2 Classification of Optimization Methods

The classification criteria presented in the previous section are summarized in the table below.

In the sequel, we do not consider semantic optimization methods further. Therefore, the main classification criteria concern the *search strategy* (bottom-up or top-down), the *objective* (rewriting or pure evaluation), and the *formalism* (logic or algebraic).

CRITERION	ALTERNATIVES	
Formalism	logic	relational algebra
Search technique	bottom-up	top-down
Traversal order	depth-first	breadth-first
Objective	rewriting	pure evaluation
Approach	syntactic	semantic
Structure	rule structure	goal structure

By combining approaches, and excluding some alternatives that do not correspond to relevant classes of methods, we obtain four classes, each including rather homogeneous methods:

1) Bottom-up evaluation methods
2) Top-down evaluation methods
3) Logic rewriting methods
4) Algebraic rewriting methods

We do not stress much the difference between logic and algebraic evaluation methods.

In Sect. 9, we first examine algebraic bottom-up evaluation methods; these methods are quite similar in nature to the algorithm INFER, the logic version of *naive* evaluation, which was presented in Chap. 7. Then, we examine top-down evaluation methods which use a mixed logic/algebraic formalism.

	BOTTOM-UP	TOP-DOWN
	Section 9.1	Section 9.2
EVALUATION METHODS	Naive (Jacobi, Gauss-Seidel) Semi-naive Henschen-Naqvi	Query-subquery (and variants)
	LOGIC	ALGEBRAIC
	Section 10.1	Section 10.2
REWRITING METHODS	Magic sets Counting Static filtering Special semi-naive	Variables reduction Constant reduction

The table above contains the methods which will be presented, together with an indication of the section in which they are considered.

The existence of various optimization methods and formalisms generates the potential for several alternative approaches to the evaluation of a *Datalog* goal, represented in Fig. 8.1.

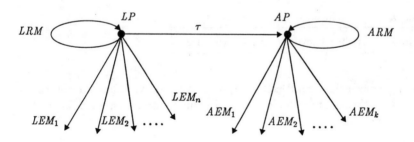

Fig. 8.1. Possible approaches to goal evaluation

Given an initial *Datalog* goal operating on a *Datalog* program LP, the following steps are feasible:

a) LP can be optimized by applying (possibly more than one) logical rewriting methods LRM.
b) LP can be directly evaluated by one of the logical evaluation methods LEM, thus producing the goal answer.

c) *LP* can be translated into equations of relational algebra, turning into an Algebraic Program *AP*.
d) *AP* can be optimized by (possibly more than one) algebraic rewriting methods *ARM*.
e) *AP* can be directly evaluated by one of the algebraic evaluation methods *AEM*, thus producing the goal answer.

Various paths along the graph represented in Fig. 8.1 are feasible; in particular:

a) The *logic approach* consists in solving the initial logic program *LP* using the methods *LRM* and *LEM* in sequence. This is achieved, for instance, by using the *magic set* or *counting* method in order to obtain a logic program LP' which exploits the structure of the goal constants, followed by a bottom-up computation using the logical version of the *naive* or *semi-naive* method.
b) The *algebraic approach* consists in applying immediately the transformation τ to the initial program *LP*, thus producing the corresponding initial program *AP*, and then solving *AP* using the methods *ARM* and *AEM* in sequence. This is achieved, for instance, by using the two algorithms for *variable* and *constant reduction*, followed by the *Gauss-Seidel* iterative method.
c) The *mixed approach* consists in applying a rewriting method *LRM* to the initial program *LP*, obtaining an optimized program LP', then applying the transformation τ to generate the corresponding algebraic program AP', and finally solving AP' using an evaluation method *AEM*.

The above three cases are represented in Fig. 8.2.

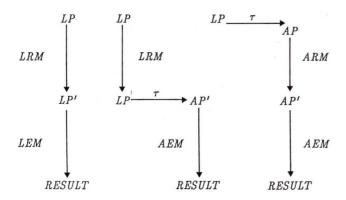

Fig. 8.2. Reasonable approaches to goal evaluation

Finally, we expect optimization methods to satisfy three important properties:

- Methods must be *sound*: they should not include in the result tuples which do not belong to it.
- Methods must be *complete*: they must produce all the tuples of the result.
- Methods must *terminate*: the computation should be performed in finite time.

All the methods we introduce satisfy these properties, although formal proofs are omitted.

8.3 Translation of Datalog into Relational Algebra

In this section, we deal formally with the translation of *Datalog* into Relational Algebra. Each clause of a *Datalog* program is translated, by a syntax-directed translation algorithm, into an inclusion relationship of Relational Algebra. The set of inclusion relationships that refer to the same predicate is then interpreted as one *equation* of *Positive Relational Algebra (RA⁺)*. Thus, a *Datalog* program gives rise to a system of algebraic equations. Each IDB predicate of the *Datalog* program corresponds to a *variable relation*; each EDB predicate of the *Datalog* program corresponds to a *constant relation*. Determining a *solution of the system* corresponds to determining the value of the variable relations which satisfy the system of equations.

Let us consider a *Datalog* clause:

$$C : p(\alpha_1, \ldots \alpha_n) \; :- \; q_1(\beta_1, \ldots, \beta_k), \ldots q_m(\beta_s, \ldots, \beta_h).$$

Let $P, Q_1, \ldots Q_m$ denote the intensional and extensional relations that correspond to predicates $p, q_1, \ldots, q_m$. Note that some of the q_i might be p itself. The translation will associate to C an inclusion relationship:

$$Expr(Q_1, \ldots, Q_m) \subseteq P.$$

The syntax-directed translation algorithm requires the definition of some notation, plus two term-rewriting functions:

- $occurs(\alpha_i, RHS)$, of type boolean, evaluates to **true** if the term α_i occurs on the right-hand side of the clause, **false** otherwise.
- $corr(i)$ denotes the function returning, for any α_i on the left-hand side, the index j of the first variable β_j on the right-hand side (if it exists) such that $\beta_j = \alpha_i$.
- $const(\alpha_i)$, type boolean, evaluates to **true** if α_i is a constant, to **false** otherwise. $const(\beta_j)$ is defined in the same way.
- $var(x_i)$, type boolean, is defined as "$\neg const(x_i)$".
- $newvar(x)$ is a procedure which, when called, returns a new name of variable.
- Let E be a string, x and y symbols. Then $E < x, y >$ is the new string obtained by replacing with y the first occurrence of x in E.
- $EQ(x, y)$ denotes a fictitious binary relation having one tuple for each constant value in the EDB, such that the first and second attributes take equal values. EQ is used by the translator to express the equality of two attributes; obviously, the *implementation* of EQ is not through an extensional relation.

The following two recursive rules apply respectively to the left- and right-hand side of C. The first rule concerns two special cases:

a) The left-hand side contains constant arguments.
b) The left-hand side contains more than one occurrence of the same variable.

Rule 1 converts each of these special cases, using appropriate string transformations, into an equivalent case, where all the left-hand side (LHS) arguments are *different variables*. Then, rule 2 is applied to the right hand side of C. This rule generates selection conditions due to the presence of either constants or replicated variables in the right-hand side (RHS). Once these cases are considered, rule 2 builds the cartesian product of all the (intensional and extensional) relations which correspond to the rule predicates. We denote by $\times_{i=1}^{n} R_i$ the cartesian product of relations $R_1, \ldots, R_n$.

After the application of the second rule, we obtain an inclusion relationship between two relational algebra expressions. The LHS of the inclusion relationship is a single variable relation; the RHS of the inclusion relationship is a relational expression, which can be transformed according to the usual algebraic transformations. For instance, *joins* can replace selections on cartesian products.

TRANSLATION ALGORITHM

    INPUT: clause $C = LHS \ : - \ RHS$
                $(= \ p(\alpha_1, \ldots \alpha_n) \ : - \ q_1(\beta_1, \ldots, \beta_k), \ldots q_m(\beta_s, \ldots, \beta_h))$.
    OUTPUT: an expression $Expr(Q_1, \ldots, Q_m)$ of the relations
                corresponding to predicates $q_1 \ldots, q_m$.
    BEGIN
        APPLY RULE 1 TO C
    END
  RULE 1: $T(C)$
    BEGIN
      IF $\exists \, i \, : \, const(\alpha_i)$ THEN   /*case a*/
      BEGIN
         $newvar(x)$; RETURN $T\Big(LHS < \alpha_i, x > \ : - \ RHS, \ EQ(x, \alpha_i)\Big)$
      END
      ELSIF $\exists \, i, j \, : \, \alpha_i = \alpha_j, \ i < j$ THEN   /*case b*/
      BEGIN
         $newvar(x)$; RETURN $T\Big(LHS < \alpha_i, x > \ : - \ RHS, \ EQ(x, \alpha_j)\Big)$
      END
      ELSE RETURN $\Pi_{corr(1), \ldots corr(n)} T'\Big(RHS\Big)$
    END
  RULE 2: $T'(C)$
    BEGIN
      IF $\exists i \, : \, const(\beta_i)$ THEN   /* case a */
      BEGIN
         $newvar(x)$; RETURN $\sigma_{i=\beta_i} T'\Big(RHS < \beta_i, x >\Big)$
      END

```
ELSIF ∃i,j : β_i = β_j, i < j    THEN    /* case b */
BEGIN
    newvar(x); RETURN  σ_{i=j}T'(RHS < β_i, x >)
END
ELSE RETURN  ×_{i=1}^m Q_i
END
```

Note that we use special EDB predicates to express arithmetic conditions (e.g., $lessequal(X,Y)$ to express terms as $X \leq Y$); the syntactic translation of these predicates requires the introduction of extensional relations to express all pairs of database constants that satisfy the arithmetic conditions (e.g., the binary relation $LESSEQUAL$). These *builtin* predicates will be dealt with in more depth in Chap. 11. Once the translation is performed, however, it is possible to express arithmetic conditions through selection predicates. Thus, for instance, the RHS of

$$p(X,Y) : -q(X,Y), lessequal(X,Y)$$

is translated by the syntax-directed algorithm into:

$$\Pi_{1,2}\sigma_{1=3 \wedge 2=4}(Q \times LESSEQUAL)$$

but can be simplified into:

$$\Pi_{1,2}\sigma_{1 \leq 2}Q.$$

Example 8.1. The following is a systematic application of the translation algorithm. Given the rule:

$$p(X,X,Z) :- s(X,Y), r(Y,a,Z).$$

by T, case (b):

$$T(p(N1,X,Z) :- s(X,Y), r(Y,a,Z), eq(N1,X))$$

by T, last recursive call:

$$\Pi_{6,1,5}T'(S(X,Y), R(Y,a,Z), EQ(N1,X))$$

by T', cases (a) and (b):

$$\Pi_{6,1,5}\sigma_{4=a \wedge 2=3 \wedge 1=7}T'(S(N2,N3), R(Y,N4,Z), EQ(N1,X))$$

by T', last recursive call:

$$\Pi_{6,1,5}\sigma_{4=a \wedge 2=3 \wedge 1=7}(S \times R \times EQ)$$

After "pushing" selection and join conditions:

$$\Pi_{6,1,5}((S \underset{2=1}{\bowtie} (\sigma_{2=a}R)) \underset{1=2}{\bowtie} EQ)$$

Final inclusion relationship:

$$\Pi_{6,1,5}((S \underset{2=1}{\bowtie} (\sigma_{2=a}R)) \underset{1=2}{\bowtie} EQ) \subseteq P$$

□

8.3 Translation of Datalog into Relational Algebra

We now transform algebraic inclusion relationships into algebraic equations. For each intensional predicate p, we collect all the inclusion relationships of the type:

$$Expr_i(Q_1,\ldots,Q_m) \subseteq P \qquad (1 \leq i \leq m_p)$$

obtained from rules T and T'. From these inclusion relationships, we generate an *algebraic equation* having P as LHS, and Q as the *union* of all the left-hand sides of the inclusion relationships:

$$P = Expr_1(Q_1,\ldots,Q_m) \cup Expr_2(Q_1,\ldots,Q_m) \ldots \cup Expr_{m_p}(Q_1,\ldots,Q_m).$$

This transformation of several disequations into one equation really captures the least Herbrand model semantics of a *Datalog program*. In fact, it expresses the fact that *we are only interested in those ground facts that are consequences of our program*, i.e., we only want to compute facts that are *defined* by some relationship contained in the program.

Example 8.2. The following *Datalog* program:

$$p(X) :- p(Y), r(X,Y).$$
$$p(X) :- s(Y,X).$$
$$q(X,Y) :- p(X), r(X,Z), s(Y,Z).$$

is translated into:

$$\Pi_2(P \underset{1=2}{\bowtie} R) \subseteq P$$

$$\Pi_2 S \subseteq P$$

$$\Pi_{1,4}((P \underset{1=1}{\bowtie} R) \underset{3=2}{\bowtie} S) \subseteq Q$$

From these three inclusion relationships, we obtain the following two equations:

$$\Pi_2(P \underset{1=2}{\bowtie} R) \cup \Pi_2 S = P$$

$$\Pi_{1,4}((P \underset{1=1}{\bowtie} R) \underset{3=2}{\bowtie} S) = Q$$

□

Notice that, by construction, the final system of equations is based on the use of the relational operations selection, projection, cartesian product, and union, but does *not* include the difference operation. Thus, the final system of equations is written in *positive relational algebra*, RA^+ (see Chap. 2).

We now translate *logic goals* into *algebraic queries*. Input *Datalog* goals are translated into projections and selections over one variable relation of the system of algebraic equations. For this translation, we extend the projection operator to include the projection over the empty set:

$$\Pi_\emptyset E = \begin{cases} \text{``no''}, & \text{if } E \text{ is the empty relation;} \\ \text{``yes''}, & \text{otherwise.} \end{cases}$$

We use the projection on the empty set to express goals with no free arguments.

Let p be the goal predicate, let argument i be bound to constant c_i in the goal, and let P denote the relational variable corresponding to p; the constant c_i is translated into a selection term $i = c_i$.

a) A *goal selection* is required when at least one of the arguments of p is bound to a constant value; it is obtained as the conjunction of the selection terms corresponding to all goal constants.

b) A *goal projection* $\Pi_\emptyset P$ is generated only when all arguments of the goal predicate are bound to constants.

In the sequel, we will use the special relational operator *composition* between two binary relations. Let R and S be binary relations. The composition (o) of R and S is defined as:

$$R \circ S = \Pi_{1,4}(R \underset{2=1}{\bowtie} S).$$

Note that composition is associative.

Example 8.3

a. The following are examples of translations of logic goals into algebraic queries.

$? - p(X). \Leftrightarrow P$
$? - p(a). \Leftrightarrow \Pi_\emptyset(\sigma_{1=a}P)$
$? - q(a,X). \Leftrightarrow \sigma_{1=a}Q$
$? - q(a,b). \Leftrightarrow \Pi_\emptyset(\sigma_{1=a \wedge 2=b}Q)$

b. The following *Datalog* program will be used as an example in the following sections:

$p_1(X,Y) :- c_1(X,Y).$
$p_1(X,Y) :- p_1(X,Z), p_3(Z,Y).$
$p_1(X,Y) :- p_2(X,Y).$
$p_2(X,Y) :- p_1(X,Z), p_3(Z,Y).$
$p_2(X,Y) :- c_3(X,Y).$
$p_3(X,Y) :- p_3(X,Z), c_2(Z,Y).$
$p_3(X,Y) :- c_4(X,Y).$

Using the composition operator, we translate this program into the following set of relational algebra inclusion relationships (where each predicate p_i corresponds to the relational variable X_i, and each database predicate c_j corresponds to the relational constant C_j).

$C_1 \subseteq X_1$
$X_1 \circ X_3 \subseteq X_1$
$X_2 \subseteq X_1$
$X_1 \circ X_3 \subseteq X_2$
$C_3 \subseteq X_2$
$X_3 \circ C_2 \subseteq X_3$
$C_4 \subseteq X_3$

This yields the system of equations:

$$X_1 = C_1 \cup (X_1 \circ X_3) \cup X_2$$
$$X_2 = (X_1 \circ X_3) \cup C_3$$
$$X_3 = (X_3 \circ C_2) \cup C_4$$

The query $? - p(a, X).$ is translated into $\sigma_{1=a} X_1$. □

Though the translation from *Datalog* to RA^+ is the more useful, it is also possible to translate equations of RA^+ into *Datalog* rules. We demonstrate this fact informally, by showing that all operations of RA^+ can be translated. "$eq(X,Y)$" denotes the equality predicate, which is satisfied iff $X = Y$.

Projection. Consider the equation $X = \Pi_{i_1,\ldots i_h} Y$, with X an intensional relation of arity h and Y an intensional or extensional relation of arity k ($h \leq k$). This equation corresponds to the following *Datalog* rule:

$$x(\alpha_{i_1}, \alpha_{i_2}, \ldots \alpha_{i_h}) : - y(\alpha_1, \alpha_2, \ldots \alpha_k).$$

Selection. Consider the equation $X = \sigma_{i_1=t_1 \wedge \ldots i_n=t_n} Y$, where X and Y have the same arity k, the i_j are attribute numbers, and the t_j are either constants or attribute numbers. This equation corresponds to the following *Datalog* rule:

$$x(\alpha_1, \alpha_2, \ldots \alpha_k) : - y(\alpha_1, \alpha_2, \ldots, \alpha_k), eq(\alpha_{i_1}, \tau_1), \ldots, eq(\alpha_{i_n}, \tau_n).$$

where

$$\tau_j = \begin{cases} t_j, & \text{if } t_j \text{ is a constant;} \\ \alpha_{t_j}, & \text{if } t_j \text{ is an attribute number.} \end{cases}$$

Cartesian product. Consider the equation $X = Y_1 \times \ldots \times Y_n$. Let k be the arity of X and k_i be the arity of Y_i, $1 \leq i \leq n$, $k_1 + k_2 + \ldots + k_n = k$. We translate this as:

$$x(\alpha_1 \ldots \alpha_k) : - y_1(\alpha_1, \ldots \alpha_{k_1}), \ldots, y_n(\alpha_{k-k_n-1+1}, \ldots \alpha_k).$$

Union. Consider the equation $X = Y_1 \cup Y_2$, X, Y_1, and Y_2 being relations with the same arity k. This equation produces the two following *Datalog* rules:

$$x(\alpha_1, \alpha_2, \ldots \alpha_k) : - y_1(\alpha_1, \alpha_2, \ldots \alpha_k).$$
$$x(\alpha_1, \alpha_2, \ldots \alpha_k) : - y_2(\alpha_1, \alpha_2, \ldots \alpha_k).$$

Note that the join is obtained by applying selection and cartesian product. We show an example of the translation of a join; consider the following equation:

$$P = Q_1 \underset{1=2}{\bowtie} Q_2 \underset{3=2}{\bowtie} Q_3 \qquad (1)$$

where $arity(P) = 6$, $arity(Q_1) = arity(Q_2) = arity(Q_3) = 2$. Then, the translation is easily obtained:

$$p(x_1, \ldots x_6) : -q_1(x_1, x_2), q_2(x_3, x_4), q_3(x_5, x_6), eq(x_1, x_4), eq(x_3, x_6).$$

Example 8.4

a. Consider the following definition of relation R:

$$R = Q_1 \cup Q_2 \cup \Pi_{1,3} Q_3$$

where $arity(R) = arity(Q_1) = arity(Q_2) = 2, arity(Q_3) = 3$. Then, the *Datalog* rules we obtain are:

$r(X, Y) : -q_1(X, Y).$
$r(X, Y) : -q_2(X, Y).$
$r(X, Z) : -q_3(X, Y, Z).$

b. Consider now the following equation:

$$P = \Pi_{1,3}((Q_1 \underset{1=2}{\bowtie} Q_2) \underset{3=2}{\bowtie} Q_3)$$

where $arity(P) = arity(Q_1) = arity(Q_2) = arity(Q_3) = 2$. This equation is obtained from equation (1) above by changing the arity of P and by adding a projection to the RHS. The translation is:

$$p(x_1, x_3) : -q_1(x_1, x_2), q_2(x_3, x_4), q_3(x_5, x_6), eq(x_1, x_4), eq(x_3, x_6).$$

This rule can be simplified into a more readable one:

$$p(x_1, x_3) : -q_1(x_1, x_2), q_2(x_3, x_1), q_3(x_5, x_3).$$

□

Thus, we have shown informally that each defining expression of RA^+ can be translated into a *Datalog* program.

8.4 Classification of Datalog Rules

We give now a classification of *Datalog* rules; this classification will be used in the following sections, where we will select the optimization methods based on the class of rules of a given *Datalog* program.

Initially, we introduce some useful notions. Let P be a *Datalog* program,

$$P : p_i(\alpha_1, \ldots \alpha_{n_i}) : - q_1(\beta_1, \ldots, \beta_k), \ldots q_m(\beta_s, \ldots, \beta_h). \qquad i = 1 \ldots r,$$

defining n IDB predicates $p_1 \ldots p_n$ ($1 \leq n \leq r$). Let X_P be the set of IDB predicates of P. Let Σ be the system of n equations equivalent to P,

$$\Sigma : P_i = E_i(P_1, \ldots P_n), \qquad i = 1 \ldots n,$$

and let X_Σ be the set of variable relations (IDB predicates) of Σ.

Definition 8.1. The *dependency graph* of P is the directed graph $G_P = <N, E>$ such that:

- $N = X_P$
- $E = \{<p_i, p_j> \iff p_j \text{ occurs in the body of a rule defining } p_i\}$

Definition 8.2. The *dependency graph* of Σ is the directed graph $G_\Sigma = <N, E>$ such that:

- $N = X_\Sigma$
- $E = \{<P_i, P_j> \iff P_j \text{ occurs in } E_i\}$

8.4 Classification of Datalog Rules

From the above definitions, the dependency graphs of P and Σ are identical up to node renaming; in the sequel, we will use dependency graphs both in the logic and algebraic formalism. Figure 8.3 shows the dependency graph of the system of Example 8.3b.

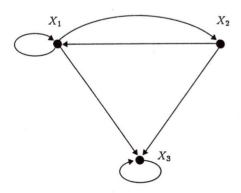

Fig. 8.3. Dependency graph for the system of Example 8.3b

Definition 8.3. Given a *Datalog* rule

$$C : p(\alpha_1, \ldots \alpha_n) :- q_1(\beta_1, \ldots, \beta_k), \ldots q_m(\beta_s, \ldots, \beta_h).$$

we say that it is *recursive* if at least one of the q_i is p. For example, the rule: $anc(X,Y) : -par(X,Z), anc(Z,Y).$ is recursive, while the rule: $aunt(X,Z) : -par(X,Y), sister(Y,Z).$ is not.

Definition 8.4. Given a *Datalog* program

$$P : p_i(\alpha_1, \ldots \alpha_{n_i}) :- q_1(\beta_1, \ldots, \beta_k), \ldots q_m(\beta_s, \ldots, \beta_h). \qquad (1 \leq i \leq r)$$

we say that P is *recursive* if the dependency graph of P contains at least *one* cycle.

A program containing recursive rules is recursive. The converse is not true: there are programs which are recursive but do not contain recursive rules.

Example 8.5. The following program, computing the couples $< X, Y >$ with Y ancestor of X, is recursive:

$$anc(X,Y) :- par(X,Y).$$

$$anc(X,Y) :- par(X,Z), anc(Z,Y).$$

In fact, the second rule is recursive. Moreover, its dependency graph shown in Fig. 8.4(a) contains a cycle.

The following program is recursive, as is shown by its dependency graph in Fig. 8.4(b), but does not contain any recursive rules.

$$p(X,Y) :- r(X,Y), t(X,X).$$
$$p(X,Y) :- w(X,Y).$$
$$r(X,Y) :- s(W,X), p(W,Y).$$
$$r(X,Y) :- t(X,Y).$$
$$w(X,Y) :- s(W,X), t(Z,Y).$$

□

Definition 8.5. Given an intensional predicate p in a *Datalog* program P, we say that *p is recursive* if the corresponding node in the dependency graph of P is contained in a cycle.

Definition 8.6. Given two intensional predicates p and q in a *Datalog* program P, we say that p and q are *mutually recursive* if the corresponding nodes in the dependency graph of P are both contained in the same cycle.

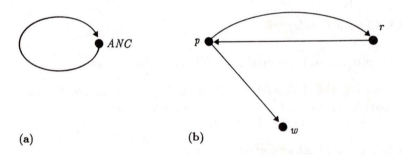

(a)   (b)

Fig. 8.4. Dependency graphs for recursive programs

Thus, predicates p and r of Example 8.5 are recursive, and in particular mutually recursive, while predicate w is not recursive.

Definition 8.7. Given a *Datalog* rule

$$R : p(\alpha_1, \ldots \alpha_n) :- q_1(\beta_1, \ldots, \beta_k), \ldots q_m(\beta_s, \ldots, \beta_h).$$

we say that *R is linear with respect to q_i* (or *in q_i*) if there is at most one occurrence of q_i in the rule body.

For example, $anc(X,Y) :- par(X,Z), anc(Z,Y).$ is linear in anc. The rule $anc(X,Y) :- anc(X,Z), anc(Z,Y).$ is not linear in anc.

Definition 8.8. Given a *Datalog* program P, containing a rule

$$R : p(\alpha_1, \ldots \alpha_n) :- q_1(\beta_1, \ldots, \beta_k), \ldots q_m(\beta_s, \ldots, \beta_h).$$

8.4 Classification of Datalog Rules

we say that rule R is *linear* if there is at most one q_i in the rule body (possibly p itself) which is mutually recursive to p in P.

Note that a rule can be *linear* and *recursive*. Of the two rules above, for predicate *anc*, the first one is linear and recursive, the second is only recursive.

Definition 8.9. Given a *Datalog* program

$$P : p_i(\alpha_1, \ldots \alpha_{n_i}) :- q_1(\beta_1, \ldots, \beta_k), \ldots q_m(\beta_s, \ldots, \beta_h). \qquad i = 1 \ldots r$$

we say that P is *linear* if all its rules are linear.

Example 8.6. Consider the program:

(1) $p_1(X, Y) :- c_1(X, Y).$
(2) $p_1(X, Y) :- p_1(X, Z), p_3(Z, Y).$
(3) $p_1(X, Y) :- p_2(X, Y).$
(4) $p_2(X, Y) :- p_1(X, Z), p_3(Z, Y).$
(5) $p_2(X, Y) :- c_3(X, Y).$
(6) $p_3(X, Y) :- p_3(X, Z), c_2(Z, Y).$
(7) $p_3(X, Y) :- p_1(X, Y).$

obtained from the program of Example 8.3b by modifying rule 7. Its dependency graph is shown in Fig. 8.5.

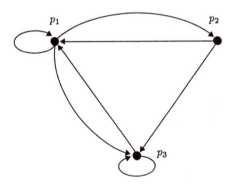

Fig. 8.5. Dependency graph of the program in Example 8.6

In this program, predicates p_1, p_2, and p_3 are all mutually recursive, and thus recursive. Moreover, we can observe that all the rules are linear with respect to all the predicates, but rules 2 and 4 are *not* linear rules. In fact, the body of rule 2 contains one occurrence of p_1 and one of p_3, which are both mutually recursive to p_1. A similar observation holds for rule 4. Thus, the program considered is not linear. □

Let us consider the algebraic counterpart of linearity:

Definition 8.10 An equation $X = E(X_1, \ldots, X_n)$ of RA^+ is *linear with respect to X_i* (or *in X_i*) iff it is derived from a set of Datalog rules which are linear in the predicate p_i corresponding to X_i.

For instance, the equation $ANC = PAR \cup PAR \circ ANC$, derived from the anc program, is linear in the variable relation ANC.

It is very easy to see that, if an equation $X = E(X_1, \ldots, X_n)$ of RA^+ is linear with respect to X_i, then, for any two relations $X_i^{(1)}$ and $X_i^{(2)}$ of the same arity as X_i, the following *mathematical linearity property* holds:

$$E(X_1, \ldots, X_i^{(1)} \cup X_i^{(2)}, \ldots, X_n) = $$
$$E(X_1, \ldots, X_i^{(1)}, \ldots, X_n) \cup E(X_1, \ldots, X_i^{(2)}, \ldots, X_n).$$

However, the converse is not true: there are non-linear equations which have the *mathematical linearity property*, which is thus a weaker condition than linearity. This more general property is useful because some optimization methods are applicable to mathematically linear equations, thus to a wider class of programs. One of these methods is *seminaive evaluation*, presented in Sect. 9.1.2.

The two notions are typically confused; this happens because they differ only in correspondence to special cases (see Example 8.7).

Example 8.7. Consider the rule:

$$p(X,Y) : -p(X,Z), a(Z,Y), p(X,W).$$

Clearly, this rule is not linear with respect to p, thus the corresponding algebraic equation

$$P = \Pi_{1,2}[(P \circ A) \underset{1=1}{\bowtie} P]$$

is not linear with respect to variable P. However, this equation has the *mathematical linearity property*:

$$\Pi_{1,2}[((P_1 \cup P_2) \circ A) \underset{1=1}{\bowtie} (P_1 \cup P_2)] = $$
$$\Pi_{1,2}[((P_1 \circ A \cup P_2 \circ A) \underset{1=1}{\bowtie} (P_1 \cup P_2)] = $$
$$\Pi_{1,2}[(P_1 \circ A) \underset{1=1}{\bowtie} P_1] \cup \Pi_{1,2}[(P_2 \circ A) \underset{1=1}{\bowtie} P_1] \cup $$
$$\Pi_{1,2}[(P_1 \circ A) \underset{1=1}{\bowtie} P_2] \cup \Pi_{1,2}[(P_2 \circ A) \underset{1=1}{\bowtie} P_2] = $$
$$(P_1 \circ A) \cup (P_2 \circ A).$$

The last equality comes from:

$$\Pi_{1,2}[(P_1 \circ A) \underset{1=1}{\bowtie} P_1] \subseteq P_1 \circ A.$$
$$\Pi_{1,2}[(P_2 \circ A) \underset{1=1}{\bowtie} P_2] \subseteq P_2 \circ A.$$
$$\Pi_{1,2}[(P_2 \circ A) \underset{1=1}{\bowtie} P_1] \subseteq \Pi_{1,2}[(P_2 \circ A) \underset{1=1}{\bowtie} P_2]$$
$$\Pi_{1,2}[(P_1 \circ A) \underset{1=1}{\bowtie} P_2] \subseteq \Pi_{1,2}[(P_1 \circ A) \underset{1=1}{\bowtie} P_1]$$

On the other hand:

$$\Pi_{1,2}[(P_1 \circ A) \underset{1=1}{\bowtie} P_1] \cup \Pi_{1,2}[(P_2 \circ A) \underset{1=1}{\bowtie} P_2] = $$
$$(P_1 \circ A) \cup (P_2 \circ A).$$

The equation is thus mathematically linear in P. □

8.4 Classification of Datalog Rules

Definition 8.11. An equation $X = E(X_1, \ldots, X_n)$ of a system Σ of RA^+ is *linear* if it is linear with respect to all the variables of Σ that derive from recursive predicates.

There is also an analogous definition for systems of relational algebra equations:

Definition 8.12. Given a system Σ of n equations:

$$\Sigma : X_i = E_i(X_1, \ldots X_n), \quad i = 1 \ldots n,$$

Σ is *linear* if and only if all its equations are linear.

Another important notion is *stability*. Let C be a linear recursive *Datalog* rule:

$$C : p(\alpha_1, \ldots \alpha_n) :- q_1(\beta_1, \ldots, \beta_k), \ldots q_m(\beta_s, \ldots, \beta_h).$$

A *chain* $q_1, \ldots q_k (1 \leq k \leq m)$ of predicate occurrences in the body of C is called a *stability chain* if the three conditions (a-c) below hold:

a) $q_1, \ldots, q_{k-1}$ are database predicates and q_k is p;
b) q_1 contains at least one variable argument x_1, called the *initial variable* of the chain, which matches one variable of p in the rule head;
c) For $1 \leq k$, $1 \leq i \leq k-1$, q_i contains a variable x_i which matches a variable x_{i+1} of q_{i+1}; x_k is called the *final variable* of the chain corresponding to the initial variable x_1.

Note that, in the definition of stability chain, the order of $q_1, \ldots, q_k$ does not necessarily coincide with the order of predicates in the rule body.

Definition 8.13. Let C be a linear recursive *Datalog* rule as above. Then C is *stable* iff for any stability chain $q_1, \ldots, q_k (1 \leq k \leq m)$ and any initial variable x_1, it happens that x_1 and the final variable x_k corresponding to x_1 have the same place in p and q_k respectively.

Example 8.8. The following are instances of stable rules:

(1) $p(X, Y) :- q_1(X, Z), p(Z, Y).$
(2) $r(X, Y, Z) :- s(X, X), r(X, Y, W), t(Z, W).$
(3) $sgc(X, Y) :- par(X, X_1), sgc(X_1, Y_1), par(Y, Y_1).$

The following rules are instead not stable:

(4) $sgc(X, Y) :- par(X, X_1), sgc(Y_1, X_1), par(Y, Y_1).$
(5) $s(X, Y, Z) :- r(X, Z), s(X, Y, W), t(Z, Y).$
(6) $p(X, Y) :- s(Z), r(X, W), p(Z, W).$

Consider for example rule (4). Here, $par(X, X_1), sgc(Y_1, X_1)$ is a stability chain, where X is the initial variable (in par), and X_1 is the final variable (in sgc). X occupies the first place in the LHS predicate sgc, while X_1 occupies the second place in the RHS predicate sgc. Thus, rule (4) is not stable. □

Stability is a very strong condition in a rule; however, it is very frequent. In practice, most programs have only stable rules.

Stability is important since some algorithms may become very simple if applied to stable rules. See for instance the algorithm *reduction of constants* described in Sect. 10.2.6.

8.5 The Expressive Power of Datalog

In Sect. 8.3 we have informally shown that each defining expression of RA^+ can be translated into a *Datalog* program. This means that *Datalog* is at least as expressive as RA^+.

It is easy to see that nonrecursive *Datalog* is exactly equivalent to RA^+; thus, full *Datalog* is strictly more expressive than RA^+, because in *Datalog* it is possible to express recursive queries, which are not expressible in RA^+. For example, the ANC relation can only be expressed by recursive rules, or by recursive RA^+ equations. This holds for most recursive *Datalog* programs.

On the other hand, there are expressions in full relational algebra that cannot be expressed by *Datalog* programs. These are the queries that make use of the *difference* operator. For example, given two binary relations R and S, there is no *Datalog* rule defining $R - S$. However, we will see in Chap. 11 that these expressions can be captured by the use of logical negation ($\neg$). Figure 8.6 represents the situation.

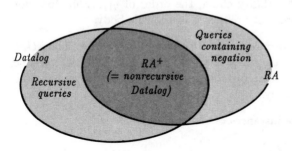

Fig. 8.6. The expressive power of Datalog

Note also that, even though *Datalog* is syntactically a subset of first-order logic, strictly speaking they are not comparable. Indeed, the semantics of *Datalog* is based on the choice of a particular model (the least Herbrand model), while first-order logic does not a priori require the choice of a particular model.

8.6 Bibliographic Notes

Semantic optimization concerns the use of additional semantic knowledge on the database in order to produce an efficient answer to a query. Obviously, the combination of the query with additional semantic information should be performed automatically. A pragmatic approach to semantic optimization, influenced by Artificial Intelligence, is proposed by King [King 81]; users' queries are composed with integrity constraints, both expressed through a simple formalism, thus enabling optimizations to be made at the logical level (e.g. determining that the answer to a query is empty because it contradicts some constraint) or at the physical level (e.g., using a better access path based on a condition inferred from a constraint). A quantitative model enables distinguishing between profitable and nonprofitable query transformations. Malley and Zdonik [Mall 86] show the usage of a knowledge base (a "library" of query optimization strategies) as a system's tool for query optimization. An approach based on standard logic is proposed by Chakravarthy, Minker, and Grant [Chak 86], [Chak 87], which show the technical problems posed by the combination of queries and constraints, both expressed through a logic formalism.

A problem related to semantic optimization, also not considered in this book, concerns the efficient evaluation of integrity constraints. Logic is an extremely powerful formalism for expressing constraints; as was shown in Chap. 1, the use of a logic programming formalism to describe queries and constraints provides a uniform environment, which is suitable for semantic query optimization. The main problem of expressing database constraints through logic concerns the efficient evaluation of the consistency of database states or transactions. Efficiency is typically achieved by assuming a consistent database state as the transaction's before-state, and by building the minimum proof tree that enables one to conclude the consistency of the transaction's after-state. Approaches to the construction of such a reduced proof tree can be driven by elementary operations of transactions (insertions or deletions), or by violated constraints, or by a combination of both. Various proposals have been presented by Decker [Deck 86], Bry and Manthey [Bry 86], Bry, Decker, and Manthey [Bry 88], Qian and Smith [Qian 87], Sheard and Steample [Shea 86], and Kowalski, Sadri, and Soper [Kowa 87]. The classic approach to constraint verification in database systems aims at the early detection of incorrect transactions in order to abort them. However, once violated constraints and violating operations are detected, it becomes possible to attempt to correct the transaction in order to produce an after-state which is consistent and reflects the user's needs. This approach has been proposed by Ceri and Garzotto in [Ceri88e], where the correction process is driven by the interaction with the user.

An interesting unifying paradigm for bottom-up vs. top-down evaluation is in [Bry 89b]. The syntax-directed translation algorithm from *Datalog* to RA^+ was presented in [Ceri 86a]. Various definitions of *linearity*, equivalent to ours, are present in [Banc 86b], [Ioan 87a], and [Ceri 86a]. The definition of *stable* rule is similar to that in [Hens 84].

Interesting analyses of the problem of the expressive power of relational query languages can be found in [Aho 79a] and [Abit 87b].

8.7 Exercises

8.1 Translate the following *Datalog* program into inclusion relationships of RA^+:

$b(X,Y) :- a(X), t(X,Z), s(Y,Z,X).$
$a(X) :- p(Y), b(Z,Y), c(X,Z).$
$a(X) :- t(Y,X), b(w,X).$

8.2 Obtain the equations of RA^+ corresponding to the inclusion relationships of Exercise 8.1.

8.3 Translate the following *Datalog* goals into relational algebra queries:

$?- s(z,X,y).$
$?- b(Y,X).$
$?- b(w,w).$

8.4
(1) Find the *Datalog* rules that correspond to the following system of algebraic equations on binary relations:

$X = A \cup Y$
$Y = Y \circ X \circ B$
$Z = X \circ B \circ A$

(2) Find the *Datalog* rules that correspond to the following system of algebraic equations. The relations are all binary.

$X_1 = \Pi_{1,6}\left((X_2 \underset{2=1}{\bowtie} X_3) \underset{2=1}{\bowtie} C_1\right)$
$X_2 = C_2 \circ X_2 \circ C_1 \circ X_4$
$X_3 = X_2 \circ X_3 \circ C_3$
$X_4 = C_4 \cup X_3 \cup C_2$

8.5 Classify all the properties of *Datalog* predicates, rules, and programs obtained as the results of Exercise 8.4.

8.6 Why are rules (5) and (6) of Example 8.7 not stable?

Chapter 9
Evaluation Methods

In this chapter, we present methods for *evaluating* a *Datalog* program, namely, for generating the actual set of tuples which satisfy a given user's goal for a given set of *Datalog* rules. We deal with *bottom-up* and *top-down* evaluation methods in two separate sections. Evaluation methods are expressed using either of the logic and algebraic formalisms introduced in the previous chapter, and in fact we do not stress the choice of formalism very much, given the equivalence between them.

9.1 Bottom-up Evaluation

In this section we present three bottom-up methods: *naive, semi-naive*, and *Henschen-Naqvi* methods. The former is the most general method, and has already been presented in Chap. 7 using a logical formalism (algorithm INFER); we present here two *algebraic* versions of the naive method, called *Jacobi* and *Gauss-Seidel* solution methods; these methods get their names from well-known algorithms for the iterative solution of systems of equations in Numerical Analysis. We then show how the *naive* evaluation can be improved, to give the so-called *semi-naive* evaluation; in particular, we show two variants of semi-naive methods which hold for linear and for general systems of relational algebra, respectively. We then briefly present a specialized method, the *Henschen-Naqvi* method, named after its designers, which applies to linear *Datalog* programs.

9.1.1 Algebraic Naive Evaluation

The algebraic naive method evaluates the solution (*fixpoint*) of a system of algebraic equations. Thus, this method is the simplest of the algebraic evaluation strategies AEM of Fig. 8.1.

Assume the following system Σ of relational equations:

$$R_i = E_i(R_1, \ldots, R_n) \quad (i = 1, \ldots, n).$$

The *Jacobi* method proceeds as follows. Initially, the variable relations R_i are set equal to the empty set. Then, the computation $R_i = E_i(R_1, \ldots, R_n)$, $i = 1, \ldots, n$ is iterated until none of the R_i changes between two consecutive iterations (namely, until the R_i reach a *fixpoint*).

ALGORITHM JACOBI
INPUT: A system of algebraic equations Σ and an Extensional Database EDB.
OUTPUT: The values of the variable relations $R_1, \ldots, R_n$.
METHOD:
FOR $i := 1$ TO n DO $R_i := \emptyset$;
REPEAT
    $cond := true$;
    FOR $i := 1$ TO n DO $S_i := R_i$;
    FOR $i := 1$ TO n DO
        BEGIN
        $R_i := E_i[S_1, \ldots, S_n]$;
        IF $R_i \neq S_i$ THEN $cond := false$;
        END;
UNTIL $cond$;
FOR $i := 1$ TO n DO OUTPUT(R_i).
ENDMETHOD

We now give a formal description of the *Jacobi* method in terms of fixpoint theory. First of all, we need to give some definitions. Let k_i be the arity of R_i, for $1 \leq i \leq n$. Let

$$Dom(R_i) = \{<a_1, \ldots, a_{k_i}> \text{ such that } a_j \in Const,\ j = 1, \ldots, k_i\}.$$

Let $\Omega = \{(X_1, \ldots, X_n) \text{ such that } X_i \subseteq Dom(R_i)\}$. We define a mapping:

$$T_\Sigma : \Omega \longrightarrow \Omega$$

such that $T_\Sigma(X_1, \ldots, X_n) = (E_1(X_1, \ldots, X_n), \ldots, E_n(X_1, \ldots, X_n))$.
The partial order $\preceq$ in Ω is defined in the following way:

$$(X_1, \ldots, X_n) \preceq (Y_1, \ldots, Y_n) \Longleftrightarrow (X_1 \subseteq Y_1 \wedge \ldots \wedge X_n \subseteq Y_n).$$

It is easy to see that:

a) $(\Omega, \preceq)$ is a complete lattice,
b) T_Σ is a monotonic transformation on $(\Omega, \preceq)$,
c) T_Σ corresponds exactly to T_S of Sect. 7.2.2. In fact, the tuples of constants contained in $Dom(R_i)$ are the relational correspondent of the elements of the Herbrand Base HB of Sect. 7.2.2.

Theorem 9.1. The result computed by the Jacobi algorithm is the least fixpoint of the transformation:

$$T_\Sigma : \Omega \longrightarrow \Omega$$

such that $T_\Sigma(X_1, \ldots, X_n) = (E_1(X_1, \ldots, X_n), \ldots, E_n(X_1, \ldots, X_n))$ induced by the system Σ.

We omit the proof of the Theorem; indeed, the structure of the Jacobi algorithm corresponds exactly to the structure of the algorithm LFP as described in Sect. 7.2.

The convergence of the *Jacobi* method can be slightly improved if, at each step k, in order to compute the new value $R_i^{(k)}$, we substitute in E_i the values of $R_j^{(k)}$, $j = 1\ldots, i-1$, that have just been computed in the same iteration instead of the old values $R_j^{(k-1)}$. This variant of the *Jacobi* method is called the *Gauss-Seidel* method; both methods are well known in numerical analysis.

ALGORITHM GAUSS-SEIDEL
INPUT: A system of algebraic equations Σ, and an Extensional Database EDB.
OUTPUT: The values of the variable relations $R_1, \ldots, R_n$.
METHOD:
FOR $i := 1$ TO n DO $R_i := \emptyset$;
REPEAT
   $cond := true$;
   FOR $i := 1$ TO n DO
      BEGIN
      $S := R_i$;
      $R_i := E_i[R_1, \ldots, R_n]$;
      IF $R_i \neq S$ THEN $cond := false$;
      END;
UNTIL $cond$;
FOR $i := 1$ TO n DO OUTPUT(R_i).
ENDMETHOD

Theorem 9.2. The Gauss-Seidel algorithm, applied to a system of algebraic equations Σ and to an extensional database EDB, produces the same result as the Jacobi algorithm.

The proof is trivial, and we leave it as an exercise to the reader (see Exercise 9.3).

Variants of the *Jacobi* or *Gauss-Seidel* methods are obtained by computing the various algebraic expressions concurrently rather than sequentially. Synchronization among computations is induced by the *dependency graph*: the computation of the value of a given relation R_i at the k-th iteration ($R_i^{(k)}$) is possible as soon as the values of all relations $R_j^{(k-1)}$ appearing in the equation $R_i = E_i(R_1, \ldots, R_n)$ have been computed. In this case, we say that $R_i^{(k)}$ is *computable*. The *Chaotic method* consists in evaluating computable relations in any order. The so-called *lazy* and *data flow* evaluations are subcases of this method, which correspond to evaluating computable relations at the latest and earliest convenience respectively.

We now give some examples of application of the Jacobi method.

Example 9.1.
a. We compute the logic program $L1$:

$$r_1: \qquad anc(X,Y) :- anc(X,Z), par(Z,Y).$$
$$r_2: \qquad anc(X,Y) :- par(X,Y).$$

with the EDB:

$PAR = \{<b,a>, <b,g>, <a,d>, <a,e>, <d,f>, <c,h>\}.$

We translate program $L1$ into the equation:

$$ANC = \Pi_{1,4}(ANC \underset{2=1}{\bowtie} PAR) \cup PAR$$

where ANC is the variable relation that corresponds to the intensional predicate anc. Note that by using the composition operator defined in Chap. 8, we can rewrite the above equation as:

$$ANC = ANC \circ PAR \cup PAR$$

Let us follow the *Jacobi* algorithm. Initially, we set $ANC^{(0)} = \emptyset$. We then enter the $REPEAT$ loop. We set $cond = true$. We initialize $S_1 = ANC = \emptyset$. Now we compute the first value of ANC. The join with an empty relation is empty, therefore:

$ANC^{(1)} = PAR = \{<b,a>, <b,g>, <a,d>, <a,e>, <d,f>, <c,h>\}$

$ANC^{(1)} \neq S_1$, thus $cond = false$

We therefore enter the second iteration. The next value of S_1 is $ANC^{(1)}$. We compute the next value of ANC as follows:

$$ANC^{(2)} = \{<b,a>, <b,g>, <a,d>, <a,e>, <d,f>,$$
$$<c,h>, <b,d>, <b,e>, <a,f>\}$$

Note that $ANC^{(2)} = ANC^{(1)} \cup \{<b,d>, <b,e>, <a,f>\}$.

$ANC^{(2)}$ is different from S_1, so we enter the third iteration, setting $S_1 = ANC^{(2)}$; we obtain:

$$ANC^{(3)} = \{<b,a>, <b,g>, <a,d>, <a,e>, <d,f>, <c,h>,$$
$$<b,d>, <b,e>, <a,f>, <b,f>\}$$

Note that $ANC^{(3)} = ANC^{(2)} \cup \{<b,f>\}$.

$ANC^{(3)}$ is different from S_1, so we enter the fourth iteration, setting $S_1 = ANC^{(3)}$; we obtain:

$$ANC^{(4)} = \{<b,a>, <b,g>, <a,d>, <a,e>, <d,f>,$$
$$<c,h>, <b,d>, <b,e>, <a,f>, <b,f>\}$$

Finally, $ANC^{(4)} = ANC^{(3)} = S_1$ and our evaluation is finished.

b. We use the same method to compute the goal:

$$? - sgc(a, Y).$$

for the logic program $L2$:

$r_1:$ $sgc(X, Y) :- eq(X, Y).$
$r_2:$ $sgc(X, Y) :- par(X, X_1), sgc(X_1, Y_1), par(Y, Y_1).$

The program $L2$ computes the same-generation cousins in a certain database. The EDB is represented by the relation

$$PAR = \{<d,g>, <e,g>, <b,d>, <a,d>, <a,h>, <c,e>\}$$

and by the relation EQ which equates each element of the DB domain to itself:

$$EQ = \{<a,a>, <b,b>, <c,c>,$$
$$<d,d>, <e,e>, <f,f>, <g,g>, <h,h>\}$$

Note that the extensional representation of EQ is introduced here for convenience, but EQ need not be stored explicitly in practice. Our goal asks for the cousins of a given constant value a. Initially, we translate the program and the goal. $L2$ is translated into:

$$SGC = \Pi_{1,5}((PAR \underset{2=1}{\bowtie} SGC) \underset{4=2}{\bowtie} PAR) \cup EQ$$

By using the composition operation, this becomes:

$$SGC = PAR \circ SGC \circ (\Pi_{2,1} PAR) \cup EQ.$$

According to Chap. 8, the goal becomes:

$$\sigma_{1=a} SGC.$$

The resulting values of SGC are the following:

$$SGC^{(0)} = \emptyset.$$

$$SGC^{(1)} = EQ = \{<a,a>, <b,b>, <c,c>, <d,d>,$$
$$<e,e>, <f,f>, <g,g>, <h,h>\}.$$

$$SGC^{(2)} = \{<a,a>, <b,b>, <c,c>, <d,d>, <e,e>, <f,f>,$$
$$<g,g>, <h,h>, <d,e>, <e,d>, <b,a>, <a,b>\}$$
$$= SGC^{(1)} \cup \{<d,e>, <e,d>, <b,a>, <a,b>\}$$

$$SGC^{(3)} = \{<a,a>, <b,b>, <c,c>, <d,d>, <e,e>, <f,f>,$$
$$<g,g>, <h,h>, <d,e>, <e,d>, <b,a>, <a,b>,$$
$$<b,c>, <a,c>, <c,b>, <c,a>\}$$
$$= SGC^{(2)} \bigcup \{<b,c>, <a,c>, <c,b>, <c,a>\}$$
$$SGC^{(4)} = \{<a,a>, <b,b>, <c,c>, <d,d>, <e,e>, <f,f>,$$
$$<g,g>, <h,h>, <d,e>, <e,d>, <b,a>, <a,b>,$$
$$<b,c>, <a,c>, <c,b>, <c,a>\} = SGC^{(3)} = S_1$$

Computation is finished, and the value returned by the *Jacobi* algorithm is $SGC^{(4)}$. In order to answer to the initial goal, we have to select the result relation $\sigma_{1=a} SGC^{(4)}$. The resulting tuples are:

$$\{<a,a>, <a,b>, <a,c>\}.$$

We see from this example that the *Jacobi* algorithm first produces the entire relation SGC, and then selects the tuples satisfying the goal. Only a subset of the tuples computed by the *Jacobi* method is really required for the answer. □

These two examples have demonstrated two sources of inefficiency.

a) Several tuples are computed many times during the iteration process. In particular, during the iterative evaluation of a relation R, tuples belonging to relations $R^{(i)}$ will also belong to all subsequent relations $R^{(j)}$, $j \geq i$, until the fixpoint is reached. This derives from the monotonicity property of the transformation T_Σ.

b) Several tuples are computed without being really required, and are eliminated by the final selection. Note that when the program is not recursive the algebraic expression that results from the translation algorithm can be easily optimized by standard relational algebra techniques, including the classical "push" of selection conditions into the expression. When the program is instead recursive, these transformations are not obvious, and more sophisticated optimization methods become necessary. Thus, for instance, the final selection in Example 9.1b cannot be "pushed" trivially.

We partially eliminate the inefficiencies due to observation a) through the *semi-naive* methods; we deal with the inefficiency due to observation b) with subsequent evaluation methods (*Henschen-Naqvi* and *Query-Subquery*) and with rewriting methods of Chap. 10.

9.1.2 Semi-naive Evaluation

Semi-naive evaluation is a bottom-up technique designed to eliminate redundancy in the evaluation of tuples at different iterations. This method does not use any information about the structure of the goal. It only exploits the structure of the program.

9.1 Bottom-up Evaluation

There are two possible settings of the semi-naive algorithm: the *(pure) semi-naive* and the *pseudo-rewriting semi-naive*. We will explain the first one here, and the second one in the next chapter, which deals with rewriting methods.

Consider the Jacobi algorithm. Let $R_i^{(k)}$ be the temporary value of relation R_i at iteration step k. The *differential* of R_i at step k of the iteration is defined as

$$D_i^{(k)} = R_i^{(k)} - R_i^{(k-1)}.$$

The differential term, expressing the new tuples of $R_i^{(k)}$ at each iteration k, is exactly what we would like to use at each iteration, instead of the entire relation $R_i^{(k)}$.

Consider the case of linearity. As remarked in Sect. 8.4, a recursive equation

$$R_i = E(R_i)$$

which is linear with respect to R_i satisfies the *mathematical linearity property*:

$$E(R_i' \cup R_i'') = E(R_i') \cup E(R_i'')$$

for any two relations R_i' and R_i'' having the same arity as R_i. As a consequence, the differential of R_i at step k of the iteration is simply the application of the expression E to the differential D_i, i.e. $E(D_i)$:

$$E(D_i^{(k)}) = E(R_i^{(k)} - R_i^{(k-1)}) = E(R_i^{(k)}) - E(R_i^{(k-1)})$$

so that one can compute $R_i^{(k+1)} = E(R_i^{(k)})$ as the union

$$E(R_i^{(k-1)}) \quad \cup \quad E(D_i^{(k)}).$$

Thus, when the whole system is linear, one can legally substitute D_i for R_i in the Jacobi or Gauss-Seidel algorithms: the result is given by the union of the newly obtained term and the old one.

Then we replace the Jacobi algorithm with the following one:

ALGORITHM SEMI-NAIVE
INPUT: A (mathematically) linear system of algebraic equations Σ, and an Extensional Database.
OUTPUT: The values of the variable relations $R_1, \ldots, R_n$.
METHOD:

FOR $i := 1$ TO n DO $D_i := \emptyset$;
FOR $i := 1$ TO n DO $R_i := \emptyset$;
REPEAT
    $cond := true$;
    FOR $i := 1$ TO n DO
        BEGIN
        $D_i := E_i[D_1, \ldots, D_n] - R_i$;

$$R_i := D_i \cup R_i;$$
IF $D_i \neq \emptyset$ THEN $cond := false$;
END;
UNTIL $cond$;
FOR $i := 1$ TO n DO OUTPUT(R_i).
ENDMETHOD

The most common case, and the easiest to deal with, is linearity. Differentials of higher order (e.g., quadratic) can also be used for nonlinear equations; however, the expressions become quite cumbersome.

Another possibility is to transform each nonlinear rule into a set of rules, each of which becomes *linear with respect to each of the original rule's variables*; then, we can use a variant of the *semi-naive* method, the *general semi-naive* method.

We start by presenting the transformation that applies to nonlinear rules; for simplicity, we consider only one rule, recursive and nonlinear, containing m occurrences of the head predicate in the body:

$$C: \quad p:-q_1,\ldots,q_h,p^{(1)},\ldots,p^{(m)}.$$

For the sake of simplicity, we have omitted the variables from the literals, and we represent each literal by its predicate symbol. This rule is equivalent to the following program:

$$p:-q_1,\ldots,q_h,p,r_1$$

$$r_1:-p,r_2$$

$$r_2:-p,r_3$$

$$\ldots$$

$$r_{m-1}:-p$$

The resulting rules are linear *with respect to all predicates*; obviously, they are not linear, because there is mutual recursion between p and each of the r_i. By applying this construction to all nonlinear recursive rules, we obtain a program, all of whose rules are linear with respect to all variables. This system can be solved by applying a variant of the (pure) *semi-naive* algorithm which is less efficient but still saves some computation. We can use the differential of each predicate only in the equation that defines that predicate, as follows.

ALGORITHM GENERAL SEMI-NAIVE
INPUT: A system of algebraic equations Σ, each linear in its LHS variable, and an extensional database EDB.
OUTPUT: The values of the variable relations $R_1,\ldots,R_n$.
METHOD:

FOR $i := 1$ TO n DO $D_i := \emptyset$;

9.1 Bottom-up Evaluation

FOR $i := 1$ TO n DO $R_i := \emptyset$;
REPEAT
   $cond := true$;
   FOR $i := 1$ TO n DO
     BEGIN .
       $D_i := E_i[R_1, \ldots, D_i, \ldots, R_n] - R_i$;
       $R_i := D_i \cup R_i$;
       IF $D_i \neq \emptyset$ THEN $cond := false$;
     END;
UNTIL $cond$;
FOR $i := 1$ TO n DO OUTPUT(R_i).
ENDMETHOD

The advantage of this method with respect to the *naive* method is that, at each iteration step, a *differential term* D_i is used in each equation instead of the whole value of R_i. Of course, this is far less efficient than the *linear semi-naive* method, where *all* the D_i were introduced in *all* the equations, thus reducing sensibly the size of the computation factors. Instead, in this case, we have to accept a much smaller reduction, i.e. *only one* differential for each equation.

Example 9.2. Let us see how the method of semi-naive evaluation improves the computation of Example 9.1(b). We want to compute the goal $\sigma_{1=a}SGC$. on the linear equation

$$SGC = \Pi_{1,5}((PAR \underset{2=1}{\bowtie} SGC) \underset{4=2}{\bowtie} PAR) \cup EQ$$

Due to linearity, we can apply the *semi-naive* method. The EDB is the same as in Example 9.1(a). We obtain:

$D^{(0)} = \emptyset$
$SGC^{(0)} = \emptyset$

$$D^{(1)} = EQ = \{<a,a>, <b,b>, <c,c>, <d,d>,$$
$$<e,e>, <f,f>, <g,g>, <h,h>\}.$$

$SGC^{(1)} = D^{(1)} \cup SGC^{(0)} =$
$$\{<a,a>, <b,b>, <c,c>, <d,d>, <e,e>, <f,f>,$$
$$<g,g>, <h,h>\}.$$

$D^{(2)} = \{<a,a>, <b,b>, <c,c>, <d,d>, <e,e>, <f,f>, <g,g>,$
$$<h,h>, <d,e>, <e,d>, <b,a>, <a,b>\} - SGC^{(1)} =$$
$$= \{<d,e>, <e,d>, <b,a>, <a,b>\}$$

$SGC^{(2)} = D^{(2)} \cup SGC^{(1)} =$

$\{<a,a>, <b,b>, <c,c>, <d,d>, <e,e>, <f,f>,$
$<g,g>, <h,h>, <d,e>, <e,d>, <b,a>, <a,b>\}$

$D^{(3)} = \{<b,c>, <a,c>, <c,b>, <c,a>\}$

$SGC^{(3)} = D^{(3)} \cup SGC^{(2)} =$

$\{<b,c>, <a,c>, <c,b>, <c,a>\} \cup SGC^{(2)}$

$= \{<a,a>, <b,b>, <c,c>, <d,d>, <e,e>, <f,f>,$
$<g,g>, <h,h>, <d,e>, <e,d>, <b,a>, <a,b>,$
$<b,c>, <a,c>, <c,b>, <c,a>\}$

$D^{(4)} = \emptyset,$ and we are finished:

$SGC^{(4)} = \{<a,a>, <b,b>, <c,c>, <d,d>, <e,e>, <f,f>,$
$<g,g>, <h,h>, <d,e>, <e,d>, <b,a>, <a,b>,$
$<b,c>, <a,c>, <c,b>, <c,a>\}$

If we now compare the result of this computation with that of Example 9.1(b), we see that the result of the two computations is the same, but the computation here is more efficient. Only the $D^{(i)}$ have been involved in the joins, while in Example 9.1 we had to compute joins for each of the temporary values $SGC^{(i)}$, which always have more tuples than the corresponding values of $D^{(i)}$. □

9.1.3 The Method of Henschen and Naqvi

One of the first pure evaluation methods proposed in the literature was developed by Henschen and Naqvi, and applies to linear *Datalog* programs with goals that contain bound arguments. It exploits the form of the goal, as well as the structure of the program.

This method produces an iterative program that evaluates the goal in several steps. Each step produces some of the answer tuples; at the same time, it computes symbolically the expression that will be evaluated at the following step. The method is based on a "functional" interpretation of predicates: any predicate p having at least two arguments is viewed as a set function from some of its arguments to the remaining ones, associating to each set S of values of its first arguments a set S' of values of its second arguments; for instance, if p is a binary predicate, we denote by f_p the functional mapping:

$$S' = f_p(S) = \{y \mid x \in S \text{ and } p(x,y)\}.$$

With this notation, one can also perform the composition of predicate functions, in the usual mathematical sense. This can be applied to the solution of linear *Datalog* programs, by considering the bound arguments as the function's domains. We exemplify the method of Henschen-Naqvi on program SGC

$r_1:$ $sgc(X,Y) :- eq(X,Y).$
$r_2:$ $sgc(X,Y) :- par(X,X1), sgc(X1,Y1), par(Y,Y1).$

with the goal $?- sgc(a,X)$.

The values of X which satisfy the goal can be obtained as the set:

$$f_{eq}(a) \ \cup \ f_{rap}(f_{eq}(f_{par}(a))) \ \cup \ f_{rap}(f_{rap}(f_{eq}(f_{par}(f_{par}(a))))) \ \cup \ \ldots$$

where f_{eq} denotes the function returning, for each constant, itself; f_{par} is the set function from the first to the second argument of the relation PAR, and f_{rap} is the set function from the second to the first argument of the relation PAR.

The iterative algorithm of Henschen and Naqvi produces this answer by computing at each step i:

$$E(a) \ = \ f_{rap}(f_{eq}(f_{par}\ldots(a)))$$

and produces symbolically the new expression to be computed, i.e.

$$f_{rap}(E(f_{par}(a))).$$

This is the first example of optimization based on the fact that one of the arguments of the goal predicate is bound to a constant. We shall not spend much time discussing this method, because it is not fully general (it requires linearity of the rules), and because we will later introduce many other methods that exploit the same knowledge for optimization. The most interesting feature of this method is that it integrates two kinds of computation: at each step, some tuples of the answer are computed, but some symbolic manipulation is also performed. The first kind of computation is typical of pure evaluation methods, the second is characteristic of rewriting methods.

9.2 Top-down Evaluation

We now introduce some efficient top-down evaluation algorithms, optimizing the behaviour of the backward-chaining methods described in Chap. 7. In Sect. 9.2.1 we present the *Query-Subquery* algorithm. In Sect. 9.2.2, a variation of this method is introduced, the so-called *Recursive Query Answering / Frozen Query Iteration*. They are both *set-oriented* methods, as they process whole relations instead of tuples. In this sense, they are appropriate for database processing.

9.2.1 Query-Subquery

Query-Subquery is a top-down, set-oriented method; it differs from Prolog, which is also a top-down method, because Prolog acts one tuple at a time. Set-orientation is quite desirable from a database viewpoint. Also, QSQ is complete, while Prolog, being based on SLD-resolution without any additional halting conditions, may not terminate in some cases. As we will see below, the QSQ algorithm uses a *dynamic halting condition* (see Sect. 7.3.1), based on the new tuples inferred and on the *new subqueries* generated. It also applies a kind of *subtree factoring* as described in Sect. 7.3.1.

QSQ is a fully general method, because it can be applied to any initial program and goal. The objective of QSQ is to be able to access the minimum number of facts that are needed to answer the query. In order to do this, the fundamental notion of *subquery* is introduced.

A subquery is a natural notion of backward reasoning: a goal, together with a program, determines a *query*; literals in the body of any of the rules defining the goal predicate are subgoals of the given goal. Thus, a subgoal, together with the program, yields a *subquery*; this definition applies recursively to subgoals of rules which are subsequently activated. The query decomposition tree is shown in Fig. 9.1.

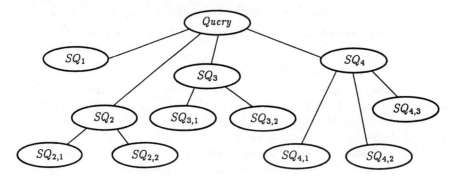

Fig. 9.1. Schema of the reduction of a query to subqueries

The method groups together all the subqueries that have the same adornment (see Chap. 4) into one *generalized query*. For instance, if during the execution two subqueries $p(a, Y)$ and $p(b, Y)$ are derived, the corresponding generalized query is $p(\{a, b\}, Y)$. We say that a rule head *matches such a generalized subquery* iff it matches one of the subqueries.

The method maintains two sets:

- The set P of *answer tuples*, which contains answers to the main goal, and answers to intermediate subqueries. P is represented by a set of *temporary relations*, one relation for each IDB predicate. At the end of the process, only the answers to the main goal are retained among those produced. However, the set of answers found in P before discarding the answers to other goals is the minimum set of answers (the minimal knowledge) required to solve the goal.
- The set Q of *current generalized subqueries* (or *subquery instances*), which contains all the subgoals that are currently under consideration. In fact, because of the recursion in the rules, the process consists of "opening" the search of

the answers to some subqueries, and during this search also generating new subqueries to be answered. The set Q keeps record of all these subqueries.

At each step of the algorithm, QSQ stores a state, represented by the current pair of these sets. The function of the algorithm is twofold: *to generate new answers* and *to generate new subqueries that must be answered*.

There are two versions of the QSQ algorithm, an iterative one (QSQI) and a recursive one (QSQR). The difference between the two concerns which of these two functions has priority over the other. QSQI privileges the production of answers; thus, when a new subquery is encountered, it is suspended until all the possible answers that can be produced up to that moment have been produced, and only then does the algorithm expand it. QSQR behaves in the opposite way: whenever a new subquery is found, it is recursively expanded and the process of finding all the partial answers is postponed to the moment when this subquery has been completely solved. In this chapter, we describe QSQR, which is more efficient.

For each rule whose head matches the initial goal, the evaluation process starts with $P = \emptyset$ (all the temporary relations are empty) and Q only containing the initial query ($Q = \{goal\}$).

At each recursion step, a new collection of answer tuples, and a new set of subquery instances to be answered are produced, and added to the respective sets. The process ends when there are no more literals to be expanded in the rule body, and no new tuples are generated.

Another important notion is that of *selection function*: it chooses the next subgoal of the rule body that will be analyzed by the algorithm. This function has exactly the same meaning that was introduced in Chap. 7. In particular, Example 7.15 gives an idea of the most common selection functions. This function is not detailed in the algorithm specification. However, the "most instantiated" selection function is the most interesting, because most of the time it yields the best optimization. As in SLD-resolution, the choice of the selection function does not affect the completeness of the algorithm. Different selection functions will yield different kinds of optimization

Since the method is rather difficult to understand, we advise the reader to follow it step by step in Example 9.3.

ALGORITHM RECURSIVE QUERY-SUBQUERY
INPUT: A (generalized) query q.
OUTPUT: Answer tuples for the query.
METHOD:
REPEAT
If the goal is the main goal, then set $Q := \{q\}$.
    FOR each rule of the program whose head matches the goal DO
      BEGIN
        REPEAT
          (a) Choose the first/next subgoal $r(\ldots)$ in the rule body

according to the selection function; this subgoal
corresponds to a predicate, and thus to a temporary
relation R.

(b) According to the available information (provided by the
bindings already known from the previously selected
and processed subgoals of this rule body)
symbolically determine the basic operations to be
performed on R (selection, join, projection, ...).

(c) Now we have a new subquery (a predicate and a set
of values). If this predicate *is not recursive*,
generate answers for it by executing
algebraically the operations specified by (b),[1]
and continue with the next subgoal. If it
is recursive, consider the set of subquery
instances (generalized subqueries) determined by (b):

(c.1) the instances contained in the current set
of instances Q are answered (partially) by
looking in the temporary relation R
to see if there are any tuples satisfying the
conditions in (b).[2]
If there are, the bindings
provided by these answers are retained (they
might be used for the next iteration of step (b))
and we continue with the next subgoal;

(c.2) for the new instances, call recursively
this procedure with these instances as input;

UNTIL there are no more subgoals in the rule body;

infer answer tuples for the head goal
by performing the required algebraic operation(s)
on the sets of tuples solving the right hand side
(for example projection).

[1,2] Note that the algebraic operations are always executed by using the current values already derived for the recursive predicates, and the (known) values of the nonrecursive ones.

        END
UNTIL no more answer tuples are generated.
ENDMETHOD

We exemplify the method in an example. For simplicity, we take the function "*next*" as selection function. In this particular case this selection function happens to correspond to the "*most instantiated*" selection function.

Example 9.3.
Let us evaluate the goal ? $- sgc(john, X)$. on program $L2$ of Example 9.1(b):
r_1: $sgc(X, Y) : - eq(X, Y)$.
r_2: $sgc(X, Y) : - par(X, X_1), sgc(X_1, Y_1), par(Y, Y_1)$.

with the relation :

PAR
$= \{<john, henry>, <george, jim>, <henry, mary>, <jim, mary>\}$

Initially, $P = \emptyset$, $Q = \{sgc(\{john\}, Y)\}$. Note that, in Q, we have the generalized query corresponding to the goal.

from r_1: $sgc(john, Y)$
(a): **select** $eq(john, Y)$
(b): the operation to be done is: $\sigma_{1=john} EQ$
(c): EQ is nonrecursive. Answer: $eq(john, john)$;
Solution: $sgc(john, john)$, $P = \{sgc(john, john)\}$.

from r_2: $sgc(john, Y)$
(a): **select** $par(john, X_1)$
(b): the operation to be done is: $\sigma_{1=john} PAR$
(c): PAR is nonrecursive. Answer: $par(john, henry)$;

(a): **select** $sgc(X_1, Y_1)$
(b): the operation to be done is a join between $\sigma_{1=john} PAR$ and SGC
(c): **subquery:** $sgc(henry, X)$
this is a **new** instance: $Q = \{sgc(\{john, henry\}, Y)\}$.
Call procedure recursively :

from r_1: $sgc(henry, Y)$
(a): **select** $eq(henry, Y)$
(b): the operation to be done is: $\sigma_{1=henry} EQ$
(c): EQ is nonrecursive. Answer: $eq(henry, henry)$;
Solution: $sgc(henry, henry)$, $P = \{sgc(john, john), sgc(henry, henry)\}$.

from r_2: $sgc(henry, Y)$
(a): **select** $par(henry, X_1)$
(b): the operation to be done is: $\sigma_{1=henry} PAR$
(c): PAR is nonrecursive. Answer: $par(henry, mary)$;

(a): **select** $sgc(X_1, Y_1)$
(b): the operation to be done is a join between $\sigma_{1=henry} PAR$ and SGC

(c): **subquery:** $sgc(mary, X)$
this is a **new** instance: $Q = \{sgc(\{john, henry, mary\}, Y)\}$.
Call procedure recursively :

from r_1: $sgc(mary, Y)$
(a): **select** $eq(mary, Y)$
(b): the operation to be done is: $\sigma_{1=mary} EQ$
(c): EQ is nonrecursive. Answer: $eq(mary, mary)$;
Solution: $sgc(mary, mary)$,
$P = \{sgc(john, john), sgc(henry, henry), sgc(mary, mary)\}$.

from r_2: $sgc(mary, Y)$
(a): **select** $par(mary, X_1)$
(b): the operation to be done is: $\sigma_{1=mary} PAR$
(c): PAR is nonrecursive. Answer: no tuples;
we end this level of recursion here. Back to the previous level:

(a): **select**(next): $par(Y, Y_1)$
(b): the operation to be done is a join between $\sigma_{1=henry} PAR \underset{2=1}{\bowtie} SGC$ and PAR
(c): PAR is nonrecursive. Answer: $par(jim, mary)$;

The subgoals of the rule body are finished at this level of recursion. Thus, we perform the required final projection: solution: $sgc(henry, jim)$.
$P = \{sgc(john, john), sgc(henry, henry), sgc(mary, mary), sgc(henry, jim)\}$.
We go up another level of recursion:

(a): **select** $par(Y, Y_1)$
(b): the operation to be done is a join between $\sigma_{1=john} PAR \underset{2=1}{\bowtie} SGC$ and PAR
(c): PAR is nonrecursive. Answers: $par(george, jim)$, $par(john, henry)$;

The subgoals of the rule body are finished at this level of recursion.
Thus, we perform the required final projection; after that, the new solution is: $sgc(john, george)$.
$P = \{sgc(john, john), sgc(henry, henry), sgc(mary, mary), sgc(henry, jim), sgc(john, george)\}$.

This was the last level of recursion still open. Yet, we are still in the initial **repeat** loop. We leave it as an exercise to the reader to show that no new tuples will be generated.

The final answer to the initial query is obtained by choosing from P only the tuples of SGC that have the first element equal to $john$. □

9.2.2 The RQA/FQI Method

We introduce here very briefly the method *Recursive Query Answering / Frozen Query Iteration* (RQA/FQI), a variant of the QSQ method. It is built from a generic algorithm similar to that of QSQ, only implemented by a two-step approach.

In the first step of an RQA/FQI evaluation, a recursive processing strategy like that of Prolog is used, expanding the search tree top-down. Answers to recursive queries are stored to be re-used later. The expansion stops whenever *subsumed queries* are encountered, or after subqueries are answered completely using basic facts and nonrecursive predicates. Up to this point, this method, apart for some implementation features, follows the same top-down expansion strategy as QSQ. Note that stopping at subsumed queries is also a kind of subtree factoring.

The second step uses a different approach. There are still branches in the search tree which are incomplete because of a subsumed query, and therefore have been stored as *frozen clauses*. These are processed using an efficient variant of Least Fixpoint iteration (see Chap. 6) over these frozen clauses, working bottom-up both from database facts and facts which have already been found as answers to recursive queries. This step is necessary in order to propagate all answers of the evaluated subsuming goal to its corresponding frozen counterparts (subsumed goals). It corresponds to plugging new answers into subsumed queries and using these answers to expand the search tree further. If new subgoals are found, they are expanded again by the top-down evaluation strategy.

9.3 Bibliographic Notes

The algebraic naive evaluation methods (named after Jacobi and Gauss Seidel) have been introduced, among others, by Ceri, Gottlob, and Lavazza in [Ceri 86a]. In the same paper, semi-naive evaluation methods for a unique equation were also presented. Generalizations of these to systems of equations can be found in [Ceri 87] and [Ioan 87a]. Many logical versions of naive and semi-naive evaluation can also be found in the literature [Banc 85, McKa 81, Chan 81, Marq 83, Marq 84, Baye 85]. In particular, a survey paper by Bancilhon and Ramakrishnan [Banc 86b] is good reference.

The method of Henschen and Naqvi [Hens 84] was one of the first pure evaluation methods proposed. We also want to mention that substantially the same functional interpretation of predicates is given by Gardarin and De Maindreville in [Gard 86], who propose a method for evaluating queries as function series, where the functions involved are just the functional interpretation of the predicates. Follow-up can be found in [Gard 87b].

The Query-Subquery algorithm was introduced by Vieille in [Viei 86a]. The version introduced in that paper was found to be incomplete. The author [Viei 86b, 87] and others [Nejd 87], [Roel 87] have subsequently provided corrections or complete versions of the algorithm. The version we present here is of course a complete one. The RQA/FQI algorithm has also been introduced in [Nejd 87].

9.4 Exercises

9.1 Use the algorithm of Jacobi to compute the goal
$$? - sgc(X, a).$$
on the logic program and extensional database of Example 9.1(b).

9.2 Use Gauss-Seidel's algorithm to compute the goal
$$q(X) : -sgc(a, X).$$
on the logic program of Example 9.1(b), but using the relation PAR of the extensional database of Example 9.1(a).

9.3 Prove Theorem 9.2 (Hint: you can prove it either by showing that algorithm Gauss-Seidel computes exactly the same result as algorithm Jacobi, or by showing that it computes *the least fixpoint* of T_Σ (which is unique ...)).

9.4 Is it possible to use the linear semi-naive evaluation method to compute the solution of Example 9.3(a)? If so, compute it. If not, say why.

9.5 Evaluate the query of Example 9.1(a) using Recursive Query-Subquery.

9.6 Write a *Prolog* or *Lisp* program that implements the QSQR algorithm as given here.

9.7 Use QSQR to evaluate the following goal:
$? - n(c, Y).$
on the program:
$n(X, Y) : -r(X, Y).$
$n(X, Y) : -p(X, Z), n(Z, W), q(W, Y).$

and the database:
$P = \{< c, b >, < b, c >, < c, d >\},$
$R = \{< d, e >\},$
$Q = \{< e, a >, < a, i >, < i, o >\}.$

Chapter 10
Rewriting Methods

In this chapter we introduce a number of rewriting methods for *Datalog* programs. Section 10.1 presents rewriting methods that apply directly to *Datalog* programs: the Magic Sets, Counting, Static Filtering and Semi-naive Evaluation by Rewriting. Section 10.2 presents a structured approach to the optimization of algebraic systems; it consists of a number of preliminary steps, after which the optimization methods of *reduction of variables* and *reduction of constants* are applied.

10.1 Logical Rewriting Methods

Let us first explain, by use of a very simple example, the rationale of logical rewriting methods. Consider the usual *ancestor* example, where the *anc* predicate is defined in terms of the EDB predicate *par*:

$r_1:$ $anc(X,Y) :- par(X,Y).$
$r_2:$ $anc(X,Y) :- anc(X,Z), par(Z,Y).$

and assume that a user wants to compute the ancestors of one particular individual a. Then, he or she issues the goal:

$$? - anc(a, X).$$

To compute bottom-up the set of answers for this goal, we evaluate the program by naive evaluation, as we did in Example 9.1, and then perform a selection $\sigma_{1=a}$ on the resulting relation ANC:

$$ANC = \{<b,a>, <b,g>, <a,d>, <a,e>, <d,f>, <c,h>,$$
$$<b,d>, <b,e>, <a,f>, <b,f>\}$$

The result of the selection is:

$$\sigma_{1=a} ANC = \{<a,d>, <a,e>, <a,f>\}.$$

This approach is certainly not efficient. As we already pointed out in Sect. 9.1.1, several irrelevant facts, such as $anc(b,d)$, or $anc(c,h)$, are derived before applying the selection; these facts have no connection with the final result. Moreover, during the process of naive evaluation, they do not play any role in the derivation of the facts contained in the final result.

Given our small database, this seems not to be such an important problem. However, imagine a large, loosely connected demographical database, containing facts about hundreds of thousands of families. The application of this method would then result in a tremendous overhead.

A "smart" programmer, for the same problem, would write a more efficient program, which would exploit the knowledge provided by the binding of the first argument of predicate anc to the constant a in the goal. He or she would then write the following program:

$r'_1:$ $\quad anc'(a,Y) :- par(a,Y).$
$r'_2:$ $\quad anc'(a,Y) :- anc'(a,Z), par(Z,Y).$

This program computes much fewer useless facts. To realize this, we first translate it into RA^+; here, we denote by ANC' the result of the selection $\sigma_{1=a}ANC$:

$$ANC' = \sigma_{1=a}PAR \cup ANC' \circ PAR$$

Then, we apply the process of naive evaluation to this program using the same EDB as in Example 9.1. The following are the results at each iteration step:

$ANC'^{(0)} := \emptyset.$
$ANC'^{(1)} = \sigma_{1=a}PAR = \{<a,d>, <a,e>\}.$
$ANC'^{(2)} = \{<a,d>, <a,e>, <a,f>\}.$
$ANC'^{(3)} = \{<a,d>, <a,e>, <a,f>\} = ANC'^{(2)}.$

Therefore the result is:

$$ANC' = \{<a,d>, <a,e>, <a,f>\}$$

In this development, the size of the relations involved in the algebraic operations was much smaller than in Example 9.1, and the number of iterative steps is one less. For instance, note that in $ANC'^{(2)}$, the tuples $<b,d>$ and $<b,e>$ are not computed, while they are computed by the Jacobi algorithm at step $ANC^{(2)}$. We conclude that the number of facts produced during the whole evaluation of this "smart" program is much less than the number of facts produced by the inefficient one.

Logical rewriting methods automatically transform inefficient programs into "smart" ones, by exploiting the argument bindings provided by goals. Given a goal G and a program P, the program P' obtained from such a transformation on P is *equivalent* to P *with respect to* G, as it produces the same result. Formally, two programs P and P' are equivalent with respect to a goal G iff $\mathcal{M}_{P,G} = \mathcal{M}_{P',G}$ (recall the definition of $\mathcal{M}_{P,G}$ in Sect. 6.2.1).

Notice that this example was very easy, and even an unexperienced programmer could have performed the transformation. However, the rewriting methods proposed in this chapter are so powerful that can be applied to very complicated programs, giving quite unintuitive transformations which may lead to a substantial increase of efficiency. Moreover, these transformations usually modify programs in a more general way than just for a single constant value. Programs are rewritten in the same way for all the goals with the same *adornment* (see Chap. 4 for the definition of adornment).

Notice also that this example was conducted using the naive evaluation method just for the sake of convenience, since we had already computed the program's result in Chap. 9; however, the program considered is linear, and thus evaluable by the semi-naive evaluation method. Nevertheless, by applying the semi-naive method, the rewriting of the program leads to a drastic improvement of efficiency.

In this section, we consider four logical rewriting methods: *Magic Sets*, *Counting*, *Static Filtering*, and *Rewriting Semi-naive*. The Magic Sets method is one of the most famous optimization methods, and is fully general. Thus, it will be explained in detail. Afterwards, we will introduce the Counting method, which is not general, though very efficient in its application domain. We will explain this method through a transformation that takes as its starting point the output of the Magic Sets method.

The Static Filtering method is also general, and applies to any *Datalog* program with a goal having a bound argument. Both Static Filtering and Magic Sets, when applied to the *ancestor* program discussed above, achieve a similar result as the program rewritten by the "smart" programmer.

The last method, Semi-naive evaluation by rewriting, does not consider the structure of the goal for the optimization, and can be applied to *Datalog* programs without goals.

10.1.1 Magic Sets

The purpose of the Magic Sets method is the optimization of *Datalog* programs with particular adornments of the goal predicates. According to the classification criteria given in Chap. 8, it is a logical rewriting method exploiting the form taken by the query.

This method is based on the idea of *sideways information passing (SIP)*. Intuitively, given a certain rule and a subgoal in the rule body with some bound argument(s), one can solve this subgoal, and so obtain bindings for uninstantiated variables in other argument positions. These bindings can be transferred to other subgoals in the rule body, and they, in their turn, transmit bindings to other variables.

This is the normal behavior of top-down evaluation methods (for instance, QSQ and, in general, resolution-based methods). In order to simulate the binding passing strategy of top-down methods, the Magic Sets method introduces constraints into the program, by means of additional subgoals added to the RHS of the original rules, and additional rules defining these goals added to the program. The additional rules constrain the program variables to satisfy other predicates also (called "magic" predicates). Thus, during bottom-up computation, the variables assume only *some* values instead of all possible ones. In most cases, this makes the new program more efficient.

Before presenting the Magic Sets transformation, we describe a few preparatory standardization steps, which have to be applied to the original input program, and we give a few preliminary definitions.

Let P be a *Datalog* program, and consider a goal on P. We can view the goal itself as being defined by a rule, e.g., the goal $?-anc(a, X)$. becomes the rule $q(X) : -anc(a, X)$. We add the goal rule to the program.

Consider a rule r of P. Given an adornment of its head predicate, an argument of a subgoal in r is said to be *distinguished* if either of the following conditions holds:

- it is a constant;
- it is bound by the adornment;
- it appears in an EDB predicate occurrence that has a distinguished argument.

Hence, an argument is distinguished if the range of possible values it can take is restricted by some constant appearing in the same predicate, or by some other variable, which also has a restricted range of values.

From this definition, variables in an EDB predicate occurrence are either all distinguished or all not distinguished. An EDB predicate occurrence with all variables distinguished is a *distinguished predicate occurrence*. We present an example.

Example 10.1. Consider the following program, which already contains the goal rule:

$r_0:$ $\quad\quad\quad\quad q(X) :- anc(a, X).$
$r_1:$ $\quad\quad\quad\quad anc(X, Y) :- par(X, Y).$
$r_2:$ $\quad\quad\quad\quad anc(X, Y) :- anc(X, Z), par(Z, Y).$

In rule r_0, a is distinguished in anc, while X is not because anc is an IDB predicate. Consider then the head predicate of rule r_1 with the same adornment b, f (this adornment will be passed from r_0 to r_1 and r_2). Then, X is distinguished in par because it corresponds to the bound variable in the head. Y is also distinguished in par because par is an EDB predicate and has another distinguished argument. So, this is a distinguished occurrence of the EDB predicate par.

Consider now rule r_2, also with the adornment (b, f). Then, X is distinguished in anc because it corresponds to the bound variable in the head. anc is an IDB predicate, so it does not transmit the "distinction" to other arguments. Thus, no argument of par is distinguished in this rule, and this is not a distinguished occurrence of the EDB predicate par. $\square$

The concept of distinguished argument formalizes the idea of using bound arguments to bind other, free, variables, thus indicating the direction of *sideways information passing*.

For each rule r of P, and for each adornment of the head predicate of r, we generate an *adorned rule* as follows. We consider all the distinguished arguments to be bound; this will generate an adornment for all the IDB predicates that are in the rule. The rule obtained by replacing all these predicates with their adorned version is an *adorned* rule. From this step we can already see the generality of the algorithm, since all goals that generate the same adornment produce the same rewritten program.

For instance, consider rule r_2 of Example 10.1, for the adornment (b, f) of its head predicate. The corresponding adorned rule is

$$r_{2(ad)}: \quad anc^{bf}(X,Y) :- anc^{bf}(X,Z), par(Z,Y).$$

We give distinct numbers to different occurrences of the same predicate p in the right-hand side of a rule. For a predicate p, we denote its i-th occurrence by "p_i". However, if there is only one occurrence of a certain predicate in the right hand side of a rule, we may omit the occurrence number in that rule.

For example, in the following rule:

$$r(X,Y) :- s(X,Z), t(Z,Z), s(W,Y).$$

we write the occurrence numbers as:

$$r(X,Y) :- s\_1(X,Z), t\_1(Z,Z), s\_2(W,Y).$$

We replace each rule of P with the corresponding set of all possible adorned rules.

Example 10.2. From the program of Example 10.1, the set of possible adorned rules is

$$q^f(X) :- anc^{bf}\_1(a, X).$$

for the adornment (b, f) of anc:

$$anc^{bf}(X,Y) :- par\_1(X,Y).$$
$$anc^{bf}(X,Y) :- anc^{bf}\_1(X,Z), par\_1(Z,Y).$$

and for the adornment (f, b) of anc:

$$anc^{fb}(X,Y) :- par\_1(X,Y).$$
$$anc^{fb}(X,Y) :- anc^{fb}\_1(X,Z), par\_1(Z,Y).$$

Note that, in this example, we have added the numbers of the predicate occurrences in the rule bodies; however, since there is only one such occurrence, we could just as well have omitted them. □

Now we have transformed P into a new set P^A of adorned rules, which are not all necessary to evaluate the goal. We eliminate all the rules that are not *reachable* for the goal from P^A. We say that an adorned rule is *reachable* for the goal iff either it is the adorned rule corresponding to the goal rule, with all the LHS predicate arguments free, or its head predicate appears, with the same adornment, in the RHS of a reachable rule.

Example 10.3. The reachable adorned system of Examples 10.1 and 10.2 is:

$$q^f(X) :- anc^{bf}(a, X).$$
$$anc^{bf}(X,Y) :- par(X,Y).$$
$$anc^{bf}(X,Y) :- anc^{bf}(X,Z), par(Z,Y).$$

Note that the last two rules of Example 10.2 are not reachable for the goal. In fact, their *LHS* adorned predicates do not appear in the *RHS* of any other reachable rule. □

The concept of reachability is not essential to the algorithm, but is useful for isolating the rules that are really involved in the goal evaluation, by restricting the computation to the "predecessors" of the goal rule.
We give now the proper Magic Sets algorithm.

ALGORITHM MAGIC SETS

INPUT:
A set of adorned rules P^A, including the goal rule, all reachable from the goal.

OUTPUT:
A new set of rules P^{magic}, equivalent to P^A with respect to the goal.

METHOD:

$P^{magic} := P^A$;
  FOR each adorned rule r, and FOR each occurrence of an intensional predicate p in the RHS of r DO
  BEGIN
    Generate one *magic rule* in the following way:
    a) Delete all other occurrences of IDB predicates in the RHS;
    b) Replace the name of p in this occurrence with $magic\_r\_p^a\_i$ where a is the adornment of p in that occurrence and i is the occurrence number;
    c) Delete all nondistinguished variables of this occurrence of p, thus possibly obtaining a predicate with fewer arguments;
    d) Delete all nondistinguished EDB predicates in r;
    e) Replace the name of the head predicate p' with $magic\_p'\ ^{a'}$ where a' is the adornment of p';
    f) Delete all nondistinguished variables of p';
    g) Exchange the places of the head magic predicate and the body magic predicate.
    Add this rule to P^{magic}.
  END

  FOR each adorned rule r in the original program DO
  BEGIN
    Generate a *modified* rule in the following way:
      FOR each occurrence of an intensional predicate p in the

        RHS of r DO
        BEGIN
            add to the RHS the predicate $magic\_r\_p^a\_i(X)$
            where a is the adornment of the occurrence of p,
            i is the occurrence number, X is the list
            of distinguished arguments in this occurrence.
                IF p is not the head predicate
                THEN the magic predicate must be inserted just
                before that occurrence;
                ELSE the magic predicate is to be inserted at the
                beginning of the rule body, before all other literals.
        END
        Replace r with its modified version in P^{magic};
    END

    FOR each IDB predicate p and FOR each adornment DO
    BEGIN
        Generate a *complementary* rule as follows:
        FOR each adorned rule r, and FOR each occurrence
        of p in the RHS of r DO
        BEGIN
            add the rule:
            $magic\_p^a(X) :- magic\_r\_p^a\_i(X).$
            where i is the considered occurrence of p, a
            is its adornment and X is the list of its
            distinguished arguments.
        END
        Add this rule to P^{magic}.
    END
ENDMETHOD

The proof that the new set P^{magic} of rules is equivalent to P with respect to the goal is given in the original paper by Bancilhon et al. (see Bibliographic Notes). We will only give here some significant examples, showing that in most cases the method produces an optimization. Moreover, in spite of the apparent complication of the algorithm, examples show that simple initial programs produce simple sets of magic rules.

The tuples of the relation *magic* defined by the magic rules form the *Magic Set*. Notice that, in general, this method produces more tuples than exactly those of the goal answer. The resulting relation must be finally selected to obtain the answer.

Example 10.4. We now give a complete example of application of this method using the program:

r_1 : $\qquad anc(X,Y) := par(X,Y)$.
r_2 : $\qquad anc(X,Y) := anc(X,Z), par(Z,Y)$.

a. We first develop it for the goal $? - anc(X,a)$. We add the goal rule to P:

r_0 : $\qquad q(X) := anc(X,a)$.
r_1 : $\qquad anc(X,Y) := par(X,Y)$.
r_2 : $\qquad anc(X,Y) := anc(X,Z), par(Z,Y)$.

The reachable adorned system P^A is:

R_0 : $\qquad q^f(X) := anc^{fb}(X,a)$.
R_1 : $\qquad anc^{fb}(X,Y) := par(X,Y)$.
R_2 : $\qquad anc^{fb}(X,Y) := anc^{fb}(X,Z), par(Z,Y)$.

We now generate the modified system:

First DO loop: Produces a magic rule for each occurrence of anc in the RHS of each adorned reachable rule:
$from\ R_0$: $\quad magic\_R_0\_anc^{fb}(a)$.
$from\ R_2$: $\quad magic\_R_2\_anc^{fb}(Z) := magic\_anc^{fb}(Z), par(Z,Y)$.

Second DO loop: Produces a modified rule for each adorned rule:
$from\ R_0$: $\quad q^f(X) := magic\_R_0\_anc^{fb}(a), anc^{fb}(X,a)$.
$from\ R_2$: $\quad anc^{fb}(X,Y) := magic\_R_2\_anc^{fb}(Z), anc^{fb}(X,Z), par(Z,Y)$.
R_1 does not generate modified rules because there are no occurrences of IDB predicates in its RHS.

Third DO loop: Produces a complementary rule for each IDB predicate and each adornment:
$from\ R_0$: $\quad magic\_anc^{fb}(X) := magic\_R_0\_anc^{fb}(X)$.
$from\ R_2$: $\quad magic\_anc^{fb}(Z) := magic\_R_2\_anc^{fb}(Z)$.

Finally, since in this case there is a unique predicate $magic\_anc^{fb}(X)$, by renaming it as $magic(X)$, the system reduces to the equivalent

$\qquad magic(a)$.
$\qquad magic(Z) := magic(Y), par(Z,Y)$.
$\qquad q^f(X) := anc^{fb}(X,a)$.
$\qquad anc^{fb}(X,Y) := par(X,Y)$.
$\qquad anc^{fb}(X,Y) := magic(Z), anc^{fb}(X,Z), par(Z,Y)$.

In this case, the Magic Set is defined by the first two rules. The final system is equivalent to the initial one with respect to goal $? - anc(X,a)$. Indeed, if we apply to this program some evaluation algorithm, such as naive or semi-naive evaluation, and then apply the selection $\sigma_{2=a}$ to the result, this will produce the same set of answers as the application of an evaluation algorithm to the initial program, and of the final selection to its result.

b. Now consider the goal $? - anc(a,Y)$. We examined it informally at the beginning of Sect. 9.1. The reader can verify as an exercise that the output of the transformation for this goal is:

10.1 Logical Rewriting Methods

$$magic\_R_0\_anc^{bf}(a).$$
$$magic\_R_2\_anc^{bf}(X) :- \quad magic\_anc^{bf}(X).$$
$$q^f(X) :- \quad magic\_R_0\_anc^{bf}(a), anc^{bf}(a,X).$$
$$anc^{bf}(X,Y) :- \quad par(X,Y).$$
$$anc^{bf}(X,Y) :- \quad magic\_R_2\_anc^{bf}(X), anc^{bf}(X,Z), par(Z,Y).$$
$$magic\_anc^{bf}(a) :- \quad magic\_R_0\_anc^{bf}(a).$$
$$magic\_anc^{bf}(X) :- \quad magic\_R_2\_anc^{bf}(X).$$

Since in this case too there is a unique predicate $magic\_anc^{bf}(X)$, by renaming it as $magic(X)$ the whole system reduces to the equivalent

$$magic(a).$$
$$q^f(X) :- \quad anc^{bf}(a,X).$$
$$anc^{bf}(X,Y) :- \quad par(X,Y).$$
$$anc^{bf}(X,Y) :- \quad magic(X), anc^{bf}(X,Z), par(Z,Y). \qquad \square$$

Note that in part (a) of Example 10.4 the magic rule acts as a constraint on an argument of the EDB predicate par, while in part (b) it restricts the values of an argument of anc.

Unfortunately, in part (a) there is no real optimization because the two magic rules in the transformed program compute the second argument of the goal answer, so that the use of the three rules that follow merely duplicates the computation.

More sophisticated techniques exist, modifying the Magic Sets method (see the bibliographic notes), that eliminate this duplication.

While the application of the Magic Sets method to Example 10.4(a) is not effective, in part (b) we have a real optimization. The resulting system is in fact equivalent to the following one:

$$q^f(X) :- \quad anc^{bf}(a,X).$$
$$anc^{bf}(X,Y) :- \quad par(X,Y).$$
$$anc^{bf}(a,Y) :- \quad anc^{bf}(a,Z), par(Z,Y).$$

Once we have eliminated the adornments, which are now useless, the two rules for predicate anc are exactly the same as the "smart" program at the beginning of Sect. 9.1.

The predicate $magic$ defined by the first two rules of Example 10.4(a.) is often called the **cone** of a, referring to the fact that, starting from a, it computes its ancestors by "fanning out" like a cone. This concept is clearer if one looks at Fig. 10.1: relatives of a form an ascending structure similar to a cone.

The same cone is obtained as Magic Set in the case of the program *same-generation cousin*. Although useless in the *ancestor* case, we will see that the Magic Sets method achieves a good optimization in the *same-generation cousin* program (Example 10.5). The first part of Example 10.5 provides the Magic Sets transformation of the standard *same-generation cousin* program, also introduced

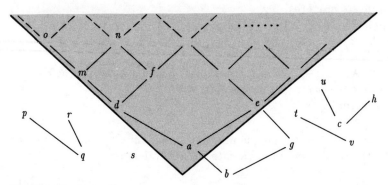

Fig. 10.1. The cone of a

in Chap. 9. In the second part, we study carefully another, syntactically different but semantically equivalent, version of the same program. We describe in full detail the application of the algorithm to this program, because it is an example of an "unintuitive" result of the rewriting process, achieving a real optimization. No programmer, even a "smart" one, can be reasonably expected to produce such an optimization for this program. Thus, this example illustrates the need for *automatic* rewriting methods.

Example 10.5

a. Consider the program

$sgc(X,Y) : -eq(X,Y).$
$sgc(X,Y) : -par(X,X1), sgc(X1,Y1), par(Y,Y1).$

which computes cousins in the same generation. Consider the goal:

$$? - sgc(a,Y).$$

After the transformations described, the program becomes:

$magic(a).$
$magic(X1) : -magic(X), par(X,X1).$
$sgc(X,Y) : -eq(X,Y).$
$sgc(X,Y) : -magic(X1), par(X,X1), sgc(X1,Y1), par(Y,Y1).$

You can see now a real optimization: only the ancestors of a are considered in the new program, so that in the execution only the tuples that are related to the genealogical tree of a are considered.

b. Consider the program

$sgc(X,Y) : -eq(X,Y).$
$sgc(X,Y) : -par(X,X1), sgc(Y1,X1), par(Y,Y1).$

10.1 Logical Rewriting Methods

which also computes cousins in the same generation. Consider the goal

$$? - sgc(a, Y).$$

We now describe a full application of the method to this program. The following is the reachable adorned system:

$R_0 \quad q^f(Y) : -sgc^{bf}(a, Y).$
$R_1 \quad sgc^{bf}(X, Y) : -par(X, X1), sgc^{fb}(Y1, X1), par(Y, Y1).$
$R_2 \quad sgc^{fb}(X, Y) : -par(X, X1), sgc^{bf}(Y1, X1), par(Y, Y1).$
$R_3 \quad sgc^{bf}(X, Y) : -eq(X, Y).$
$R_4 \quad sgc^{fb}(X, Y) : -eq(X, Y).$

First DO loop:

from $R_0 \quad magic\_R_0\_sgc^{bf}(a).$
from $R_1 \quad magic\_R_1\_sgc^{fb}(X1) : -par(X, X1), magic\_sgc^{bf}(X).$
from $R_2 \quad magic\_R_2\_sgc^{bf}(Y1) : -par(Y, Y1), magic\_sgc^{fb}(Y).$

Second DO loop:

from $R_1 \quad sgc^{bf}(X, Y) : -par(X, X1), magic\_R_1\_sgc^{fb}(X1),$
$\qquad sgc^{fb}(Y1, X1), par(Y, Y1).$
from $R_2 \quad sgc^{fb}(X, Y) : -par(X, X1), magic\_R_2\_sgc^{bf}(X1),$
$\qquad sgc^{bf}(Y1, X1), par(Y, Y1).$

Third DO loop:

from $R_0 \quad magic\_sgc^{bf}(X) : -magic\_R_0\_sgc^{bf}(X).$
from $R_1 \quad magic\_sgc^{fb}(X) : -magic\_R_1\_sgc^{fb}(X).$
from $R_2 \quad magic\_sgc^{bf}(X) : -magic\_R_2\_sgc^{bf}(X).$

The resulting rewritten program is:

$q^f(Y) : -sgc^{bf}(a, Y).$
$magic\_R_0\_sgc^{bf}(a).$
$magic\_R_1\_sgc^{fb}(X1) : -par(X, X1), magic\_sgc^{bf}(X).$
$magic\_R_2\_sgc^{bf}(Y1) : -par(Y, Y1), magic\_sgc^{fb}(Y).$
$magic\_sgc^{bf}(X) : -magic\_R_0\_sgc^{bf}(X).$
$magic\_sgc^{fb}(X) : -magic\_R_1\_sgc^{fb}(X).$
$magic\_sgc^{bf}(X) : -magic\_R_2\_sgc^{bf}(X).$
$sgc^{bf}(X, Y) : -par(X, X1), magic\_R_1\_sgc^{fb}(X1),$
$\qquad sgc^{fb}(Y1, X1), par(Y, Y1).$
$sgc^{fb}(X, Y) : -par(X, X1), magic\_R_2\_sgc^{bf}(Y1),$
$\qquad sgc^{bf}(Y1, X1), par(Y, Y1).$
$sgc^{bf}(X, Y) : -eq(X, Y).$
$sgc^{fb}(X, Y) : -eq(X, Y).$

which, by simple transformation and renaming, may be written as:

$q^f(Y) : -sgc^{bf}(a, Y).$
$magic^{bf}(a).$

$$magic^{fb}(X1) :- par(X, X1), magic^{bf}(X).$$
$$magic^{bf}(Y1) :- par(Y, Y1), magic^{fb}(Y).$$
$$sgc^{bf}(X, Y) :- par(X, X1), magic^{fb}(X1),$$
$$sgc^{fb}(Y1, X1), par(Y, Y1).$$
$$sgc^{fb}(X, Y) :- par(X, X1), magic^{bf}(Y1),$$
$$sgc^{bf}(Y1, X1), par(Y, Y1).$$
$$sgc^{bf}(X, Y) :- eq(X, Y).$$
$$sgc^{fb}(X, Y) :- eq(X, Y).$$ □

10.1.2 The Counting Method

The *Counting method* is a rewriting method based on the knowledge of the goal bindings; the method includes the computation of the Magic Set, but each element of the Magic Set is complemented by additional information expressing its "distance" from the goal constant.

Consider again the *same-generation cousin* program of Example 10.5(a). In this case, the Magic Sets method restricts the computation to the ancestors of a; in the graph of Fig. 10.1, these ancestors are those elements that are linked to a by upwards paths formed by edges belonging to the relation PAR.

The Counting method maintains information about whether an element is one of a's parents (distance 1), a's grandparents (distance 2), a's great-grandparents (distance 3), etc. The rewritten program contains the computation of these distances. At this point, computation may be restricted to the children of a's parents, to the grandchildren of a's grandparents, to the great-grandchildren of a's great-grandparents, etc. Figure 10.2 , which introduces levels in the graph of Fig. 10.1, explains this idea better.

Rather than giving the full algorithm for transforming the set of rules, we sketch how to pass from the set of rules already transformed by the Magic Sets algorithm to rules for the Counting method, using the output of Example 10.5(a). The magic rules are transformed into counting rules, defining a counting predicate.

We start from the *magic* program:

$magic(a).$
$magic(X1) :- magic(X), par(X, X1).$
$sgc(X, Y) :- eq(X, Y).$
$sgc(X, Y) :- magic(X1), par(X, X1), sgc(X1, Y1), par(Y, Y1).$

for the goal:

$$?- sgc(a, Y).$$

Initially, we change the name of the magic predicate to *counting*. Then, we augment each magic predicate with a new argument, which is a variable expressing the *distance* of each element of the Magic Set from a. In the first rule, the distance must be 0, because the element is a itself:

$$counting(a, 0).$$

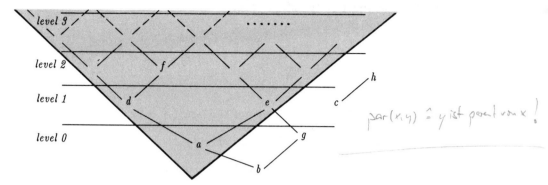

Fig. 10.2. The cone of a with the indication of levels

In the second magic rule, let I be the distance of the head predicate argument $X1$ from a. Then, since we know $par(X, X1)$, the distance between $X1$ and a must be one more than that between X and a.

Thus, if J is the distance between X and a, which is used in the body counting predicate, we can set $I = J + 1$. Note that, in pure *Datalog*, there is no possibility of expressing arithmetical operations. With some liberality, we use built-in predicates for arithmetic and comparison operators that will be discussed in Chap. 11. We will see in Sect. 10.1.4 another rewriting method that uses non-pure *Datalog* constructs in its output. For the moment, we use them in their infix notations (e.g. $I = J + 1$ or $X = Y$). The second rule becomes:

$$counting(X1, I) :- counting(X, J), par(X, X1), I = J + 1.$$

In this way, the two counting rules have associated a *level* to each element of the magic (counting) set. Now we change the predicate sgc into sgc', also by augmenting it with a new argument: the distance between a and its two original arguments. The third rule becomes:

$$sgc'(X, Y, I) :- X = Y, integer(I).$$

Note that we have added the predicate *integer* in order to obtain a *Datalog* rule: in fact, I would otherwise have been present in the head but not in the body.

Consider now the fourth rule: this has to be transformed in order to compute the steps "downwards". Let I be the distance between a and the head predicate arguments (X, Y). Let J denote the distance between a and the pair $(X1, Y1)$ in the body:

$$sgc'(X, Y, I) :- counting(X, I),$$
$$par(X, X1), sgc'(X1, Y1, J), par(Y, Y1), I = J - 1.$$

The counting predicate restricts X to take only the elements of the counting set that have distance $I = J - 1$ from a as its values. In this way, these are the only elements involved in the computation at each step.

Now the goal must be transformed into

$$?-\ sgc'(a, Y, 0).$$

and the final program is:

$counting(a, 0).$

$counting(X1, I) \ :-\ counting(X, J), par(X, X1), I = J + 1.$

$sgc'(X, Y, I) \ :-\ X = Y, integer(I).$

$sgc'(X, Y, I) \ :-\ counting(X, I), par(X, X1),$

$$sgc'(X1, Y1, J), par(Y, Y1), I = J - 1.$$

Observe the characteristics of the final program: the counting predicate "counts" generation levels from a upwards, marking the level of each element of the Magic Set. The sgc' predicate goes the opposite way, coming down again. The goal chooses only the sgc pairs of level 0, i.e., those at the same level as a. At each step, among the tuples of the Magic Set, the program only uses the tuples of appropriate distance from a.

Now we can look again at Fig. 10.2, which represents the cone of a. It is exactly the same as Fig. 10.1, but with an indication of levels for each generation. At each step the computation is limited to those elements that are placed together between the same two horizontal lines. This means that different counting sets are involved at different steps of iteration, each counting set being only a very small part of the whole magic set. The application of the Counting method allows more precise selections during the computation, but it creates a problem when the path between two elements of the database is not unique. Consider the following graph representation of a generic binary relation R. We build a directed graph $\Gamma_R =< N_R, E_R >$ defined as follows: N_R is the set of constants present in R; $< n_1, n_2 >\in E_R$ iff there is a tuple $< n_1, n_2 >$ in R.

In general, the set of all the possible paths between two nodes n_1 and n_2 contains more than one element; thus the length of the path between the two nodes is not unique. Thus, such a concept of distance is not well defined.

This problem, which in general makes the rewritten program less efficient, affects the correctness of the Counting method in the case of *cyclic* databases. A cyclic database is a database whose Γ_R graph is cyclic. In this case, since the set of possible paths is *infinite*, each element is assigned an infinite set of distances from the binding element. In cases like this, the rewritten program would keep counting edges without realizing that it is counting the same edge several times.

Thus, for the Counting Method to be *sound and complete*, and to *terminate*, it must be hypothesized that the underlying database is acyclic. A further hypothesis is that the program is linear, and there is at most one recursive rule for each predicate.

10.1.3 The Static Filtering Method

The Static Filtering method is also a rewriting strategy, which assumes the existence of a successive bottom-up evaluation. Its input is a goal on a program with just one IDB predicate.

In this method, the process of bottom-up evaluation is viewed as a *flow* of tuples through a graph derived from the program, called a *relation-axiom graph*.

As in the Magic Set method, some preliminary modifications are performed on the original *Datalog* program; then the corresponding graph is built, and only at that point are we ready to perform the real transformation, which consists of a sequence of modifications performed on the graph itself.

Initially, we add the goal rule to the program, in the same way as we did in the section on Magic Sets. We introduce the *AC-notation*: each rule in the program

$$r: \quad p(\alpha_1, \ldots, \alpha_n) :- q_1(\beta_1, \ldots, \beta_k), \ldots, q_m(\beta_s, \ldots, \beta_h).$$

where p and the q_i are not necessarily different, is rewritten in the AC-notation as:

$$r^{AC}: \quad p(A_1, \ldots, A_m) : -q_1(B_{11}^{i_1}, \ldots, B_{1h}^{i_1}), \ldots, q_n(B_{n1}^{i_n}, \ldots, B_{nk}^{i_n}), Cond_r. \quad (1)$$

The superscript to the arguments is needed to distinguish the same argument in different occurrences of the same predicate. As in the Magic Sets method, it is only added if there is more than one occurrence of the same predicate in the RHS of the rule. However, note that this time the occurrence number is attached to the arguments and not to the predicates. We have:

1) Renamed all the predicate arguments, so that they are all different from one another.
2) Introduced $Cond_r$, which is a condition related to rule r, that expresses the relationships among arguments in the rule. In particular, it restores the original equalities among arguments that are now all different. It has the form

$$(v_{i1}\theta_1 v_{j1}) \wedge \ldots \wedge (v_{it}\theta_t v_{jt})$$

where the θ_j's are comparison operators $(\leq, <, >, \geq, =, \neq)$. Note that, after translating standard *Datalog* rules, $Cond_r$ only contains equalities.

The relation-axiom graph of a program has two types of nodes: *relation nodes*, associated to predicates, each labeled with the corresponding predicate symbol, and *axiom nodes*, associated to rules, each labeled with the rule name. The name "axiom node" derives from the fact that rules are viewed as axioms of a logical theory. For each rule r as above, there is a directed edge from node r to node p, and one from each of the q_i to r. The goal evaluation is done on the relation-axiom graph of the program augmented with the goal rule.

Example 10.6 Given the program

$r_1:$ $anc(X,Y) :- par(X,Y).$
$r_2:$ $anc(X,Y) :- par(X,Z), anc(Z,Y).$

with the goal

$? - anc(X, a).$

We first add the goal rule to the program:

$r_0:$ $\quad q(X) :- anc(X, a).$
$r_1:$ $\quad anc(X, Y) :- par(X, Y).$
$r_2:$ $\quad anc(X, Y) :- par(X, Z), anc(Z, Y).$

In AC-notation this is transformed into:

$r_0:$ $\quad q(Q1) :- anc(A1, A2), Q1 = A1, A2 = a.$
$r_1:$ $\quad anc(A1, A2) :- par(P1, P2), A1 = P1, A2 = P2.$
$r_2:$ $\quad anc(A1^1, A2^1) :- par(P1, P2), anc(A1^2, A2^2),\ A1^1 = P1,$
$\hspace{5cm} A2^1 = A2^2, P2 = A1^2.$

Note that we have numbered the predicate arguments only when we had more than one occurrence of the same predicate inside a rule, namely only in rule r_2, where we had two occurrences of anc. Fig. 10.3 shows the relation-axiom graph for the $ANCESTOR$ program. □

The bottom-up evaluation of a goal can be considered as a flow of tuples through the graph, from the relation nodes corresponding to the EDB predicates, following the graph arrows, to the relation node corresponding to the goal node. Computation is ideally performed inside axiom nodes. Edges that go out from a node are called the node's "output ports", while incoming edges are called "input ports". Tuples enter each axiom node through its input ports, and contribute to a computation that produces new tuples. These are transmitted, through the output port, to the next relation node. The flow of tuples in the graph ends when

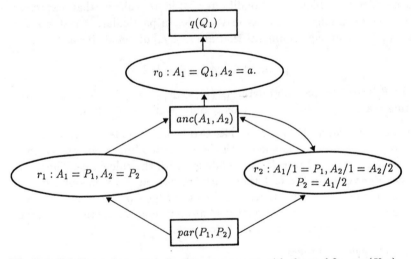

Fig. 10.3. Relation-axiom graph for the anc program with the goal $? - anc(X, a)$

no new tuples can be generated by any axiom node. This is the termination condition of the evaluation process.

The process as it has been described, with no other conditions, is also called "simplistic evaluation", a version of *naive evaluation*.

Note that the goal binding(s), expressing *constraints* that must be satisfied by the tuples of the solution, are among the conditions of the goal node. Consider all the tuples accumulated in the goal node at the end of the computation. Many of these will not satisfy the binding condition, so they will eventually be cut off in this node. The idea of the Static Filtering optimization technique is to cut off useless tuples from the computation as soon as possible, at an earlier stage of their flow towards the goal node.

This is achieved by imposing *filters* on the output ports of each relation node, which are also input ports of the axiom nodes that *follow them* in the order imposed by the graph edges. These filters are sets of conditions on predicate arguments, and should, in fact, *filter out* tuples that are irrelevant to the computation. The filters are derived from the conditions attached to each axiom node, including the goal bindings. In order to make the computation as efficient as possible, it is convenient to impose the strongest possible conditions. Thus an optimization algorithm must find the best filter for each output port of each node, i.e., the most restrictive condition.

Propagation of filters to all the output ports of the relation nodes starts from the goal node. The initial filters are the goal bindings, imposed on the input port of the goal axiom node, i.e., the output port of the relation node of the queried predicate. In the last program of Example 10.6, the initial filter is thus "$A2 = a$" coming from the body of rule r_0.

The propagation of filters to the rest of the graph is achieved by *pushing* the conditions backwards with respect to the graph orientation, as far as possible. In the absence of recursion, the graph is acyclic, so that this is done very easily, and corresponds to pushing selection conditions inside a conventional algebraic expression.

Consider the rule in AC-notation introduced in (1). Fig. 10.4 shows a subgraph representing this rule.

Suppose relation node p has k output ports with filters $F_1, \ldots, F_k$ (see Fig. 10.4). The axiom node r has n input ports coming from relation nodes $q_1, \ldots, q_n$. The *push* operation consists of finding filters $F_{r,1}, \ldots, F_{r,n}$ for the input ports of node r, by deducing them from information about the filters $F_1 \ldots F_k$.

This corresponds to deriving new, restrictive conditions for the arguments of $q_1, \ldots, q_n$ in r^{AC} from known information about restrictive conditions on the arguments of p. This can be better understood by following the operation on Example 10.7 below.

In the case of recursive rules, the problem of pushing filters is rather complicated. In Fig. 10.5 we see that, in order to impose filters on the input ports of a recursive rule r, one also needs to know the filters on the output ports of the recursive predicate defined by that rule, and this information also comes from the input ports of rule r! This problem is solved by building filters incrementally on the ports, until they are stabilized. Before giving the optimization algorithm, we define the *push* operation more formally. This operation can be followed in Fig. 10.4.

INPUT:

A relation-axiom graph G; a specific relation node p; a conjunction of filters $F = F_1 \wedge \ldots \wedge F_k$ on the output ports of p (i.e. a conjunction of the conditions $Cond$ attached to the rules that contain p in the RHS); the axiom node for rule r defining the predicate p; "old" (temporary, already existing) filters $F_{r,i}$, ($i = 1 \ldots n$) for the input ports of node r, coming from relation nodes $q_1, \ldots, q_n$.

OUTPUT:

A new graph G', in particular **new** temporary filters $F'_{r,i}$ ($i = 1 \ldots n$) for the input ports of node r.

PUSH OPERATION:

For each $i = 1 \ldots n$ do:
    change filter $F_{r,i}$ to:

$$F'_{r,i} = F_{r,i} \vee (Cond_r \wedge F).$$

Modify $F'_{r,i}$ to the conjunction of all transitive consequences of $F'_{r,i}$, where all the conjunctive terms only containing arguments not belonging to q_i have been deleted.

ENDPUSH

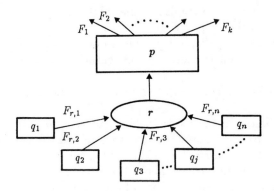

Fig. 10.4. Push operation of filters from node p to nodes $q_1, \ldots q_n$

As we will see from the full algorithm below, filters are built incrementally by iterating the PUSH operation, starting from the condition "*false*", and adding new disjunctive terms until they are complete.

Example 10.7. Consider the program of Example 10.6. Consider rule r_1 represented in Fig. 10.3. We want to push the filters of the output ports of node *anc* through the axiom node r_1 towards the filters of the output port of node *par*, which is in the RHS of rule r_1. Suppose the old temporary filters on the output ports of *par* were "*false*" at this stage of the computation.

Suppose that, at this point of the computation, the filter F (i.e. the current condition on anc) is "$A2 = a$". Then we must take

$$F'_{r_1} = F_{r_1} \vee (Cond_{r_1} \wedge \text{``}A2 = a\text{''})$$

($Cond_{r_1} \Leftrightarrow$ "$A1 = P1 \wedge A2 = P2$") as new filter for the output port of *par*. Thus,

$$F'_{r_1} = \text{``}P2 = a\text{''}$$

which is the consequence of the condition "$A2 = a$" expressed on anc. □

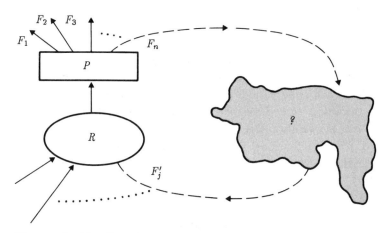

Fig. 10.5. Pushing filters in the recursive case

After the application of each *push* operation, the new graph G' only differs from G in the values of the filters $F_{r,i}$. The optimized graph is obtained by iterating this transformation until all filters stabilize. To keep track of filters' propagation, the rules are progressively marked and unmarked during iterations. Termination is ensured by the finiteness of the number of constants, variables, and nodes in the graph.

Now we give the complete optimization algorithm. Again, we advise the reader to follow the algorithm using Example 10.8.

ALGORITHM STATIC FILTERING
INPUT
A relation-axiom graph G.
OUTPUT
An optimized graph G^*.
METHOD
Initialize all filters of the graph to *false*;

FOR each axiom node r defining a goal q DO
  BEGIN
    Perform a PUSH operation of the filter *true*
    to the input ports of r;
    FOR each relation node n which has an output port
    towards r and whose filters were changed by this
    PUSH operation DO
    IF n is an intensional relation THEN mark n;
  END;
WHILE there are marked relation nodes n DO
  BEGIN
    Unmark n;
    FOR each axiom node s defining n DO
      BEGIN
        perform a PUSH operation of the output
        filters of n to the input ports of s;
        FOR each relation node p which has an output port
        towards s and whose filters were changed by this
        PUSH operation DO
        IF p is an intensional relation THEN mark p;
      END;
  END;
ENDMETHOD

It can be shown that the optimization procedure terminates, and that it is complete.

Example 10.8. Consider the ANCESTOR program, and the goal in Example 10.6. In that example, we performed the initial program transformation. Follow the optimization algorithm on the graph of Fig. 10.3.

Initially, $\forall i$, $F_{r_i} = \text{"}false\text{"}$. We start from axiom node r_0. We push the filter $\text{"}true\text{"}$ through the input port of q:

$$F'_{r_0} = F_{r_0} \vee (Cond_{r_0} \wedge \text{"}true\text{"}) \Leftrightarrow Cond_{r_0} \Leftrightarrow \text{"}A2 = a\text{"}$$

Then, we mark node *anc*. The termination condition is not verified, because there are marked nodes in the graph, so we go on.

We unmark node *anc*. Consider the two axiom nodes r_1 and r_2 defining *anc*. First we push F'_{r_0} to F_{r_1}:

$$F'_{r_1} = F_{r_1} \vee (Cond_{r_1} \wedge \text{"}A2 = a\text{"}) \Leftrightarrow \text{"}P2 = a\text{"}.$$

Now we push filters through rule r_2:

$$F'^{(par)}_{r_2} = F^{(par)}_{r_2} \vee (Cond_{r_2} \wedge \text{"}A2 = a\text{"}) \Leftrightarrow Cond_{r_2}.$$

$$F_{r_2}^{\prime(anc)} = F_{r_2}^{(anc)} \vee (Cond_{r_2} \wedge \text{``}A2 = a\text{''}) \Leftrightarrow \text{``}A2^1 = a\text{''}.$$

As the filter of rule r_2 changes, we mark node anc. Therefore, termination is not achieved. We unmark anc and proceed to the next iteration. We first consider $F_{r_1}^{\prime\prime}$:

$$F_{r_1}^{\prime\prime} = F_{r_1}^{\prime} \vee (Cond_{r_1} \wedge \text{``}A2^1 = a\text{''}) \Leftrightarrow \text{``}P2 = a\text{''}(= F_{r_1}^{\prime}).$$

This time the filter was not affected by the push operation. Then we consider $F_{r_2}^{\prime\prime}$.

$$F_{r_2}^{\prime\prime(par)} = F_{r_2}^{\prime(par)} \vee (Cond_{r_2} \wedge \text{``}A2^1 = a\text{''}) \Leftrightarrow Cond_{r_2}(= F_{r_2}^{\prime(par)}).$$

Also this filter was not changed by the push operation.

$$F_{r_2}^{\prime\prime(anc)} = F_{r_2}^{\prime(anc)} \vee (Cond_{r_2} \wedge \text{``}A2^1 = a\text{''}) \Leftrightarrow \text{``}A2^1 = a\text{''}.$$

As the filters are not affected, we do not mark any relation, thus the algorithm is terminated.

The resulting rules are:

r_0: $\quad q(Q1) :- anc(A1, A2), Q1 = A1, A2 = a.$
r_1: $\quad anc(A1, A2) :- par(P1, P2), A1 = P1, A2 = P2, P2 = a.$
r_2: $\quad anc(A1^1, A2^1) :- par(P1, P2), anc(A1^2, A2^2), A1^1 = P1,$
$\qquad\qquad A2^1 = A2^2, P2 = A1^2, A2^1 = a.$

By renaming variables, and by reintroducing conditions on predicate arguments, we obtain:

r_0: $\quad q(X) :- anc(X, a).$
r_1: $\quad anc(X, a) :- par(X, a).$
r_2: $\quad anc(X, a) :- par(X, Z), anc(Z, a).$ $\qquad\square$

It is evident that the result of the optimization of Example 10.8 is really more efficient than the original program. Observe that all the occurrences of the predicate anc in this program have the second argument bound. We will discuss the various similarities among this and other methods of optimization at the end of this chapter.

10.1.4 Semi-naive Evaluation by Rewriting

Recall from the previous chapter that the idea of *semi-naive evaluation* is modifying the traditional iterative evaluation methods by forcing them to produce only *really new* tuples at each iteration step. Unfortunately, the application of this approach to nonlinear algebraic expressions is quite difficult, as the form of the *differential* may be very complicated. Such complicated calculations for the evaluation of the differential would reduce substantially the benefit produced by reducing the size of the relations involved.

Here we introduce a rewriting method which computes the differential expression in a different way, thus eliminating the source of difficulty. The drawback of this method is that it uses constructs which are not proper of *Datalog*. Indeed, the rewritten program contains negated literals in rule bodies, which are not considered part of the language. However, in Chap. 11 we will introduce the concept of *locally stratified* program. The output of the *semi-naive rewriting method* is, in fact, a *locally stratified* program, and it thus has a well-defined solution in the new context provided by the addition of negation to a *Datalog* program.

We recall, in logical terms, what we said in Chap. 9 about semi-naive evaluation. We express in logical terms the *differential* of a predicate p, at step k of the iteration, with respect to a rule $p : -E(p)$.:

$$D^{(k)} = p^{(k)} - p^{(k-1)} = E(p^{(k-1)}) - p^{(k-1)}.$$

Expressing this differential only in terms of $p^{(k-1)}$ is rather cumbersome if the rules are nonlinear. With this method, rules are rewritten in order to express the differential *directly* in terms of $p^{(k)}$.

Let us consider the nonlinear program for the computation of ancestors (Chap. 8):

$$anc(X, Y) : -anc(X, Z), anc(Z, Y).$$

$$anc(X, Y) : -par(X, Y).$$

The rewriting achieved for the program is:

$$anc(I, X, Y) : -anc(J, X, Z), anc(J_1, Z, Y), J_1 \geq 0,$$
$$J \geq J_1, I = J + 1, \neg anc(W, X, Y).$$

$$anc(I, X, Y) : -anc(J_1, X, Z), anc(J, Z, Y), J_1 \geq 0,$$
$$J > J_1, I = J + 1, \neg anc(W, X, Y).$$

$$anc(0, X, Y) : -par(X, Y).$$

Note the use of the built-in predicates for inequalities and summation; the latter was already used for the counting method. Note also the use of negation.

Let us follow the evaluation of the first rule. Initially, it produces the second generation ancestors, with $I = 2$, $J = 1$, $J1 = 1$, and the negated occurrence of $anc(W, X, Y)$ eliminates the ancestors of first generation from the result. The second rule can now compute the third generation ancestors. Here, $I = 3$, $J = 2$, $J1 = 1$. The first rule works again for the fourth generation, and so on, until some rule produces the empty set.

Note that each rule application creates only new tuples, since those already produced are eliminated by the negated literal. Thus, we manage to express the differential of the *anc* predicate.

In general, if there are k occurrences of the recursive predicate in the rule, then the rewriting consists of replicating this rule k times, each time decrementing by 1 the upper bound of the inequality that appears in the new rule body.

In order to obtain the solution to the original program, one must then apply the naive evaluation method to the rewritten program, taking care of the new situation produced by the presence of the negated literals in the *RHS* (see situation produced by the presence of the negated literals in the RHS (see Chap. 11). The resulting value of predicate anc is the union of all the nonempty sets of pairs (X, Y) resulting from all the values of $anc(J, X, Y)$ found by the application of naive evaluation.

10.2 Rewriting of Algebraic Systems

In this section, we turn our attention to the algebraic formalism. We introduce a structured approach for rewriting systems of equations of relational algebra. This optimization is developed in a sequence of steps: each step is implemented by a particular algorithm.

A *Datalog* program, defining more than one intensional relation, yields a *system* of equations of relational algebra. Each equation defines a *variable relation*, so that, from now on, we will deal with *systems of n equations and n variables*, with the following standard form:

$$\Sigma : \quad X_i = E_i(X_1, \ldots X_n, C_1, \ldots C_m) = \bigcup_{j=1}^{n_i} T_{ij} \quad , \quad i = 1 \ldots n \quad (1)$$

where the X_i are relational variables (IDB predicates), and the C_j are relational constants. Each term $T_{i,j}$ has the form $T_{ij} = \Pi_{L_{ij}} \sigma_{P_{ij}} CP_{ij}$, where L_{ij} is a list of attribute numbers, P_{ij} is a conjunction of simple predicates corresponding to the selection and join conditions, and each CP_{ij} is a cartesian product of variable and constant relations.

Making an analogy between the translated goals with the form of relational algebra queries, we also will refer to goals as *queries*. X_P, the query variable, will be called *principal variable* of the system.

In the sequel, we will often use the system E_Σ, introduced in Example 8.3.

$$X_1 = C_1 \cup (X_1 \circ X_3) \cup X_2$$
$$X_2 = (X_1 \circ X_3) \cup C_3$$
$$X_3 = (X_3 \circ C_2) \cup C_4$$

10.2.1 Reduction to Union-Join Normal Form

For each equation E_i in (1), let $n_i > 1$. Then, for each term T_{ij}, we introduce a new variable N_{ij} into the system, and we rewrite E_i in the following way:

$$X_i = \bigcup_{j=1}^{n_i} N_{ij}$$
$$N_{ij} = T_{ij} = \Pi_{L_{ij}} \sigma_{P_{ij}} CP_{ij}$$

As a consequence, our equations are of one of these possible forms:

a) **Union**-type equations (*U-equations*), which are fully characterized by a pair

$$< X_i, U_i >,$$

where X_i is the relational variable in the *LHS* of the equation, U_i is the set of relational variables and constants which appear in the union in the *RHS*.

b) **Join**-type equations (*J-equations*), which are fully characterized by a quadruple

$$< X_i, L_i, P_i, J_i >,$$

where X_i is the *LHS* of the equation, L_i is a list of column numbers that appear in the projection operator, P_i is a predicate, and J_i is the set of variables and constants that are in the cartesian product on the *RHS*.

The resulting system is in *Union-Join normal form* (UJNF).

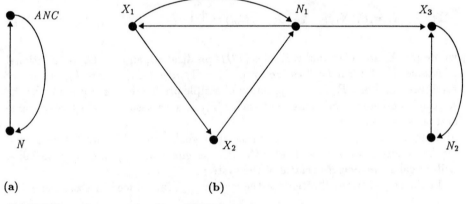

(a)   (b)

Fig. 10.6. Dependency graphs of two UJNF systems

Example 10.9.

a. Consider the system that computes ancestors (Example 9.1 (a)), namely:

$$ANC = ANC \circ PAR \cup PAR.$$

Its UJNF is:

$$N = ANC \circ PAR$$
$$ANC = N \cup PAR.$$

Its dependency graph is shown in Fig. 10.6(a).

b. The UJNF of system E_Σ is:

$$X_1 = C_1 \cup N_1 \cup X_2$$
$$X_2 = N_1 \cup C_3$$
$$X_3 = N_2 \cup C_4$$
$$N_1 = X_1 \circ X_3$$
$$N_2 = X_3 \circ C_2.$$

Its dependency graph is in Fig. 10.6(b). □

10.2.2 Determination of Common Subexpressions

Determining common subexpressions of algebraic expressions may cause a reduction in computation, as some operations may be factored. The reduction into UJNF makes this determination much easier: we may look for common subexpressions separately in U-equations or in J-equations.

a) A common subexpression of two U-equations for X_i and X_j corresponds to the (maximum) common subset U_C of U_i and U_j. In the system,

$$\{< X_i, U_i >, \; < X_j, U_j >\}$$

is replaced with

$$\{< X_i, U_i - U_C \cup \{X_C\} >, < X_j, U_j - U_C \cup \{X_C\} >, < X_C, U_C >\}$$

where X_C is a new variable. The search for common elements is done by inspection.

b) A common subexpression of two J-equations for X_i and X_j corresponds to a subset J_C which is common to J_i and J_j; notice, however, that we are interested in common subexpressions such that the relations in J_C have the same join conditions in E_i and E_j. This makes the search for common subexpressions in J-equations nontrivial.

Consider two J-equations $< X_i, L_i, P_i, J_i >$ and $< X_j, L_j, P_j, J_j >$. Let $J_i = \{J_{i1}, \ldots J_{im}\}$ and $J_j = \{J_{j1}, \ldots J_{jl}\}$ be such that $\exists r, \; 1 \leq r \leq min\{m, l\}$: $\forall k = 1 \ldots r, J_{ik} = J_{jk}$. Let $P_i = p_{i1} \wedge \ldots p_{iu}$ and $P_j = p_{j1} \wedge \ldots p_{jv}$ be such that $\exists s, \; 1 \leq s \leq min\{u, v\}$: $\forall h = 1 \ldots s, p_{ih} = p_{jh}$. Then, the common subexpression is:

$$\sigma_{p_{i1} \wedge \ldots \wedge p_{is}}(J_{i1} \times \ldots \times J_{ir}).$$

We introduce a new equation for X_C and modify the equations for X_i and X_j in the obvious way. This construction is quite tedious, and intuitive, so we omit it.

Finding common subexpressions within $n \; (> 2)$ J-equations is computationally hard, and iterating this construction in different ways does not necessarily yield the same result.

Example 10.10.

a. Consider the following UJNF system:

$X_1 = C_1 \cup C_2 \cup X_3$
$X_2 = C_2 \cup X_3 \cup C_3$
$X_3 = X_1 \circ X_2 \circ X_4$
$X_4 = C_4 \circ X_1 \circ X_2$

After determining common subexpressions, we have:

$X_1 = C_1 \cup X_5$
$X_2 = X_5 \cup C_3$
$X_3 = X_6 \circ X_4$
$X_4 = C_4 \circ X_6$
$X_5 = C_2 \cup X_3$
$X_6 = X_1 \circ X_2$

b. Given the UJNF system:

$X = A \cup Y$
$Y = Y \circ X \circ B$
$Z = X \circ B \circ A$

this can be rewritten as:

$X = A \cup Y$
$Y = Y \circ W$
$Z = W \circ A$
$W = X \circ B$

c. Given the system:

$X_1 = \Pi_{1,6}\left((X_2 \underset{2=1}{\bowtie} X_3) \underset{2=1}{\bowtie} C_1\right)$
$X_2 = C_2 \circ X_2 \circ C_1 \circ X_4$
$X_3 = X_2 \circ X_3 \circ C_3$
$X_4 = C_4 \cup X_3 \cup C_2$

we can determine common subexpressions in two alternative ways. The first alternative is:

$X_1 = \Pi_{1,6}(W \underset{2=1}{\bowtie} C_1)$
$X_2 = C_2 \circ X_2 \circ C_1 \circ X_4$
$X_3 = (\Pi_{1,4}W) \circ C_3$
$X_4 = C_4 \cup X_3 \cup C_2$
$W = X_2 \underset{2=1}{\bowtie} X_3$

The second alternative is:

$X_1 = \Pi_{1,6}(Z \underset{2=1}{\bowtie} X_3)$
$X_2 = C_2 \circ (\Pi_{1,4}Z) \circ X_4$
$X_3 = X_2 \circ X_3 \circ C_3$

$$X_4 = C_4 \cup X_3 \cup C_2$$
$$Z = X_2 \underset{2=1}{\bowtie} C_1$$

Note that evaluating the relative convenience of the two alternatives, in order to select one of them, is quite hard. □

10.2.3 Query Subsetting and Strong Components

Given a system Σ and a goal Q on X_P, the principal variable of Σ, we determine the subset $D(X_P) \subseteq X_\Sigma$ of variables whose computation is required for evaluating the goal Q.

Let $G(\Sigma)$ be the dependency graph of Σ. $D(X_P)$ is defined recursively as follows:

1) X_P is in $D(X_P)$.
2) If $< X_i, X_j >$ is in $E(G(\Sigma))$ and X_i is in $D(X_P)$, then also X_j belongs to $D(X_P)$.

Let Σ_P ($\subseteq \Sigma$) be the system of equations for the variables in $D(X_P)$. Then, Q can be evaluated on Σ_P. Query subsetting is equivalent to determining the subprogram reachable from a given goal (Sect. 10.1.1).

We are now interested in determining the decomposition of Σ_P into subsystems such that each subsystem can be solved independently. By construction, the graph $G(\Sigma_P)$ is connected. Let us identify strongly connnected components of Σ_P; each strongly connected component of Σ_P corresponds to a subset of mutually recursive equations. Thus, each strongly connected component corresponds to an algebraic subsystem which must be solved through an independent application of solution methods (Jacobi, Gauss-Seidel, Semi-Naive).

Finally, we determine the order of evaluation of these subsystems. Let SC_i and SC_j be two strongly connected components of $G(\Sigma_P)$, and let $< X_i, X_j >$ be an edge connecting SC_i to SC_j (and obviously not vice-versa). Then, SC_j must be evaluated before SC_i. This rule defines an order relationship among components:

$$SC_j < SC_i \Leftrightarrow \Big(< X_i, X_j > \in E\big(G(\Sigma_P)\big) \land (X_i \in SC_i) \land (X_j \in SC_j)\Big).$$

By construction of $G(\Sigma_P)$, there exists only one connected component which is maximal according to the partial order induced by this order relationship. This component, which is called *top component*, contains the variable X_P. If $SC_j < SC_i$, then SC_j must be solved before SC_i; this is due to the fact that, after solving SC_j, the variable relations of SC_j become known, and can be used as constants in the solution of SC_i.

The process of finding strongly connected components in a graph has polynomial time complexity.

Example 10.11.
a. Let us consider the system E_Σ of our current example. The system can be separated into two strongly connected components, i.e. the subsystems:

E_{Σ_1} :
$$X_1 = C_1 \cup N_1 \cup X_2$$
$$X_2 = N_1 \cup C_3$$
$$N_1 = X_1 \circ X_3$$
E_{Σ_2} :
$$X_3 = N_2 \cup C_4$$
$$N_2 = X_3 \circ C_2$$

If the principal variable is X_1, X_2, or N_1, then the *query subsetting* step does not have any effect on the system, i.e., $E_{\Sigma_P} = E_\Sigma$.

If the principal variable of the system is X_3 or N_2, then the query subsetting step is effective, and we have $E_{\Sigma_P} = E_{\Sigma_2}$.

Consider the dependency graph of Fig. 10.7 with 5 connected components. Let X_P belong to the connected component SC_1. Then, query subsetting is effective, and eliminates components SC_4 and SC_5 from the system.

The ordering induced by our rule is:

$$SC_3 < SC_2 < SC_1.$$

□

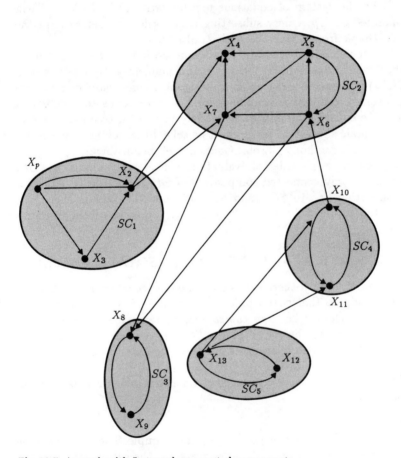

Fig. 10.7. A graph with 5 strongly connected components

A particular case is when, by appropriate substitution, it is possible to solve each equation of the system independently of the others, because each of them becomes a connected component. Systems like these are called *reducible by substitution*.

Consider a system Σ and its dependency graph G_Σ. It is easy to see that the procedure of evaluating variable X_i, and then substituting it in Σ as a constant, is equivalent to eliminating node X_i from G_Σ, and then linking all the predecessor nodes of X_i to all its successors; we will say that we have *substituted* node X_i in G_Σ.

We say that a graph G is *reducible by substitution* iff it can be reduced to a graph with a single node X_R by an appropriate sequence of node substitutions; a graph with a single node corresponds to a single equation in the variable X_R, which can be solved independently; then, other variables are computed starting from substitution equations.

A system is *reducible by substitution* if and only if its dependency graph is reducible by substitution. The problem of finding the right sequence of substitutions (if any) to solve a system of equations is reduced to a problem of graph theory. There are a number of sufficient conditions that guarantee graph reducibility.

The following are some significant results on this subject:

a) Any acyclic graph G is reducible by substitution.
b) A strongly connected graph G is reducible by substitution iff it contains a node K such that $G - \{K\}$ is acyclic.
c) A graph G is reducible by substitution if and only if all its connected components are reducible by substitution.

The constructive proofs of these theorems yield an algorithm which determines whether a given graph G is reducible by substitution; if so, it provides a sequence of substitutions for G (see the Bibliographic Notes).

Example 10.12. Consider the system

$$P = \Pi_2(Q \underset{1=2}{\bowtie} R) \cup \Pi_2 S$$
$$Q = \Pi_{1,4}((P \underset{1=1}{\bowtie} R) \underset{3=2}{\bowtie} S)$$

where R and S are extensional relations. This is clearly reducible by substitutions, yielding

$$P = \Pi_2((\Pi_{1,4}((P \underset{1=1}{\bowtie} R) \underset{3=2}{\bowtie} S)) \underset{1=2}{\bowtie} R) \cup \Pi_2 S \qquad \square$$

10.2.4 Marking of Variables

In Sects. 10.2.4 through 10.2.6 we are concerned with rewriting methods which transform subsystems of equations; each subsystem corresponds to a strong component of Σ_P. The optimization steps are driven by the knowledge of goal bindings (or adornment); thus, they have the same generality as the logic rewriting methods presented in Sect. 10.1.

Marking relations denotes the propagation of the query selection predicate to the various equations of the system, according to the following *marking rules*:

a) Propagation for U-equations: let X_i be marked ($X_i : m$), and consider the equation $< X_i, U_i >$; then, all the variables X and constants C of U_i must be marked m ($X : m$ or $C : m$).

b) Propagation for J-equations: let X_i be marked ($X_i : m$), and consider the equation $< X_i, L_i, P_i, J_i >$; using the projection list L_i, transform m into the corresponding column(s) n of variables X or constants C in E_i, and give mark n to them, yielding $(X : n)$ or $(C : n)$. Moreover, if that column is also used in some equi-joins with other columns q of other variables Y or constants K in the equation, give mark q to them as well, yielding: $(Y : q)$ or $(K : q)$.

Marked variables correspond to new variable relations obtained as selections over the original variables. In fact, $(X : k)$ corresponds to a selection over the k-th column of X : $(X : k) = \sigma_{k=const} X$. This reduction is justified by the following general equation:

$$\sigma_p X_i = \sigma_p E_i.$$

The correctness of the above rules derives from distributivity rules of selection with respect to union, commutativity rules of selection with projection and selection, and distributivity rules for selection with respect to cartesian product.

The following marking algorithm takes as input a system Σ of N equations and generates a new system Σ', equivalent to Σ with respect to the considered goal (see Sect. 10.1). Let a_p be the arity of X_P; let the query predicate be an equality on column i_P of X_P.

a) Initially, give mark i_P to variable $X_P : (X_P : i_P)$.
b) Apply recursively the marking rules to mark all possible variables and constants of Σ; consider the marked variables as new variables of Σ';
c) Include recursively in Σ' all the equations $X_i = E_i$ such that X_i is mentioned in at least one of the equations of Σ' previously generated.

Example 10.13.

a. Let Q_1 : $\sigma_{1=a} X_2$ on the subsystem E_{Σ_1} of E_Σ. We obtain:

$(X_1 : 1) = (C_1 : 1) \cup (N_1 : 1) \cup (X_2 : 1)$
$(X_2 : 1) = (N_1 : 1) \cup (C_3 : 1)$
$(N_1 : 1) = (X_1 : 1) \circ X_3$

b. Let Q_2 : $\sigma_{2=a} X_3$ on the component E_{Σ_2} of E_Σ. We obtain:

$(X_3 : 2) = (N_2 : 2) \cup (C_4 : 2)$
$(N_2 : 2) = X_3 \circ (C_2 : 2)$
$X_3 = N_2 \cup C_4$
$N_2 = X_3 \circ C_2$

□

10.2.5 Reduction of Variables

Let us compare system Σ with the system Σ' obtained from the marking algorithm. The initial system is strongly connected by construction, every variable X of Σ is present in Σ' at least once, and can be either marked or unmarked. It is also possible that Σ' contains more than one occurrence of the same variable X, possibly marked on different columns. Actually, each variable X of Σ corresponds in Σ' to either of the following:

1) One unmarked variable, and no marked variable.
2) One or more marked variables, and no unmarked variable.
3) One or more marked variables, and one unmarked variable.

We call the variables described in point 2 *reduced variables*.

After executing the marking algorithm, we face two alternatives: either we replace system Σ with system Σ', and work with this last one, or we keep system Σ. Examples 10.13(a) and 10.13(b) represent two possible cases. In Example (a), $|\Sigma| = |\Sigma'|$. Since the three variables of Σ' are reduced, Σ' is certainly more efficient than Σ. Indeed, selection conditions propagate to reduced variables. Because there are no more unmarked occurrences of X in Σ', one can use $(X : k) = \sigma_{k=const}X$ instead of X. Notice that the gain in efficiency derives from the fact that selections have been propagated to the constant relations contained in the equations of reduced variables, so that, at computation time, we need only retrieve from the database these selected (and thus smaller) constant relations.

Example 10.14. By renaming variables, the system of Example 10.13(a) is reduced to:

$$V_1 = \sigma_{1=a}C_1 \cup V_3 \cup V_2$$
$$V_2 = V_3 \cup \sigma_{1=a}C_3$$
$$V_3 = V_1 \circ X_3$$

and the goal evaluation is reduced to the evaluation of variable V_2. □

In Example 10.13(b), there are no reduced variables, because all the marked variables also have some unmarked occurrence. Thus, $\Sigma \subseteq \Sigma'$, and of course the original system Σ is more efficient than Σ'.

It is easy to realize that, since Σ is a strongly connected system, whenever a variable is not reduced, neither is any of the others. Thus, it is convenient to replace Σ with Σ' only if no unmarked variable is present in Σ'.

10.2.6 Reduction of Constants

We now consider *marked* occurrences of constant relations, that appear in equations of a system Σ' after the failure of the step of reduction of variables. Our purpose is to reduce the size of a certain *occurrence* of the constant *before* solving the system, by using the information that the constant is marked, i.e., that it has some relationship with the selection condition of the goal.

The reduction can either be successful or fail. In the first case, we say that the constant is *reducible*; otherwise, it is called *irreducible*.

This reduction requires, in the most general case, the application of a rather complex algorithm. We provide here a simplified version of the general algorithm, referenced in the Bibliographic Notes. This version deals with equations obtained from the translation of linear, stable *Datalog* rules, which altogether define only one, recursive IDB predicate. If there are other nonrecursive intensional predicates, we are able to solve them before this reduction, by standard relational algebra operations and optimizations. Therefore, we may assume as input to this transformation a system only containing one variable relation, and several constant relations.

Let Σ be the system, in the following form:

$$X = N_1 \cup N_2 \cup \ldots \cup N_n$$

$$N_1 = E_1(X)$$

$$N_2 = E_2(X)$$

$$\ldots$$

$$N_n = E_n(X)$$

The system contains only one IDB predicate, corresponding to variable X. The reduction to UJNF yields only one U-equation, defining X, and a number of J-equations. Each of these may contain at most one occurrence of X, because the program is linear. Since we only have one variable X, we assume that our goal be represented by the following selection on X: $\sigma_{m=a}X$.

After the step of *marking*, the system has the following structure:

$$(X : m) = (N_1 : m) \cup (N_2 : m) \cup \ldots \cup (N_n : m)$$

$$(N_1 : m) = E_1'(X)$$

$$(N_2 : m) = E_2'(X)$$

$$\ldots$$

$$(N_n : m) = E_n'(X)$$

Each of the resulting expressions E_i' may assume one of the following three forms:

- E_i' contains an unmarked occurrence of X. In this case, X may or may not be joined with constant relations.
- E_i' contains a marked occurrence of X. In this case, the occurrence is marked on column m by the stability hypothesis.
- E_i' contains no occurrences of X.

Thus, the dependency graph of our system has the form shown in Fig. 10.8: the configuration is star-like. Some of the N_i have outgoing edges towards X. These

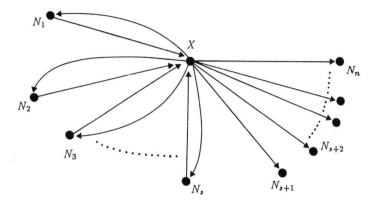

Fig. 10.8. Dependency graph before constant reduction

correspond to the E'_i that contain the occurrence (marked or unmarked) of X. Let there be s of these.

We reorder the equations in the system so that the first s J-equations are the ones containing X (so that they are $N_1, \ldots, N_s$).

Given a generic J-equation

$$Y = F(Y_1, \ldots, Y_n) = \Pi_{L_{ij}} \sigma_{P_{ij}} C P_{ij}$$

(as at the beginning of Sect. 10.2) we say that position l of Y *corresponds* to position m of Y_i if the attribute m of Y_i is projected into the l-th attribute of Y according to the list of projections L_{ij}.

Let K be a constant contained in the equation $(N_i : m) = E'_i(X)$, and marked on position k. We want to reduce K. If K has more than one mark $(k_1, \ldots k_l)$, then we collect all its marks and construct the selection predicate $sp \Leftrightarrow (k_1 = a \wedge \ldots \wedge k_l = a)$. We now provide the full algorithm. In order to understand it better, it is advisable to follow Example 10.15(b).

ALGORITHM REDUCTION OF CONSTANTS (K)

INPUT: A marked system Σ for a unique recursive variable X deriving from a program only containing linear stable rules. A marked constant K in the equation $(N_k : m) = E'_k(X)$. A selection predicate sp.

OUTPUT: An algebraic expression that reduces the size of constant K, or the word "irreducible".

METHOD:
$K' := \sigma_{sp} K$ /* *symbolic expression* */
  FOR $j := 1$ TO s DO
    BEGIN
    $K'' := \emptyset$
    IF X is unmarked in $E'_j(X)$ THEN
      BEGIN
        $K'' := K$

```
            FOR all constant relations in E'_j(X) DO
                IF E'_j contains a join X ⋈_{x_r=c_r} C_{j_r}
                such that position x_r of N_k corresponds to
                position r of K in (N_k : m) = E'_k(X)
                    THEN K'' := K'' ⋉_{r=c_r} C_{j_r}    /* symbolic expression */
                IF K'' = K THEN
                    BEGIN
                    OUTPUT "irreducible"
                    EXIT METHOD
                    END
            END
        K' := K' ∪ K''     /* symbolic expression */
    END
```

IF any of the constants C in the final expression of K' is reduced THEN substitute it with C'.

OUTPUT K'

ENDMETHOD

Note the comment "symbolic expression" added to some statements of the algorithm. This is to remind the reader that the algorithm is just a symbolic transformation that yields as output an expression, and thus no real operation is performed on the relations K, C, etc., by its application.

We explain the rationale of the algorithm from an intuitive viewpoint. All the equations and constant relations of our system contribute to the value of X. This is true in particular for the constant K. In fact, via some algebraic operations, some of the columns of K will eventually "produce" some columns of the relation X. These columns are exactly those transferred from K to N_k via the projection list $L_{i,j}$ of the equation $N_k = E_k(X)$. Suppose there is such a column x in N_k, corresponding (via the projection list) to column r in K. Then, using just $X \ltimes_{x=c} C$ instead of the whole X in this join leaves the result unchanged; we can thus propagate this semijoin to K and reduce it to: $K \ltimes_{r=c} C$.

Note that each application of the algorithm reduces one occurrence of a constant K. In order to obtain the reduction of the whole constant in all its occurrences, we should take the union of all the reduced values, and substitute it for all the instances of the constant.
Alternatively, a different reduced constant is to be substituted for each corresponding occurrence. These reduced constants may have some tuples in common.

Note also that, if the initial selection predicate is of the type "$\sigma_{m=a \wedge n=b} X$", the algorithm must be run twice, in sequence. In this case, a constant may be reduced by two applications of the algorithm.

Example 10.15. We show the behavior of the reduction algorithm using examples of growing complexity. All the extensional relations involved are binary. The reader may verify as an exercise that these systems correspond to programs with stable rules. Note that Example (a) corresponds to a general form of *ANC*.

a. We use the algorithm for the reduction of constants to optimize the system of Example 10.13(b), where we could not reduce variables. Consider the goal $\sigma_{2=a} X_3$ on system E_{Σ_2}. The initial system is:

$$X_3 = N_2 \cup C_4$$
$$N_2 = X_3 \circ C_2$$

We rewrite it in a form which makes the application of the algorithm clearer:

$$X_3 = N_2 \cup N_4$$
$$N_2 = X_3 \circ C_2$$
$$N_4 = C_4$$

The marked system is:

$$(X_3 : 2) = (N_2 : 2) \cup (N_4 : 2)$$
$$(N_2 : 2) = X_3 \circ (C_2 : 2)$$
$$(N_4 : 2) = (C_4 : 2)$$
$$X_3 = N_2 \cup N_4$$
$$N_2 = X_3 \circ C_2$$
$$N_4 = C_4$$

We can reduce constants C_4 and C_2:

$$C'_4 = \sigma_{2=a} C_4 \cup C_4 \underset{2=1}{\ltimes} C'_2$$
$$C'_2 = \sigma_{2=a} C_2 \cup C_2 \underset{2=1}{\ltimes} C'_2.$$

Notice that C'_2 itself is defined by an equation, which should be solved in order to determine the reduced relation C'_2. Once computed, this value is substituted in the equation for C'_4.

b. Consider the equation:

$$P = UP \circ P \circ DOWN \cup FLAT$$

which is a generalized version of the classical SGC problem. This yields the system

$$P = N_1 \cup N_2$$
$$N_1 = UP \circ P \circ DOWN$$
$$N_2 = FLAT$$

and the goal $\sigma_{1=a} P$, producing the marking

$$(P : 1) = (N_1 : 1) \cup (N_2 : 1)$$
$$(N_1 : 1) = (UP : 1) \circ P \circ DOWN$$
$$(N_2 : 1) = (FLAT : 1)$$
$$P = N_1 \cup N_2$$
$$N_1 = UP \circ P \circ DOWN$$
$$N_2 = FLAT$$

The reduction of variables fails, but we can reduce constants, obtaining

$UP' = \sigma_{1=a}UP \cup UP \underset{1=2}{\bowtie} UP'$

$FLAT'' = \sigma_{1=a}FLAT \cup ((FLAT \underset{1=2}{\bowtie} UP') \underset{2=1}{\bowtie} DOWN)$

Notice that the reduction of the constant UP corresponds, in this case, to the *cone* of a of the Magic Sets method. The above expressions for UP' and $FLAT'$ are produced in the following sequence of step-by-step transformations:

Reduction of FLAT:

Mark of $FLAT$ is 1.
$FLAT' := \sigma_{1=a}FLAT$;
Enter DO loop (only for N_1):
$FLAT'' := \emptyset$
P is not marked in E_1, thus:
$FLAT'' := FLAT$;
Enter DO loop:
P is joined in E_1 with UP, on 1st attribute of P and 2nd attribute of UP. 1st attribute of N_2 corresponds to 1st attribute of $FLAT$ in E_2 via the projection list, thus:
$FLAT'' := FLAT \underset{1=2}{\bowtie} UP$;
P is joined in E_1 with $DOWN$ on 2nd attribute of P and 1st attribute of $DOWN$. 2nd attribute of N_2 corresponds to 2nd attribute of $FLAT$ in E_2 via the projection list, thus:
$FLAT'' := (FLAT \underset{1=2}{\bowtie} UP) \underset{2=1}{\bowtie} DOWN$;
There are no more constants joined with P in E_1, we go out of the DO loop.
$FLAT'' \neq FLAT$, so we do not output "irreducible", and obtain the final result:
$FLAT' := FLAT' \cup FLAT''(= \sigma_{1=a}FLAT \cup ((FLAT \underset{1=2}{\bowtie} UP) \underset{2=1}{\bowtie} DOWN))$;
We have finished the reduction of constant $FLAT$. However, if any of the constants involved in $FLAT'$ is reducible, we can substitute its new expression in the expression for $FLAT'$.

Reduction of UP:

Mark of UP is 1.
$UP' := \sigma_{1=a}UP$;
Enter DO loop (only for N_1):
$UP'' := \emptyset$
P is not marked in E_1, thus:
$UP'' := UP$;
Enter DO loop:
P is joined in E_1 with UP, on 1st attribute of P and 2nd attribute of UP. 1st attribute of N_1 corresponds to 1st attribute of UP in E_1 via the projection list, thus:
$UP'' := UP \underset{1=2}{\bowtie} UP$;
P is joined in E_1 with $DOWN$ on 2nd attribute of P and 1st attribute of $DOWN$. 2nd attribute of N_1 does not correspond to any attribute of UP in E_1

via the projection list, thus we have finished the constants joined with P in N_1, so we go out of the DO loop. $UP'' \neq UP$, so we do not output "irreducible", and obtain the final result:
$UP' := UP' \cup UP'' (= \sigma_{1=a} UP \cup (UP \underset{1=2}{\bowtie} UP))$;
The reduction of constant UP is thus complete. However, we note that this constant is involved in the expressions for $FLAT'$ and UP' so we can substitute its new expression in those. The final system is:

$P = N_1 \cup N_2$
$N_1 = UP' \circ P \circ DOWN$
$N_2 = FLAT'$
$FLAT' := \sigma_{1=a} FLAT \cup ((FLAT \underset{1=2}{\bowtie} UP') \underset{2=1}{\bowtie} DOWN)$
$UP' := \sigma_{1=a} UP \cup (UP \underset{1=2}{\bowtie} UP')$

which is obviously equivalent to:

$P = N_1 \cup FLAT'$
$N_1 = UP' \circ P \circ DOWN$
$FLAT' := \sigma_{1=a} FLAT \cup ((FLAT \underset{1=2}{\bowtie} UP') \underset{2=1}{\bowtie} DOWN)$
$UP' := \sigma_{1=a} UP \cup (UP \underset{1=2}{\bowtie} UP')$

The dependency graph for this system is shown in Fig. 10.9.

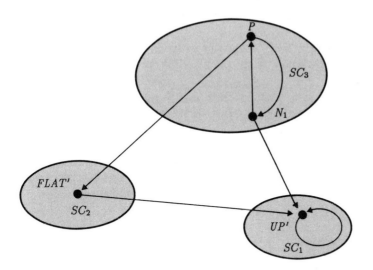

Fig. 10.9. Dependency graph of Example 10.15(b) after constant reduction

The order of evaluation of strong components is :

$$SC_1 < SC_2 < SC_3.$$

Thus, UP' has to be evaluated first, followed by $FLAT'$. □

10.2.7 Summary of the Algebraic Approach

We summarize here the transformations and optimizations which apply to systems of algebraic equations. The first optimization step concerns determining common subexpressions (Sect. 10.2.2); common subexpression analysis can in fact be applied to an entire system of equations, thus being goal-independent, or instead be applied after query subsetting over a reduced system, thus being goal-specific.

Given the principal variable of the system, query subsetting is performed and strongly connected components are identified, together with the partial ordering of subcomponents (Sect. 10.2.3).

Then, each subsystem is considered independently, in a sequence which is consistent with the order relationship. For each subsystem, we attempt the reduction of variables and the reduction of constants. Reduction of variables (Sect. 10.2.5) may introduce restrictions in the original equations; reduction of constants (Sect. 10.2.6) may produce new equations with respect to the original subsystem, whose purpose is to restrict the base relations to the relevant tuples; new equations are solved before the original subsystem.

Finally, we have to apply solution methods and produce the tuples which satisfy the goal. The following alternatives are possible:

a) The system has only one equation (or can be reduced, into one equation with a single variable, through a sequence of substitutions). Then two cases are given:
   1) The equation is nonrecursive. Then, the equation is solved by plain relational algebra computation, using the classical optimization methods of relational algebra.
   2) The equation is recursive. Then, we can use the Jacobi or Gauss-Seidel algorithms applied to a unique equation. If the equation is linear, we can use linear semi-naive evaluation.
b) The system contains *mutually dependent* equations. Then, we can use the Jacobi or Gauss-Seidel algorithms applied to a proper system of equations. If each equation is linear, we can use linear semi-naive evaluation. Otherwise, we can transform equations so as to use the general semi-naive evaluation.

10.3 A General View of Optimization

A precise comparison between all the optimization methods discussed is very difficult, because the evaluation is influenced by a great number of factors, and also because methods are very different. In this section, we attempt just two kinds of consideration. First, we observe the various methods from a *qualitative* viewpoint, searching for the basic optimization principles that they share. Then, we report performance evaluation studies that have been made on some special classes of *Datalog* programs.

Among the qualitative considerations, let us start by highlighting the common features between rewriting and pure evaluation methods. We have seen that *sideways information passing* is the basic principle of the Magic Set method. To be clear, let us concentrate on the usual *same-generation* program, considered in Example 9.3 and Example 10.5. Sideways information passing allows the selection of a particular set of values (the so-called *cone of a*, see Fig. 10.1), which is the information needed to compute the goal. This same set of values is computed by the Query-subquery method, and by most of the top-down techniques. It is possible to see that the final set of answers computed by QSQR contains the tuples whose second element is an element of the magic set. Indeed, the derivation trees obtained by the two methods in the presence of constant arguments in the goal are exactly the same.

We observe that top-down methods also manage to achieve optimization in the presence of different kinds of bindings, for instance when there are several occurrences of the same variable in the rule (literals such as $p(X, X, X, Z)$). This is not provided by the conventional Magic Sets method; however, a modification of the Magic Sets algorithm has recently been proposed that also achieves this optimization (see the bibliographic notes).

We can also notice some analogy between the Magic Sets method and the reductions of variables and constants. Consider, for instance, the linear ANC program, from Example 10.4. Suppose that the first argument of predicate anc is bound in the goal. Then, the Magic Sets method reaches the same result as the reduction of variables. Indeed, the new program obtained from the magic set transformation is equivalent to a program where the recursive IDB predicate to be evaluated is the set of tuples of ANC that have a as first element. This corresponds, in the algebraic context, to reducing the variable ANC.

In case of binding on the second argument, the reduction of variables fails, but it is possible to reduce constants. Observe Example 10.15a. We have already said that this can be viewed as a generalization of the program ANC, where C_2 takes the place of the first occurrence of PAR, X_3 represents ANC and C_4 represents the second occurrence of PAR. Note then that the reduction of constant C_4 is exactly the set of tuples whose second element is in the *cone of a*.

In more complicated cases, however, the analogy between the results of the Magic Sets method and the reductions of variables and constants is more difficult to detect. We can only notice, intuitively, that the reduction of constants also implements a kind of sideways information passing. Indeed, the semijoin reduction achieved on the reduced constant is a *generalized cone*, limiting the tuples to be used only to those related to the binding argument. This relation is established precisely by a cascade of semijoins, starting from the selection constant and "fanning out" to the other attributes of constant relations.

The characteristics of the *static filtering method* are also comparable to the reduction of variables; the main difference is that static filtering takes as input a logic program defining only one IDB predicate. Thus, programs with mutually recursive rules must first be transformed into equivalent programs for only one predicate; this transformation is quite cumbersome.

The similarity between the two methods becomes evident if we interpret *filters* imposed on the graph output ports as selection conditions which are propagated "backwards"; the same propagation occurs in the marking algorithm.

Let us see a comparison of the two methods in the same example. It is a nonlinear program proposed as an example in the paper presenting the Static Filtering method (see Bibliographic Notes [Kife 86]). Consider the following *Datalog* rules:

$R(x,y,z) :- B(x,y,z).$
$R(x,y,z) :- A(x,u,v), R(u,y,z,), R(v,z,y).$

with the goal ? $- R(x,y,a)$. They are translated into:

$$R = B \cup \Pi_{1,5,6}((A \underset{2=1}{\bowtie} R) \underset{5=3}{\bowtie_{6=2}} R).$$

with the goal $Q: \quad \sigma_{3=a} R.$

The reduction of variables gives:

$(R:3) = (B:3) \cup \Pi_{1,5,6}((A \underset{2=1}{\bowtie} (R:3)) \underset{5=3}{\bowtie_{6=2}} (R:2))$
$(R:2) = (B:2) \cup \Pi_{1,5,6}((A \underset{2=1}{\bowtie} (R:2)) \underset{5=3}{\bowtie_{6=2}} (R:3))$

which can be written as:

$(R:2 \vee 3) = (B:2 \vee 3) \cup \Pi_{1,5,6}((A \underset{2=1}{\bowtie} (R:2 \vee 3)) \underset{5=3}{\bowtie_{6=2}} (R:2 \vee 3))$

By interpreting a marking over the disjunction of two columns as a selection with a disjunctive predicate, this is precisely the filtering condition that would be obtained by the Static Filtering method.

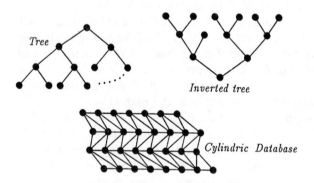

Fig. 10.10. Graph representation of database structures

We turn now to *performance* evaluation. In the following, we compare the performance of some of the methods of Chaps. 9 and 10. These results are based on a study performed by Bancilhon and Ramakrishnan (see Bibliographic Notes [Banc 86c]), using three programs as benchmarks: linear *ancestor*, nonlinear *ancestor*, and the *up-flat-down* version of the *same-generation* program (see Example 10.15(b)). Two goals are considered for the linear *ancestor* program, which bind, respectively, the first and second arguments of the rule head. On the second and third sample programs, only one binding is considered, because the other is completely symmetrical.

Some of the methods considered are evaluation methods, like QSQR, semi-naive evaluation, and the method of Henschen and Naqvi. However, rewriting methods are also considered, and the subsequent evaluation strategy assumed is semi-naive evaluation.

The study of the performance of the methods is based on three types of *EDB*. Recall the graph representation of binary relations we introduced in Section 10.1.2, when speaking about the counting method. In the sample *EDB*, the tuples are arranged either in a tree structure, or in an inverted tree structure, or in a cylinder, as shown in Fig. 10.10. Thus, cyclic data are not considered.

The cost metric chosen is the *number of successful inferences*. This is equivalent to measuring the size of each intermediate result of each operation performed on the *DB*, *before* duplicate elimination. Thus, the measure of complexity of *join*, *cartesian product*, and *selection* is the size of the result, the measure of complexity of *union* and *projection* is the size of the operand(s).

Three main factors were singled out as significant in the determination of the performance of these algorithms.

a) The amount of *work duplication*. This concerns the strategies which do not detect common inferences; for instance, naive evaluation suffers from a large amount of work duplication, while semi-naive evaluation or QSQR (which "remembers" the previous work) are more efficient.
b) The size of the set of *relevant facts*. Relevant facts are tuples effectively involved in the computation of a goal's answers. Strategies like Magic Sets or QSQR concentrate just on the selection of the relevant facts. Due to this feature, when the magic set transformation is really effective, it performs much better than Semi-naive evaluation. The Counting method manages to restrict relevant tuples even further and therefore presents a greater advantage than Magic Sets.
c) The use of either *unary* or *binary* intermediate relations. In graph terms, the difference amounts to considering either sets of nodes or sets of edges. For instance, compare the magic set to the reduction of constants: while the magic set produces a set of elements, i.e., a unary relation, constant reduction produces a binary relation. Thus, the former has to be preferred to the latter from this viewpoint.

The schema of Fig. 10.11 synthesizes the results of the study. Note that the comparison is limited to logic rewriting or evaluation methods, and does not include the algebraic rewriting methods (variable and constant reductions). We use the symbol "<<" to denote the fact that a method performs an order of magnitude better than another.

In short, in the linear *ancestor* program and in the *up-flat-down* program, the Counting method and the method of Henschen and Naqvi perform orders of magnitude better than the others. The performance of the Magic Set method

Query 1 ($anc^{b,f}$)
all data structures:
HN,C << MS, QSQR << QSQI << SN, SF << J, GS

Query 2 ($anc^{f,b}$)
all data structures:
HN,C << MS, QSQR, SF << QSQI << SN << J, GS

Query 3 ($anc^{b,f}$, non linear)
all data structures:
QSQR << QSQI << SN, MS, SF << J, GS
Note that Henschen-Naqvi and Counting cannot be applied to this query.

Query 4 ($sgc^{b,f}$, or *UP-FLAT-DOWN* program)
tree:
C << HN, MS, QSQR << QSQI << SN, SF << J, GS

inverted tree:
C << HN << MS, QSQR << QSQI << SN, SF << J, GS

cylinder:
C << HN, MS, QSQR << QSQI << SN, SF << J, GS

Where:
HN=Henschen-Naqvi; C=Counting; MS=Magic Sets; QSQR=Recursive Query-Subquery; QSQI=Iterative Query-Subquery; SN=Semi-Naive; SF=Static Filtering; J=Jacobi; GS=Gauss-Seidel.

Fig. 10.11. Synthesis of the performance evaluation study

on the linear *ancestor* program is quite similar to that of Static Filtering and QSQR. In the nonlinear case, QSQR performs definitely better than Magic Sets. In fact, in this case the magic set transformation does not improve the situation, and the transformed program behaves like the original program with semi-naive evaluation.

The Static Filtering method behaves similarly to Magic Sets. Finally, note also that, in the *up-flat-down* program, the Counting method performs even better than the method of Henschen and Naqvi. The Static Filtering method performs very poorly in this case.

We also have to note that the Static Filtering method behaves much worse when the goal binding does not "fall directly" on some argument of the goal predicate in the rule body. This same problem is present with the reduction of variables, which we know to be very similar to static filtering.

In conclusion, a general result of this study is that, for a given program and goal, strategies behave differently, and the difference amounts to *orders of magnitude*. This means that the choice of the right strategy greatly affects the performance.

10.4 Bibliographic Notes

The Magic Sets algorithm was introduced informally in [Banc 86a], and more formally in [Banc 86d] where the proofs of completeness and validity were provided. The idea of "sideways information passing" is formally described in [Beer 87b]. The method has been extended by Sacca' and Zaniolo to a class of queries to logic programs that contain function symbols [Sacc 86b].

A more sophisticated technique called "Supplementary Magic Sets" has been introduced by Beeri and others [Beer 87b]; its virtue is to eliminate some repeated computations (recall Example 10.4(b)). Intermediate results are stored in special predicates, called "supplementary magic predicates", which are potentially useful later in the computation, so that they are produced only once.

One of the most recent improvements of the Magic Set method is proposed by R. Ramakrishnan in [Rama 88b], where top-down evaluation is completely mimicked by bottom-up, using a semi-naive evaluation of rewritten rules. Rewriting uses initial goal bindings in a very sophisticated way; for instance, goals like $p(X, X)$ give rise to the binding of the two occurrences of the variables X to each other. This is not immediately covered in the Magic Sets method.

An idea similar to Magic Sets is exhibited in the "Alexander Method", published in [Rohm 86]; the method consists in rewriting the rules for a recursive predicate as problems and solutions for the predicate: the new rules obtained are all evaluated bottom-up, but the evaluation of the rules for problems simulates top-down evaluation and allows binding passing among subgoals. In general, the Alexander method produces more tuples than exactly those of the query answer, in a similar fashion to the Magic Sets method; the resulting relation must be finally selected to obtain the answer.

The Counting method was first presented in [Banc 86a]. This method suffers from the same drawback as the Magic Sets in that, sometimes, it entirely duplicates the computation (see the "ancestor" program). The Supplementary Counting method, introduced in [Beer 87b], eliminates this problem at the cost of a slightly more complicated transformation. Sacca' and Zaniolo have also introduced [Sacc 87a] the "Magic Counting method", which automatically decides whether to apply the Counting method or the Magic Set method.

The Static Filtering method was introduced by Kifer and Lozinskii in [Kife 86]. A similar approach was presented by Devanbu and Agrawal in [Deva 86], but restricting the application to linear rules with only one occurrence of the recursive predicate in the RHS. They consider a subcase, but they produce an algorithm which is less complex than the Kifer-Lozinskii one. Another interesting (pure evaluation) algorithm (the Apex method) was previously introduced by Lozinskii in [Lozi 85]. The semi-naive optimization by rewriting is explained in [Zani 87b].

The structured approach to the algebraic optimization of systems of equations derived from logic programs is taken from [Ceri 87] and [Tanc 88]; the reduction of systems of equations by substitution is discussed in [Ceri 86a].

The first method for pushing selections ahead of the closure was proposed by Aho and Ullman in a seminal paper [Aho 79a]. The method for variable reduction, presented in this chapter, is a generalization of the Aho-Ullman algorithm because it deals with a system of equations and a generic number of variable occurrences, while the Aho-Ullman method is concerned with one equation and one variable occurrence.

There are also other possible approaches to optimization, which, though not fully explored yet, appear to be promising. One of them is Sagiv's method for redundancy elimination: in [Sagi 87] an algorithm is presented for minimizing the *size* (in terms of number of rules in the program and of number of atoms in a rule) of a *Datalog* program under a decidable condition called *uniform equivalence*. Another kind of optimization is achieved by Ramakrishnan and others, in [Rama 88a], where the objective of optimization is *pushing projections*, rather than selections, in the body of a *Datalog* rule. This means deleting some argument positions in the literals of the rule body, which also has, sometimes, the effect of making some rules redundant for the computation.

The evaluation of performance of the various methods are based on a well-known paper by Bancilhon and Ramakrishnan [Banc 86c]. The study uses a benchmark which includes only few, conventional programs; it excludes cyclic databases, which are quite common. However, the paper remains one of the few quantitative approaches to the comparison of optimization methods.

10.5 Exercises

10.1 Perform the Magic Sets transformation on the following programs:

a.
$$sgc(X, X) :- h(X).$$
$$sgc(X, Y) :- par(X, X1), sgc(X1, Y1), par(Y, Y1).$$

which computes cousins in the same generation, with the goal

$$? - sgc(X, b).$$

b.
$$p_1(X, Y) :- c_1(X, Y).$$
$$p_1(X, Y) :- p_1(X, Z), p_3(Z, Y).$$
$$p_1(X, Y) :- p_2(X, Y).$$
$$p_2(X, Y) :- p_1(X, Z), p_3(Z, Y).$$
$$p_2(X, Y) :- c_3(X, Y).$$
$$p_3(X, Y) :- p_3(X, Z), c_2(Z, Y).$$
$$p_3(X, Y) :- c_4(X, Y).$$

with the goal

$$? - p_1(X, b).$$

10.2 Translate into Positive Relational Algebra and reduce to UJNF the program:

$$sgc(X, X) :- h(X).$$

$$sgc(X,Y) : -par(X,X1), sgc(Y1,X1), par(Y,Y1).$$

Note that this is the version where we have an inversion of arguments in the recursive predicate. Show both the dependency graphs of the translation result and of the UJNF system.

10.3 Consider the three systems of Example 10.10(c). Draw their dependency graphs, then separate and order their strong components.

10.4 Consider the systems of Example 10.10(c). Mark the second system with respect to query $Q = \sigma_{1=a} X_4$, and the third system with respect to the query $Q = \sigma_{1=a} X_1$.

10.5 Try to reduce variables in the system of Exercise 10.3, with both the markings. If it is not possible, try to reduce constants, if the appropriate conditions (stability, etc.) hold.

10.6 Write a Prolog or Lisp program that implements the algorithm for reduction of constants.

Chapter 11
Extensions of Pure Datalog

The *Datalog* syntax we have been considering so far corresponds to a very restricted subset of first order logic and is often referred to as *pure Datalog*. Several extensions of pure *Datalog* have been proposed in the literature or are currently under investigation. The most important of these extensions are:

- BUILT-IN PREDICATES It is often necessary or useful to express comparison relationships such as "$<$", "$\leq$", "$>$", etc., or equalities ("$=$") and disequalities ("$\neq$") between variables or constants occurring in the body of a *Datalog* rule. These special symbols can be conceived of as "built-ins" which are not defined in the program, whose semantics is implicitly known by the system, and which are automatically evaluated. It is important to note, however, that an unsafe use of these built-ins may cause the answer to a *Datalog* program to be infinite. Hence, particular *safety conditions* must be observed by the programmer whenever he or she uses such special predicates. The matter will be discussed in more detail below.
- NEGATION The negation sign "$\neg$" is an important logical connective. Its incorporation into *Datalog*, however, is associated with several problems of both semantic and computational nature. In particular, the Closed World Assumption (CWA), which is suitably adopted in the context of pure *Datalog*, leads to inconsistencies when negations are allowed to appear in rule bodies. Nevertheless, for a large class of programs, the so-called *stratified Datalog programs*, there is a satisfactory way of handling negation. This class will be described with care below. We will also discuss *local stratification*, a refinement of stratification, and we will present an alternative approach to defining the meaning of *Datalog* programs containing negated literals, based on the so-called *inflationary semantics*.
- COMPLEX OBJECTS The "objects" handled by pure *Datalog* programs correspond to the tuples of data relations which, in turn, are made up of attribute values. Each attribute value is atomic, i.e., not composed of subobjects; thus the underlying data model consists of relations in first normal form. This model has the advantage of being both mathematically simple and easy to implement. On the other hand, several new application areas (such as Computer-Aided Design, Office Automation, and Knowledge Representation) require the storage and manipulation of (deeply nested) structured objects of high complexity. Such complex objects cannot be represented as atomic entities in the normalized relational model but are broken into several autonomous

objects. This implies a number of severe problems of a conceptual and technical nature. For this reason, the relational model has been extended in several ways to allow the compact representation of complex objects. In Sect. 11.3 we show how *Datalog* can be extended accordingly. The main features we add to pure *Datalog* are *function symbols* as a glue for composing objects from subobjects and *set constructors* for being able to build objects which are collections of other objects. We will also add a number of predefined functions for manipulating sets and elements of sets to the standard vocabulary of *Datalog*. Our exposition follows mainly the approach undertaken in the LDL project (see Chap. 12). Other approaches are referenced in the Bibliographic Notes at the end of this chapter.

- UPDATES *Datalog* was originally conceived as a query language. However, it is possible to extend *Datalog* by programming constructs which allow to perform updates on the EDB relations. We will not deal with updates in this chapter, but we defer this topic to Sect. 12.1, where we show how updates are done in LDL, a concrete implementation of an extended version of *Datalog*.
- AGGREGATE FUNCTIONS AND NONDETERMINISM Aggregate functions can be efficiently implemented by use of a built-in predicate simulating nondeterminism. Such a predicate, called *choice*, exists in the LDL language. Again, we defer our discussion to the next chapter. In Sect. 12.1 we show how the cardinality of a set can be computed efficiently by use of the *choice* predicate.
- UNCERTAINTY There are several models of *approximate reasoning* which are used in Expert Systems and which can be embodied in *Datalog* (and, more generally, in Logic Programming). In such models, *certainty factors* are specified for both facts and rules. If a new fact is inferred, a certainty factor for this new fact is computed by combining the certainty factors of the underlying premises and the certainty factors of the rules used in the derivation. Different models of approximate reasoning correspond to different ways of combining certainty factors. We will not further discuss uncertainty in this book, but we will give a pointer to the relevant literature in the Bibliographic Notes at the end of this chapter.
- OTHER EXTENSIONS Finally, one may consider features from other fields, such as *typing* of variables (from programming language theory) and *null values* (from the field of databases) to be profitably incorporated into *Datalog*. In this chapter we will discuss none of these extensions, since little has been published on these topics. We wish, however, to acknowledge these items as important and appealing research problems.

In the rest of this chapter we will discuss some of these extensions of *Datalog* in more detail. Our style of presentation, however, will remain rather informal. All proofs of Theorems are omitted. Readers interested in a more thorough theoretical treatment of the topics may find useful references in the Bibliographic Notes at the end of this chapter.

11.1 Using Built-in Predicates in Datalog

Built-in predicates are denoted by special predicate symbols such as $>$, $<$, $\geq$, $\leq$, $=$, $\neq$ with a predefined meaning. These symbols can occur in the right-hand side of a *Datalog* rule; they are usually written in infix notation.

Example 11.1. Consider the following program P_1 consisting of a single rule where *parent* is a database predicate:

$$P_1 : \quad sibling(X,Y) : - \; parent(Z,X), parent(Z,Y), X \neq Y.$$

The meaning of this program is obvious. By use of the inequality built-in predicate we avoid a person being considered his own sibling. □

From a formal point of view, built-ins can be considered EDB predicates with a different physical realization from ordinary EDB predicates: they are not explicitly stored in the EDB but are implemented as procedures which are evaluated during the execution of a *Datalog* program. However, built-ins in most cases correspond to *infinite* relations and this may endanger the *safety* of *Datalog* programs.

A *Datalog* program should always have a finite output, i.e., the intensional relations defined by a *Datalog* program must be finite. In Example 11.1 the finiteness of the computed pairs of siblings is guaranteed because the variables X and Y both occur in the *parent* predicate and are thus each bound to a finite choice of constants. However, as the following example shows, it is quite easy to construct unsafe *Datalog* programs by using built-in predicates.

Example 11.2. Consider the following program P_2:

$$P_2 : \quad ordinary - citizen(X) : - \; president(Y), X \neq Y.$$

The intended meaning of this one-rule program is that everybody except the president is an ordinary citizen. Here we assume that *president* is an EDB predicate whose extension consists of exactly one ground fact. Although this program respects the safety condition of Sect. 6.1.2 requiring that each variable occurring in the head of a rule occurs as well in the body of this rule, we may infer an infinite number of facts from P_2: for each constant $\alpha \in Const$ different from the one constant denoting the president we can infer $ordinary\text{-}citizen(\alpha)$. □

The above example clearly shows that we need an additional safety condition for programs containing built-in predicates. It is easy to see that it is sufficient to require that each variable occurring as argument of a built-in predicate in a rule body must also occur in an ordinary predicate of the same rule body or must be bound by an equality (or a sequence of equalities) to a variable of such an ordinary predicate or to a constant. Here, by "ordinary predicate", we mean an ordinary EDB predicate or an IDB predicate.

Example 11.2 (continuation). Let *citizen* be an EDB predicate. The following one-rule program P_2' defines the "ordinary" citizens in a *safe* manner:

P_2': ordinary-citizen(X) :- citizen(X), president(Y), X≠Y.

Also the following, somewhat clumsy program P_2'' is safe:

P_2'': ordinary-citizen(X) :- citizen(X), president(Y), X=Z, Y=V, Z≠V.

□

During the evaluation of a *Datalog* rule with built-in predicates, the following principle has to be observed: defer the evaluation of a built-in predicate until all arguments of this predicate are bound to constants.

An exception to this principle can (sometimes must) be made for the equality predicate. An equality can be evaluated as soon as *one* of its two arguments is a constant or is bound to a constant. For instance in the program P_2'' of Example 11.2, after the evaluation of *citizen(X)* and *president(Y)*, we must evaluate X=Z and Y=V although Z and V are not bound. Only then we can evaluate Z≠V.

Sometimes it is even convenient to evaluate an equality before evaluating an ordinary predicate from the same rule body in order to exploit sideways information passing (see Chap. 10):

Example 11.3. Consider the following rule, where *p* and *q* are supposed to be ordinary predicates:

$$r(X,T) \text{ :- } p(a,X), q(Z,T), X=Z.$$

Here it is obvious that after the evaluation of $p(a, X)$ it is convenient to evaluate X=Z first, providing a binding (or a set of values) for Z, and then to evaluate q(Z,T). □

In a similar way, *arithmetic built-in predicates* can be used. For instance, a predicate *plus(X,Y,Z)* may be used for expressing $X+Y = Z$, where the variables X, Y, and Z are supposed to range over a numeric domain. During the evaluation of a rule body, such a predicate can be evaluated as soon as bindings for its "input variables" (here X and Y) are provided.

Finally, let us remark that when *Datalog* rules are transformed into algebraic equations, then built-in comparison predicates can be expressed through join conditions.

Example 11.4. The program P_1 of Example 11.1 is translated into the following equation of relational algebra:

$$SIB = \prod_{2,4}(PAR \bowtie_{1=1 \land 2 \neq 2} PAR)$$

where SIB and PAR denote the relations corresponding to the predicates *sibling* and *parent* respectively. □

11.2 Incorporating Negation into Datalog

In this section we first consider the Closed World Assumption and show that this assumption can be used to infer negative information from *Datalog* programs.

However, it turns out that this assumption is too restrictive for operating with negative facts, i.e., for using negative facts in a deduction process. Consequently, we describe two other approaches for incorporating negation into *Datalog*. The first one, referred to as *stratified Datalog* is a very natural extension of pure *Datalog*. It follows the intuition that predicates of a *Datalog* program should be evaluated in a certain order. The second approach is based on the concept of *inflationary semantics*. It is somewhat less intuitive than the first approach, but it is more powerful from a computational point of view.

11.2.1 Negation and the Closed World Assumption

In pure *Datalog*, the negation sign "$\neg$" is not allowed to appear. However, by adopting the Closed World Assumption (CWA), we may infer negative facts from a set of pure *Datalog* clauses.

Note that the CWA is not a universally valid logical rule, but just a principle that one may use in a particular situation. In our case, we may use the CWA to enrich the semantics of *Datalog* programs.

In the context of *Datalog*, the CWA can be formulated as follows:

CWA: If a fact does not logically follow from a set of *Datalog* clauses, then we conclude that the negation of this fact is valid.

Note that our choice of defining the semantics of a set S of *Datalog* clauses as the least Herbrand model of S nicely meets the CWA: in this model, all logical consequences of S are true and all other facts are false. However, until now, we only considered positive facts and did not worry about negative consequences of a set S of *Datalog* clauses.

Negative facts are positive ground literals preceded by the negation sign, for instance, $\neg loves(tom, mary)$. If F denotes a negative ground fact, then $|F|$ denotes its positive counterpart. For example: $|\neg loves(tom, mary)| = loves(tom, mary)$.

We can give a more formal definition of the CWA by extending our notion of logical consequence (denoted by "$\models$") to the concept of *consequence under the CWA* (denoted by "$\models^{cwa}$").

Let S be a set of *Datalog* clauses and let F be a positive or negative ground fact. We then define:

$$S \models^{cwa} F \text{ iff } \begin{cases} F \text{ is a positive fact and } S \models F, \text{ or} \\ F \text{ is a negative fact and } S \not\models |F|. \end{cases}$$

Accordingly, we may define the set $cons^{cwa}(S)$ by:

$$cons^{cwa}(S) = \{F \mid F \text{ is a pos. or neg. ground fact and } S \models^{cwa} F\}.$$

Example 11.5. Let S consist of the following three *Datalog* clauses:

$p(X, Y) \colon -p(Y, X)$
$p(a, b)$
$p(a, c)$

We then have the following relationships:

$S \models^{cwa} p(a, b),$

$S \models^{cwa} p(a,c)$,
$S \models^{cwa} p(b,a)$,
$S \models^{cwa} p(c,a)$,
$S \models^{cwa} \neg p(\alpha,\beta)$ if
$<\alpha,\beta> \notin \{<a,b>,<a,c>,<b,a>,<c,a>\}$,
$S \models^{cwa} \neg \psi(c_1,\ldots,c_n)$ for each n-ary predicate
symbol $\psi \neq p$ and arbitrary constants $c_1,\ldots c_n$.
For any other fact F, $S \not\models^{cwa} F$. □

For S as given in the above example, $cons^{cwa}(S)$ is infinite, since it contains an infinite number of negative facts. Note that $cons^{cwa}(S)$ is always infinite, when S is a finite set of *Datalog* clauses. This means that we cannot explicitly represent $cons^{cwa}(S)$. However, when S is finite, we are always able to decide whether a given positive or negative fact F belongs to $cons^{cwa}(S)$ or not.

The CWA applied to pure *Datalog* allows us to deduce some negative facts from a set S of *Datalog* clauses. It does not, however, allow us to use these negative facts in order to deduce some further facts. In real life it is often necessary to express rules whose premises contain negative information, for instance: "*if X is a student and X is not a graduate student then X is an undergraduate student*". In pure *Datalog*, there is no way to represent such a rule.

Note that in Relational Algebra an expression corresponding to the above rule can be formulated with ease by use of the set-difference operator "−". Assume that a one-column relation $STUD$ contains the names of all students and another one-column relation $GRAD$ contains the names of all graduate students. Then we obtain the relation UND of all undergraduate students by simply subtracting $GRAD$ from $STUD$, thus

$$UND = STUD - GRAD.$$

Our intention is now to extend pure *Datalog* by allowing negated literals in rule bodies. Assume that the unary predicate symbols *stud*, *und*, and *grad* represent the properties of being a student, an undergraduate, and a graduate respectively. Our rule could then be formulated as follows:

$$und(X) :- stud(X), \neg grad(X).$$

More formally, let us define *Datalog¬* as the language whose syntax is the same as *Datalog*'s except that negated literals are allowed in rule bodies. Accordingly, a *Datalog¬* clause is either a positive (ground) fact or a rule where negative literals are allowed to appear in the body.

In order to discuss the semantics of *Datalog¬* programs, we first generalize the notion of *Herbrand Model* (see Sect. 6.2) to cover negation in rule bodies.

Let $\Im$ be a Herbrand Interpretation, i.e., a subset of the Herbrand Base HB. Let F denote a positive or negative fact (ground literal).

F is satisfied in $\Im$ iff $\begin{cases} F \text{ is a positive fact and } F \in \Im, \text{or} \\ F \text{ is a negative fact and } |F| \notin \Im. \end{cases}$

Now, let R be a $Datalog^\neg$ rule of the form $L_0 :- L_1, \ldots, L_n$ and let $\Im$ be a Herbrand interpretation. R is satisfied in $\Im$ iff for each ground substitution θ for R, whenever it holds that for all $1 \leq i \leq n$, $L_i\theta$ is satisfied in $\Im$, then it also holds that $L_0\theta$ is satisfied in $\Im$. (Note that $L_0\theta$ is satisfied in $\Im$ iff $L_0\theta \in \Im$, since $L_0\theta$ is positive.)

Let S be a set of $Datalog^\neg$ clauses. A Herbrand interpretation $\Im$ is a Herbrand model of S iff all facts and rules of S are satisfied in $\Im$.

Example 11.6. Consider the following $Datalog^\neg$ program P, where *stud* and *grad* are EDB predicates:

$$und(X) :- stud(X), \neg\, grad(X)$$

and consider the following EDB E:

$$E = \{stud(a), stud(b), stud(c), stud(d), stud(e), grad(a), grad(c)\}.$$

Two different Herbrand models of $P \cup E$ are:

$$H_1 = E \cup \{und(b), und(d), und(e)\} \text{ and}$$
$$H_2 = E \cup \{und(b), und(d), und(e), und(f)\}.$$

Note that H_1 is minimal but H_2 is not, because we can eliminate $und(f)$ from H_2 obtaining a smaller Herbrand model. □

In analogy to pure $Datalog$, we require that the set of all positive facts derivable from a set S of $Datalog^\neg$ clauses be a minimal model of S. But, as we will see later, S may have several different minimal Herbrand models and does not, in general, have a unique least Herbrand model. This entails difficulties in defining the semantics of $Datalog^\neg$ programs: which of the different minimal Herbrand models should be chosen?

However, before discussing this question, we shall deal with two other critical problems which arise when negation is allowed to appear in rule bodies. The first problem is about the safety of $Datalog^\neg$ programs and the second problem concerns the CWA.

It is easy to see that an unrestricted use of the negation sign in rule bodies often leads to unsafe programs.

Example 11.7. Consider the following one-rule program P, where *interesting* is an EDB predicate describing a (finite) set of interesting objects:

$$P : boring(X) :- \neg interesting(X).$$

It is easy to see that each Herbrand Model of $P \cup E$, where E is an arbitrary EDB, is infinite: we can substitute infinitely many constants for X, such that $\neg interesting(X)$ is satisfied. Thus, whatever Herbrand model we choose to define the semantics of our program, it will always contain infinitely many facts of the type $boring(\alpha)$. This is a similar problem to the one that we have already encountered with built-in predicates. The way out of this problem is also similar: we just need to restrict the range of those variables which occur only in negated predicates. In our example, for instance, we may introduce an EDB predicate *object(X)* which limits the range of the variable X to a finite set of constants. Using such an additional predicate, we get the safe program:

boring(X) :- object(X), ¬interesting(X).   □

In general, we impose the following safety condition on *Datalog* programs with negation and built-in predicates:

Let the *special literals* of a rule body be the negated ones and those with built-in predicates, and let the *ordinary literals* be all others. We require that each variable occurring in a special literal also occur in an ordinary literal or be bound by a chain of equalities to a constant or to a variable occurring in an ordinary literal.

Let us now turn our attention to the CWA. As the following example shows, *Datalog* extended by negation is *inconsistent with the CWA*.

Example 11.8. Consider the following set S containing a single *Datalog¬* clause:

$$S = \{\ q(b) :- \neg p(a)\ \}.$$

This rule corresponds to the logical implication $\neg p(a) \Rightarrow q(b)$ or, equivalently, to the disjunction $p(a) \vee q(b)$. Obviously, neither the fact $p(a)$ nor the fact $q(b)$ is a logical consequence of S. Hence $S \not\models^{cwa} \neg p(a)$ and $S \not\models^{cwa} \neg q(b)$. On the other hand, there cannot exist any Herbrand model of S which satisfies both $\neg p(a)$ and $\neg q(b)$ because then the only rule of S would not be satisfied. Indeed, it is not hard to see that each Herbrand model of S must contain either $p(a)$ or $q(b)$.
□

For this reason we shall abandon the CWA and seek for a more appropriate definition of the semantics of *Datalog¬* programs. In the next section we present a semantics which associates a particular minimal Herbrand model to a set of *Datalog¬* clauses. This approach is still very close to the CWA, in other words, it approximates the CWA. We will also briefly discuss a different semantics which does not require model-minimality. That approach, called *inflationary semantics*, is somewhat less related to the CWA.

11.2.2 Stratified Datalog

The main problem is – as we have already mentioned – that in general several minimal models for a set S of *Datalog¬* clauses exist. For instance, the set S of Example 11.8 has two minimal Herbrand models: $H_1 = \{p(a)\}$ and $H_2 = \{q(b)\}$. Which one should be chosen ? Of course we would like to select the most "natural" from the different minimal Herbrand models of a set S.

In the following we describe a policy which is commonly referred to as *stratified evaluation of Datalog programs*, or simply as *stratified Datalog*. This policy permits a distinguished minimal Herbrand model to be selected in a very natural and intuitive way. However, as we will see later, this method does not apply to all *Datalog¬* programs, but only to a particular subclass, the so-called *stratified* programs. Note also that this technique can be used as well in the more general context of logic programming with negation.

Our policy of choosing a particular Herbrand model, and thus of determining the semantics of a $Datalog^{\neg}$ program is guided by the following intuition: When evaluating a rule with one or more negative literals in the body, we first compute the set of all answer-facts to the predicates which occur negatively in the rule body (if these predicates are EDB predicates then their answer-facts are already known and need not be computed). Then we proceed by computing the answers to the predicate in the head of the rule.

Consider, for instance, the rule of Example 11.8. First we compute all answer facts for the predicate p. Since S contains no facts of the form $p(\alpha)$ and no rule for p, this set is empty. This means in particular that $p(a)$ is not satisfied and thus $\neg p(a)$ is satisfied. Hence, by applying the rule, we conclude $q(b)$. Thus, the selected Herbrand model is $\{q(b)\}$.

In this way, we introduce a certain ordering among the predicates which reflects the sense of the implication sign (i.e., from right to left). In a certain sense it is very natural to first compute the premises (at least the ones corresponding to negative literals) and only then to compute the conclusion. However, in doing so, we introduce a certain procedurality, given that our choice of the minimal Herbrand model of a $Datalog^{\neg}$ program depends on the *syntax* of the program and not solely on its logical semantics. In pure $Datalog$ the following holds: sets of clauses which correspond to equivalent logical formulas have an identical least Herbrand model. The following example shows that this does not hold in $Datalog^{\neg}$.

Example 11.9. Consider the following set S' containing a single $Datalog^{\neg}$ clause:

$$S' = \{ \ p(a) :- \neg q(b) \}.$$

The logical formula corresponding to this set is $\neg q(b) \Rightarrow p(a)$ which is equivalent to the formula $\neg p(a) \Rightarrow q(b)$ of the set S given in Example 11.8. Indeed, both formulas are logically equivalent to $p(a) \lor q(b)$. Note that the sets of minimal Herbrand models of S and S' are the same, but the *chosen* minimal Herbrand models are different: by applying our policy, it is easy to see that the chosen Herbrand model for S' is $\{p(a)\}$. □

Note that up to here we have given a rather informal and incomplete description of our policy for choosing minimal Herbrand models. In particular, we tacitly restricted ourselves to considering one-rule programs. Thus we shall extend our method to cover $Datalog^{\neg}$ programs with several rules. Furthermore, we will see that our policy does not apply to all programs. Hence, we shall identify a sufficiently large class of program, the so-called *stratified* programs, for which our method works.

When several rules occur in a $Datalog^{\neg}$ program, then the evaluation of a rule body may engender the evaluation of subsequent rules. These rules may contain in turn negative literals in their bodies and so on. Our policy consists in requiring that before evaluating a predicate in a rule head, we have already completely evaluated all the predicates which occur negatively in the rule body or in the bodies of some subsequent rules and all those predicates which are needed in order to evaluate these negative predicates.

In order to describe this policy more formally, let us define the *Extended Dependency Graph* $EDG(P)$ of a $Datalog^{\neg}$ program P. The nodes of $EDG(P)$ consist of the IDB predicate symbols occurring in P. There is a (directed) edge $<p,q>$ in $EDG(P)$ iff the predicate symbol q occurs positively or negatively in a body of a rule whose head predicate is p. Furthermore, the edge $<p,q>$ is marked with a "$\neg$" sign iff there exists at least one rule R with head predicate p such that q occurs negatively in the body of R. Notice that for a pure *Datalog* program P, $EDG(P)$ coincides with the Dependency Graph G_P of P as defined in Chap. 8.

Our policy can now be defined as follows: If p is connected to q by a path in $EDG(P)$ such that at least one of the edges of this path is marked with $\neg$, then we require that q be evaluated prior to p.

Example 11.10. Consider a railroad network represented by the EDB predicate $link(X,Y)$ stating that there is a direct link between station X and station Y with no intermediate station. On the basis of the EDB predicate $link$, we define several IDB predicates with the following associated intuitive meanings:

cutpoint(X,A,B): each connection (path) from station A to station B goes through station X.

connected(A,B): there is a connection between A and B involving one or more links;

circumvent(X,A,B): there is a connection between station A and station B which does not pass through station X;

existscutpoint(A,B): there is at least one cutpoint on the way from A to B.

safely-connected(A,B): A and B are connected, but there is no cutpoint for A and B. This means that if a station or link between A and B is disabled (e.g. by an accident), then there is still another way of reaching B from A.

linked(A,B): we assume that our network is an undirected graph, i.e., if A is linked to B then B is linked to A. However, the EDB predicate $link$ is not necessarily symmetric since for expressing the fact that there is a link between A and B, it is sufficient to state either $link(A,B)$ or $link(B,A)$ but not necessarily both. The IDB predicate $linked$ consists of the symmetric closure of the $link$ predicate.

station(X): X is a railway station.

Let P be the $Datalog^{\neg}$ program consisting of the following rules:

R_1 :   *cutpoint(X,A,B)* :- *connected(A,B)*, *station(X)*, $\neg$*circumvent(X,A,B)*.
R_2 :   *circumvent(X,A,B)* :- *linked(A,B)*, $X{\neq}A$, $X{\neq}B$.
R_3 :   *circumvent(X,A,B)* :- *circumvent(X,A,C)*, *circumvent(X,C,B)*.
R_4 :   *linked(A,B)* :- *link(A,B)* .
R_5 :   *linked(A,B)* :- *link(B,A)* .
R_6 :   *connected(A,B)* :- *linked(A,B)*.
R_7 :   *connected(A,B)* :- *connected(A,C)*, *linked(C,B)*.

R_8 : $existscutpoint(A,B)$:- $station(X)$, $cutpoint(X,A,B)$.
R_9 : $safely\text{-}connected(A,B)$:- $connected(A,B)$, $\neg existscutpoint(A,B)$.
R_{10} : $station(X)$:- $linked(X,Y)$.

The extended dependency graph of P is depicted in Fig. 11.1.

From this graph it is easy to see that, according to our policy, before evaluating the goal *safely-connected* we must completely evaluate *existscutpoint* and thus also *cutpoint* and *station*. In turn, before being able to evaluate *cutpoint*, we must have completely evaluated *circumvent* and thus also *linked*.   □

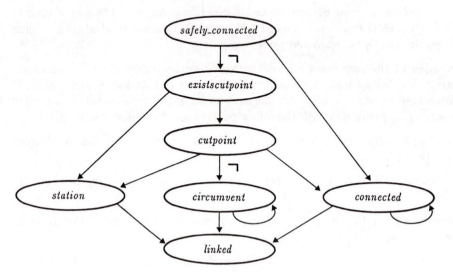

Fig. 11.1. Extended dependency graph for the "railway" program

A *stratification* of a program P is a partition of the set of IDB predicate symbols of P into subsets $P_1, \ldots, P_n$ such that the following conditions (a) and (b) are satisfied:

a) if $p \in P_i$ and $q \in P_j$ and $<p,q>$ is an edge of $EDG(P)$ then $i \geq j$.
b) if $p \in P_i$ and $q \in P_j$ and $<p,q>$ is an edge of $EDG(P)$ labeled "$\neg$", then $i > j$.

A stratification specifies an order of evaluation for the predicates in P. First we evaluate the predicates in P_1, for instance by an LFP iteration. Once we have evaluated these predicates, we may consider them as EDB predicates. Then we continue by evaluating P_2 and so on. Notice that the above conditions (a) and (b) ensure that this order of evaluation corresponds to our policy. The sets $P_1, \ldots, P_n$ are called the *strata* of the stratification.

In the literature, stratifications are often defined by using sets of clauses (subsets of the program P) instead of sets of predicate symbols. This definition is equivalent to ours. We just need to replace each predicate symbol of each stratum by all the clauses which define the predicate (see Example 11.11).

A program P is called *stratified* iff it admits a stratification. In general, as the following example shows, a stratified program has several stratifications.

Example 11.11. The following are some stratifications of the "railroad" program of Example 11.10:

Stratification s_a:

$P_1 = \{linked, station, circumvent, connected\}$
$P_2 = \{cutpoint, existscutpoint\}$
$P_3 = \{safely\text{-}connected\}$

Stratification s_b:

$P_1 = \{linked, circumvent\ \}$
$P_2 = \{cutpoint, existscutpoint, station, connected\ \}$
$P_3 = \{safely\text{-}connected\}$

Stratification s_c:

$P_1 = \{linked\ \}$
$P_2 = \{circumvent\ \}$
$P_3 = \{station\ \}$
$P_4 = \{connected\ \}$
$P_5 = \{cutpoint\ \}$
$P_6 = \{existscutpoint\ \}$
$P_7 = \{safely\text{-}connected\ \}$

The above stratifications can equivalently be indicated by sets of rules, for example:

Stratification s_a:

$P_1 = \{R_2, R_3, R_4, R_5, R_6, R_7, R_{10}\}$
$P_2 = \{R_1, R_8\}$
$P_3 = \{R_9\}$

□

Notice that if a program is stratified, then all negative literals in rule bodies of the first stratum correspond to EDB predicates.

The following Theorem states a necessary and sufficient condition for a $Datalog^\neg$ program to be stratified:

Theorem 11.1. A $Datalog^\neg$ program P is stratified iff $EDG(P)$ does not contain any cycle involving an edge labeled with "¬".

The next example displays two programs which are not stratified.

Example 11.12. Consider the program P' consisting of the single rule

$p(X,Y) :\text{-} \neg p(Y,X), q(X,Y)$

and the program P'' defined by the two rules

$p(X,Y) :\text{-} \neg s(Y,X), h(X,Y).$

$s(X,Y) \text{:- } p(X,Z), p(Z,Y).$

Obviously, neither of these two programs is stratified. □

There is a very simple method for finding a stratification for a stratified program P:

ALGORITHM STRATIFY

INPUT: a stratified $Datalog^{\neg}$ program P; $EDG(P)$.
OUTPUT: a stratification for P.

METHOD

PERFORM THE FOLLOWING STEPS:

1) Construct the graph $EDG^*(P)$ from $EDG(P)$ as follows. For each pair of vertices p, q in $EDG(P)$ do: if there is a path from p to q containing an edge labeled with "$\neg$" in $EDG(P)$, then add an edge $<p,q>$ labeled with "$\neg$" to the graph (if such an edge does not already exist).
2) $i := 1$;
3) Identify the set K of all those vertices of $EDG^*(P)$ which have no outgoing edges labeled with $\neg$. Output this set as stratum i. Delete all vertices of K in $EDG^*(P)$ and delete all corresponding edges.
4) If there are any vertices left in $EDG^*(P)$, then set $i := i + 1$ and go to step 3, otherwise stop.

ENDMETHOD

The mechanism of this algorithm is easy to understand. The first stratum consists of all predicates (vertices) whose evaluation does not depend on any negative IDB predicate. Of course, all these predicates can be evaluated immediately. The second stratum consists of all predicates whose evaluation does not depend on any negative predicate of any stratum higher than 1, and so on.

It can be seen that this algorithm always produces a stratification with a minimum number of strata. For instance, if the algorithm is applied to the program of Example 11.10, then the stratification s_a of Example 11.11 is obtained. This stratification has three strata. It is easy to see that for this program there is no stratification with fewer strata.

Let us now turn our attention to the problem of evaluating stratified programs against an Extensional Database. We will describe a bottom-up evaluation method which is quite similar to the standard bottom-up method for $Datalog$ programs except for two differences: negative information is used in the forward chaining process and the order of evaluation is in accordance with the strata (from the lowest to the highest).

In order to treat negated literals in rule bodies, we extend our "Elementary Production Rule" EP (see Sect. 7.1.1) to a more general rule $EP^{\neg}$ as follows:

Let S be a set of $Datalog^{\neg}$ clauses (i.e., $Datalog^{\neg}$ rules and $Datalog$ facts), let R be a $Datalog^{\neg}$ rule of the form $L_0 : -L_1, \ldots, L_n$ and let θ be a ground substitution for R. Of course, $FACTS(S)$, the set of all facts of S, is a Herbrand

Interpretation. If for each i with $1 \leq i \leq n$ $L_i\theta$ is satisfied in $FACTS(S)$, then the fact $L_0\theta$ is *inferred in one step by* $EP^\neg$ *from* S.

Recall that, according to our previous definition, $L_i\theta$ is satisfied in $FACTS(S)$ iff either $L_i\theta$ is positive and $L_i\theta \in FACTS(S)$ or $L_i\theta$ is negative and $|L_i\theta| \notin FACTS(S)$.

We denote by $infer1^\neg(S)$ the set of all facts inferrable in exactly one step by $EP^\neg$ from S.

In analogy to pure *Datalog* (see Sect. 7.2.2), we can now define two transformations T_S and T'_S to each set S of *Datalog$^\neg$* clauses, such that for each subset W of the Herbrand base HB:

- $T_S(W) = W \cup FACTS(S) \cup infer1^\neg(RULES(S) \cup W)$
- $T'_S(W) = FACTS(S) \cup infer1^\neg(RULES(S) \cup W)$.

Notice that, due to the term W in the outer union, for each set S of *Datalog$^\neg$* clauses T_S has the following property:

$$\forall W \subseteq HB : \quad W \subseteq T_S(W)$$

We say that a transformation having this property is a *growing* transformation.

It can be easily shown that the fact that T_S is growing is a sufficient condition for T_S to have a least fixpoint. Thus, T_S has a least fixpoint $lfp(T_S)$ for each set S of *Datalog$^\neg$* clauses.

For pure *Datalog*, this fixpoint is identical to the least Herbrand model of S and thus characterizes the semantics of S. Unfortunately, for *Datalog$^\neg$* this is not the case. As the following example shows, $lfp(T_S)$ is not necessarily a minimal Herbrand model of S.

Example 11.13. Consider a set S consisting of the following two rules R_1 and R_2:

$R_1:\ q(b) :\!\!- \neg p(a)$
$R_2:\ r(c) :\!\!- \neg q(b)$

Since there are no facts in S, both $\neg p(a)$ and $\neg q(b)$ are satisfied and hence $infer1^\neg(S) = \{q(b), r(c)\}$. Since the next step of LFP-iteration does not produce any new facts, this set is already the least fixpoint of T_S. We thus have $lfp(T_S) = \{q(b), r(c)\}$. Of course, this set is a Herbrand model for S, but it is not a minimal one: by eliminating $r(c)$ from $lfp(T_S)$ we get the smaller set $\{q(b)\}$ which is also a Herbrand model of S. □

This example shows that we have to reject T_S as a model-generator if we want to obtain a minimal Herbrand model for S.

Of course one could abandon the requirement of model minimality and just define the semantics of S to be the distinguished model $lfp(T_S)$. This approach is called *inflationary semantics* because of the rather large models that are associated with *Datalog$^\neg$* programs or clause sets. We defer a discussion of this approach to Sect. 11.2.4. In the present section we are interested in model minimality because we wish to maintain the spirit of the CWA (though, as we have shown, the CWA itself is too strong to be used).

Let us now consider the second transformation T'_S. In the context of pure *Datalog*, T_S and T'_S both have the same least fixpoint. In the context of *Datalog*¬ this is not so. T'_S is in general nonmonotonic and may have no fixpoint at all.

Example 11.14. Consider a set S consisting of a single rule

$$R:\ p(a) :\text{-}\ \neg p(a).$$

Assume that $M = lfp(T'_S)$. If $p(a) \in M$ then $p(a) \notin T'_S(M)$ and thus M would not be a fixpoint. Thus it must hold that $p(a) \notin M$. But then it follows that $p(a)$ is in $T'_S(M)$ and thus M is not a fixpoint of T_S. It follows that T'_S does not have any fixpoint. Notice that if we apply the LFP algorithm to T'_S, then we get an infinite sequence of steps which oscillates between the two values $p(a)$ and $\emptyset$. This illustrates the nonmonotonicity of T'_S. □

However, there are sets S of (true) *Datalog*¬ clauses such that T'_S is monotonic and has a least fixpoint. A sufficient condition for T'_S to have a least fixpoint is expressed by the following Theorem.

Theorem 11.2. Let S be a set of *Datalog*¬ clauses such that for each negative literal L in a rule body, $|L|$ does not unify with the head of any rule of S (but maybe with some facts of S). Then S has a least Herbrand model $M_S = lfp(T'_S)$.

Though we do not give a formal proof of this Theorem, let us make some helpful observations. Let S be a set of *Datalog*¬ clauses meeting the condition of Theorem 11.2. Every new fact produced from S by one or more applications of EP¬ is an instance of a rule head of some rule of S. Let L be a negative literal in a rule body of S. Since $|L|$ does not match with any rule head, it also holds that $|L|$ does not unify with any new fact generated by use of EP¬. It follows that the validity of any negative literal L is never changed during the computation process. From here it is not hard to prove the monotonicity of T'_S and to show that $lfp(T'_S)$ is the least Herbrand model of S.

We will now use the transformation T' for evaluating stratified *Datalog*¬ programs. Let P be such a program with stratification $P_1, \ldots, P_n$. For $1 \leq i \leq n$, let P_i^* denote the set of all clauses of P corresponding to stratum P_i. Let EDB be an extensional database. We want to define the outcome of the evaluation of P against EDB; this outcome should be a minimal model $M_{P \cup EDB}$ of $P \cup EDB$. For this purpose we define a sequence of sets $M_1, M_2, \ldots, M_n = M_S$ as follows:

$$M_1 = lfp(T'_{P_1^* \cup EDB})$$
$$M_2 = lfp(T'_{P_2^* \cup M_1})$$
$$\ldots\ldots$$
$$M_i = lfp(T'_{P_i^* \cup M_{i-1}})$$
$$\ldots\ldots$$
$$M_n = lfp(T'_{P_n^* \cup M_{n-1}}) = M_{P \cup EDB}$$

In other words, we first compute the least fixpoint M_1 corresponding to the first stratum of the program united to EDB. The existence of this fixpoint is guaranteed by Theorem 11.2. Indeed, all negative literals occurring in rule bodies of P_1^* correspond to EDB predicates, hence they do not match with any rule head of P_1^*; thus the transformation $T'_{P_1^* \cup EDB}$ satisfies the condition of Theorem 11.2. Once we have computed this least fixpoint, we take M_1 as new extensional fact base and compute the least fixpoint of the transformation $T'_{P_2^* \cup M_1}$ corresponding to the second stratum united to the set of facts M_1. In general, we compute M_{i+1} by taking the least fixpoint of the transformation $T'_{P_{i+1}^* \cup M_i}$. This fixpoint exists because due to the stratification of P, all negative literals of stratum i correspond to predicates of lower strata and thus do not unify with any rule head of stratum i. Finally the computation terminates with result $M_n = M_{P \cup EDB}$.

Theorem 11.3. If P is a stratified $Datalog^\neg$ program and EDB an extensional database, then $M_{P \cup EDB}$ is a minimal Herbrand model of $P \cup EDB$.

The following example illustrates a computation of $M_{P \cup EDB}$.

Example 11.15. Consider the program P consisting of the rules R_1 and R_2 as given in Example 11.13:

R_1: $q(b) :- \neg p(a)$
R_2: $r(c) :- \neg q(b)$

where p is an EDB-predicate and q and r are IDB-predicates. The program has a unique stratification consisting of the two strata $P_1 = \{q\}$ and $P_2 = \{r\}$. Assume that this program is run against an empty $EDB\ E = \emptyset$. We then have:

$M_1 = lfp(T'_{P_1^* \cup E}) = lfp(T'_{P_1^*}) = \{q(b)\}$ and
$M_2 = lfp(T'_{P_2^* \cup M_1}) = lfp(T''_{P_2^* \cup \{q(b)\}}) = \{q(b)\} = M_P$

It is easy to see that $\{q(b)\}$ is indeed a minimal model of P. □

Notice that our definition of M_S involves the choice of a particular underlying stratification. Since, in general, there are several different stratifications of a $Datalog^\neg$ program, the question of the uniqueness of M_S arises. This question is answered positively by the following important Theorem.

Theorem 11.4. Let S be a stratified set of $Datalog^\neg$ clauses. The minimal model M_S is independent of the choice of the underlying stratification.

The reader is encouraged to solve Exercise 11.5, where $M_{P \cup E}$ has to be computed for the "railroad program" P of Example 11.10 and an appropriate extensional database E according to the three different stratifications indicated in Example 11.11.

The semantics of a stratified $Datalog^\neg$ program P can be represented in a similar way as for pure $Datalog$ programs (see Sect. 6.2.1) as a mapping $\mathcal{M}_P$ from the powerset of EHB to the powerset of IHB such that for each finite set $\subseteq EHB$, $\mathcal{M}_P(V) = M_{P \cup V} \cap IHB$.

When a goal "? – H" is specified together with P then the output consists of all elements C of $V \cup \mathcal{M}_P(V)$ which are subsumed by H (denoted by $H \triangleright C$). Thus, in this case, the semantics of P with goal "? – H" is defined by a mapping $\mathcal{M}_{P,H}$ from the powerset of EHB to the powerset of HB such that for each finite set $V \subseteq EHB$:

$$\mathcal{M}_{P,H}(V) = \{C \mid C \in HB \wedge C \in V \cup \mathcal{M}_P(V) \wedge H \triangleright C\}.$$

11.2.3 Perfect Models and Local Stratification

In the last subsection we defined the semantics of stratified $Datalog^\neg$ programs using a procedural approach. In the present subsection we briefly outline how stratified $Datalog^\neg$ programs can be characterized equivalently through a purely *declarative* semantics. Furthermore we present the concept of *local stratification*, an interesting generalization of stratification.

If R is a $Datalog^\neg$ rule of the form $L_0 :- L_1, \ldots, L_n$ and θ is a substitution such that for $0 \leq i \leq n$, $L_i\theta$ does not contain any variable, then the rule $R\theta$: $L_0\theta :- L_1\theta, \ldots, L_n\theta$ is called a *ground instance* of R. Note that if $R\theta$ is a ground instance of R, then for $0 \leq i \leq n$, $L_i\theta$ is either a positive literal which is an element of the Herbrand base HB, or a negative literal such that $|L_i\theta|$ is an element of HB.

For each set S of $Datalog^\neg$ clauses, we define the *ground expansion* S^o of S as follows: S^o contains all facts of S and all ground instances of all rules of S.

Each set S induces a relation "$\hookrightarrow$" on HB defined as follows: for each $K \in HB$ and $L \in HB$, $K \hookrightarrow L$ iff there is a rule R in S^o such that K occurs positively or negatively in the body of R and L is the head of R.

The relation "$\hookrightarrow$" is used to define a second relation "$>$" on HB: for each $K \in HB$ and $L \in HB$, $K > L$ iff there is a finite sequence of relationships of the form $K = N_0 \hookrightarrow N_1, N_1 \hookrightarrow N_2, \ldots, N_{k-1} \hookrightarrow N_k = L$ where all N_j are elements of HB and there exists at least one i with $0 \leq i < k$ such that N_i occurs *negatively* in a rule of S^o whose head is equal to N_{i+1}.

Intuitively, if $K > L$ then the ground atom K should be considered before L in the evaluation process because a derivation attempt of L requires the prior knowledge of whether K (or $\neg K$) is derivable from S. Note, however, that our definition of the ">" relation is completely nonprocedural. Note also that the relation " > " on HB is transitive but, in general, not antisymmetric and not reflexive. Hence, in general, " > " is not a partial ordering on HB.

If $K > L$ then we say that K has *higher priority* than L.

We will use the relation ">" on HB for comparing different Herbrand models of the underlying set S of $Datalog^\neg$ clauses. Let $\Im_1$ and $\Im_2$ be two Herbrand models of S. We say that $\Im_1$ is *preferable* to $\Im_2$ iff for each element $L \in \Im_1 - \Im_2$ there exists an element $K \in \Im_2 - \Im_1$ such that $K > L$.

Intuitively, $\Im_1$ is preferable to $\Im_2$ iff $\Im_1$ can be obtained from $\Im_2$ by removing one or more higher priority ground atoms and by adding zero or more lower priority ground atoms.

A Herbrand model $\Im$ of S is *perfect* iff there are no other Herbrand models of S preferable to $\Im$.

Loosely speaking, $\Im$ is perfect iff it contains as few ground atoms of high priority as possible and not more lower priority ground atoms than necessary.

Of course, each perfect model $\Im$ of S is also a minimal model of S. Assume $\Im$ is not a minimal model; then we can remove some atoms from $\Im$, getting a preferable model $\Im'$; this is in contradiction to the perfectness of $\Im$.

Not all sets S of $Datalog^{\neg}$ clauses have a perfect model. However, as the next Theorem states, stratified $Datalog^{\neg}$ programs do.

Theorem 11.5. Let P be a stratified $Datalog^{\neg}$ program and let E be an arbitrary extensional database. The set $S = P \cup E$ has a unique perfect model; this model is precisely the model M_S as defined in Sect. 11.2.2.

Let us illustrate this Theorem by means of two short examples.

Example 11.16. Reconsider the program P made up of the rules R_1 and R_2 as given in Examples 11.13 and 11.15:

$$R_1: q(b) :\!\!- \neg p(a)$$
$$R_2: r(c) :\!\!- \neg q(b)$$

where p is an EDB-predicate and q and r are IDB-predicates. Assume that this program is run against an empty EDB $E = \emptyset$. It is easy to see that $P = P \cup E$ has only two minimal models: $\Im_1 = \{p(a), r(c)\}$ and $\Im_2 = \{q(b)\}$. Since $p(a) > q(b)$, $\Im_2$ is preferable to $\Im_1$. Thus $\Im_2$ is the perfect model of P. Notice that this is exactly the model M_P computed in Example 11.15. □

Example 11.17. Let P be the program

$$R_1: q(b) :\!\!- \neg p(a)$$
$$R_2: r(c) :\!\!- \neg p(a)$$

Obviously P has two minimal models: $\Im_1 = \{p(a)\}$ and $\Im_2 = \{q(b), r(c)\}$. $\Im_2$ can be obtained from $\Im_1$ by eliminating the higher priority atom $p(a)$ and by adding the lower priority atoms $q(b)$ and $r(c)$. Thus $\Im_2$ is preferable to $\Im_1$ and hence $\Im_2$ is the perfect model of P. This example shows that perfect models sometimes have a higher cardinality than other minimal models. □

Note that stratification is a sufficient but not a necessary condition for the existence of a unique perfect model. Stratification is a property defined by using an ordering on the set of *predicate symbols* of a program P. By using an ordering on the more refined set HB of possible *ground atoms* we are able to state a much more general sufficient condition for the existence of a unique perfect model. Programs fulfilling this condition are called *locally stratified*. In the sequel we briefly discuss this feature.

A set S of $Datalog^{\neg}$ clauses is *locally stratified* iff the Herbrand base HB can be partitioned into a finite number of strata $HB_1 \ldots HB_n$ such that for all ground atoms K and L of HB, the following two conditions are satisfied:

- If $K \hookrightarrow L$, then the stratum of K is at most as high as the stratum of L.
- If $K > L$ then K belongs to a lower stratum than L.

It is very easy to see that each stratified program is also locally stratified. The converse, however, does not hold (see Example 11.18). The notion of local stratification can be extended to apply to $Datalog^\neg$ programs with function symbols. However, in that case, a local stratification may consist of an infinite number of strata.

Theorem 11.6. If S is locally stratified then it has a unique perfect model.

Example 11.18. Let P be a program consisting of the following rules:

$R_1: \; p(b) \mathbin{:\!-} \neg p(a)$
$R_2: \; p(c) \mathbin{:\!-} \neg p(b)$

Obviously P is not stratified. However, P is locally stratified by the following strata: $HB_1 = \{p(a)\}$, $HB_2 = \{p(b)\}$, $HB_3 = HB - (HB_1 \cup HB_2)$. The unique perfect model of P is $\{p(b)\}$. Note that this model can be computed by following the order of strata from the lowest to the highest (in analogy to simple stratification). □

Although local stratification is more general than simple stratification, most existing prototypes of $Datalog^\neg$ interpreters are based on simple stratification (see Chap. 12). For the time being, local stratification is still an open research topic; in particular, the efficiency of stratification algorithms and evaluation algorithms for locally stratified programs has to be investigated. It is likely that these algorithms are much less efficient than the corresponding ones for simple stratification.

11.2.4 Inflationary Semantics and Expressive Power

In this subsection we briefly compare different $Datalog$ formalisms with respect to their expressive power. The results will be exposed in a very succinct survey style because a thorough treatment of the material would require an introduction of theoretical notions which are outside the scope of this book. The interested reader, however, is encouraged to study the original research papers on the subject which are referenced in the Bibliographic Notes at the end of this chapter.

Let us first recall some notions of Sect. 11.2.2 and settle our definition of the inflationary semantics of a $Datalog^\neg$ program:

Let P be a $Datalog^\neg$ program. The transformation T_P is defined by

$$\forall W \subseteq HB: \; T_P(W) = W \cup FACTS(P) \cup infer1^\neg(RULES(P) \cup W).$$

The inflationary semantics of P can be described as a mapping $\mathcal{M}_P^i$ from the powerset of EHB to the powerset of IHB such that for each finite set $V \subseteq EHB$: $\mathcal{M}_P^i(V) = lfp(T_{P \cup V}) \cap IHB$.

When a goal "$?-H$" is specified together with P then the output of P applied to an EDB V consists of all elements of $V \cup \mathcal{M}_P^i(V)$ which are instances of H. Thus, in this case, the inflationary semantics of P with goal $?-H$ is defined by the mapping $\mathcal{M}_{P,H}^i$ from the powerset of EHB to the powerset of HB such that for each finite set $V \subseteq EHB$:

$$\mathcal{M}_{P,H}^i(V) = \{C \mid C \in HB \land C \in V \cup \mathcal{M}_P^i(V) \land H \triangleright C\}.$$

Note that $\mathcal{M}_P^i$ and $\mathcal{M}_{P,H}^i$ are defined for each *Datalog¬* program while $\mathcal{M}_P$ and $\mathcal{M}_{P,H}$ are defined only for stratified programs.

The next Theorem states that *Datalog¬* under the inflationary semantics is computationally *as least as expressive* as stratified *Datalog¬*.

Theorem 11.7. For each stratified *Datalog¬* program P there exists a *Datalog¬* program P' such that for each goal H corresponding to a predicate symbol occurring in P, $\mathcal{M}_{P,H} = \mathcal{M}_{P',H}^i$.

Thus, the behavior, with respect to a goal H, of each stratified *Datalog¬* program P, which is processed stratum by stratum as explained in Sect. 11.2.2 can be simulated by a *Datalog¬* program P', which is processed according to the inflationary semantics.

On the other hand, it is possible to find some (quite complicated) queries which are expressible in inflationary *Datalog¬* but which cannot be formulated in stratified *Datalog¬*. This means that inflationary *Datalog¬* is *strictly more expressive* than stratified *Datalog¬*.

Furthermore, it can be seen that inflationary *Datalog¬* has the same expressive power as *Fixpoint Logic on Finite Structures*, a well known formalism obtained by extending First Order Logic with a least fixpoint operator for positive first-order formulas.

Of course, stratified *Datalog¬* is strictly more expressive than pure *Datalog*. The following is an example of a query expressible in stratified *Datalog¬* but not in pure *Datalog*.

Example 11.19. Let r and s be *EDB* predicates and let p and q be *IDB* predicates. Consider the following stratified *Datalog¬* program P:

$q(X,Y) \coloneq s(X,Y), \neg p(X,Y).$
$p(X,Z) \coloneq r(X,Y), p(Y,Z).$
$p(X,Y) \coloneq r(X,Y).$

Consider in particular the query ?- $q(X,Y)$ associated with this program. This query is issued against a generic *EDB* E where the predicates r and s are defined extensionally. It can be shown that such a query cannot be expressed by a pure *Datalog* program (even if the built-in $\neq$ is used). □

In turn, we noted in Chap. 8 that pure *Datalog* is strictly more expressive than nonrecursive *Datalog* which is equivalent to positive relational algebra.

We thus have a hierarchy of *Datalog* versions starting with nonrecursive *Datalog* as the least expressive formalism, and ending with inflationary *Datalog¬* as the most expressive formalism. This hierarchy is depicted in Fig. 11.2.

Note that almost all research prototypes of *Datalog* implementations use stratified negation, though inflationary negation is more expressive. This is not only due to historical reasons (the expressive power of inflationary negation was discovered more recently). It seems that stratified *Datalog¬* is a more natural and easier query language than inflationary *Datalog¬*. Formulating queries in inflationary *Datalog¬* may be quite complicated. For instance, it is not trivial to find an inflationary program P' which is equivalent to the stratified *Datalog¬* program

P of Example 11.19. Furthermore, it seems that those queries which are expressible under the inflationary semantics, but not under the stratification semantics are not very relevant to (even advanced) database applications. However, more research is needed to achieve a better understanding of this subject.

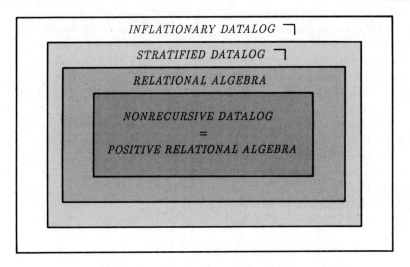

Fig. 11.2. Hierarchy of expressiveness of different versions of Datalog

11.3 Representation and Manipulation of Complex Objects

There exist several different approaches for incorporating the concept of complex structured objects into the formalism of *Datalog*. We have chosen to follow the approach of the LDL (Logic Data Language) Project because this project is well known and because the underlying language is among the most powerful ones. The LDL project is being carried out at MCC in Austin, Texas. In this section we describe some general features of LDL which are related to the concept of structured objects. Other more specific features of LDL as well as some implementation details are reported in Chap. 12.

We will treat some basic elements of LDL in Sect. 11.3.1. A more formal description of the language is given in Sect. 11.3.2. Notice that LDL also allows the use of stratified negation as described in Sect. 11.2.2. Finally, in Sect. 11.3.3, we make some brief comments on advanced relational data models which allow complex objects to be represented. Some of these models are described in more detail in Chap. 12.

11.3.1 Basic Features of LDL

The basic features we add to pure *Datalog* are (uninterpreted) function symbols, sets, and a number of special built-in functions and predicates for set manipulation. Let us treat these features in more detail.

In logic programming and, more generally, in several formalizations of first-order logic, terms are allowed to be complex structures built from function symbols, constants, and variables. The use of function symbols allows objects from the real world to be represented in a coherent and compact manner.

The function symbols in question are *uninterpreted* (or *freely interpreted*) because they have no a priori meaning. The formal semantics of each function symbol depends exclusively on the semantics of the program in which this function symbol occurs. Uninterpreted function symbols should not be confused with *evaluable* function symbols which correspond to a precise predefined data function and are automatically evaluated during program execution (one particular evaluable function *scons* used in LDL is explained later in this section).

The incorporation of uninterpreted function symbols into *Datalog* is feasible without major semantical difficulties. The resulting formalism is close to that of General Logic Programming. An important issue, however, must be taken into consideration: even from a finite vocabulary of constants and function symbols, it is possible to generate an infinite number of nested ground terms. As a consequence, the computed answers to a *Datalog* query based on a program with function symbols may be an infinite set. Thus, as in Prolog, the finiteness of the answer-set to a query remains under the responsibility of the programmer and cannot be guaranteed a priori unless the system imposes some particular safety conditions.

LDL function symbols are denoted by strings of lower case letters, in the same way as constant symbols and predicate symbols. From function symbols, variables, and constants one can construct in the obvious way nested *terms* of any depth, hich may appear as arguments of predicates. Thus, facts and rules involving complex structures can be built. By using general unification which takes into account function symbols and nested terms it is possible to generate new facts by applying rules to known facts in a similar way as in the context of pure *Datalog*.

Example 11.20. For storing information about cars and their owners, we may consider facts of the following structure:

car(vehicle(PLATE, MAKE, YEAR), owner(name(FIRST,LAST), got_from(OWNER))).

Note that facts of different levels of nesting may correspond to this scheme. In particular, the *OWNER* field is supposed to contain the list of the current owner and all previous owners of a given car. This list has the following structure:

owner(name(first1,last1), got_from(
    owner(name(first2,last2), got_from(
        . . . .

owner(name(firstn,lastn), got_from(dealer))..)).

Imagine an *EDB* containing the following facts:

car(vehicle(ny-858778, chevy-vega, 1983),
      owner(name(arthur,keller),
          got_from(dealer))).

car(vehicle(ca-baby53, oldsmobile, 1953),
      owner(name(arthur,keller),
          got_from(owner(name(gio,wiederhold),
                  got_from(owner(name(arthur,samuel),
                          got_from(dealer))))))).

car(vehicle(ny-358735, volkswagen, 1956),
      owner(name(arthur,keller),
          got_from(owner(name(georg,kreisel),
                  got_from(owner(name(kurt,gödel),
                          got_from(owner(name(norbert,wiener),
                                  got_from(dealer))))))))).

car(vehicle(ca-933473, volvo, 1988),
      owner(name(gio,wiederhold),
          got_from(dealer))).

The following rules define the *IDB* predicate *possessed* with the following meaning: *possessed(PERSON,VEHICLE)* is true iff the *VEHICLE* is currently owned or has been owned at some time by the *PERSON* in question.

$R1$: possessed(PERSON, VEHICLE):-
           car(VEHICLE,owner(PERSON,_)).

$R2$: possessed(PERSON, VEHICLE):-
           car(VEHICLE,OWNER), antecedent(PERSON,OWNER).

$R3$: antecedent(PERSON, owner(_,got_from(owner(PERSON,_)))).

$R4$: antecedent(PERSON, owner(_,got_from(OWNER1))):-
           antecedent(PERSON,OWNER1).

Rule $R1$ states that whenever a *VEHICLE* is currently owned by a *PERSON*, then we can derive the fact *possessed(PERSON,VEHICLE)*. Note that underscores can be used as in *Prolog* to denote nameless variables. Rule R_2 states that when a *PERSON* is an antecedent in the chain of all owners of the current owner of a *VEHICLE* then we can infer *possessed(PERSON,VEHICLE)*.

The notion of *antecedent* is defined by rules $R3$ and $R4$. Note that R_3 has an empty body, in other words, $R3$ is a nonground fact. Such facts are not allowed in pure *Datalog*. Note also that rules R_3 and R_4 are not safe. An infinite number of facts of the type *antecedent(X,Y)* can be derived from these rules. Nevertheless, the processing of any goal of the form *possessed(PERSON,VEHICLE)* requires the derivation of a finite number of antecedents only. In fact, the LDL compiler uses a magic-set-like technique for avoiding blind forward chaining.

In the case of our example, the *antecedent* subgoal is called in the body of rule R_2, where its second argument is bound to a ground term (i.e., a term with no variables). Note that this binding comes from the first literal of the body of R_2 and is transmitted by sideways information passing to the second literal in the body of R_2. This binding is then propagated to the rules R_3 and R_4 and enforces a safe processing.

An example of a fact generated by application of rules R_1-R_4 to our *EDB* is:

possessed(name(arthur,samuel), vehicle(ca-baby53, oldsmobile, 1953)).

□

A particularly useful structuring concept customary both in common life and in mathematics is the concept of *set*. Sets allow us to aggregate objects having the same properties without explicitly enumerating these properties for each single object. It is possible to define sets at several levels of nesting, i.e., one can consider a set whose elements are sets and so on.

In LDL sets can be defined either by enumeration of their elements (see Example 11.21) or by grouping (see Example 11.22). Two sets match iff they both contain the same elements, independently of the order in which the elements are given.

Example 11.21. We may define a predicate *hobbies_of(FIRST_NAME, SET_OF_HOBBIES)* by enumerating explicitly the set of all hobbies of a number of individuals. Consider, for instance, the following *EDB*:

hobbies_of(kurt, {gardening, physics, spiritism})
hobbies_of(john, {gardening, painting, shopping})
hobbies_of(emanuel, {surfing, ufo_watching})
hobbies_of(robert, {physics, spiritism, gardening}).

Now consider the rule:

same_hobbies(X,Y) :- hobbies_of(X,Z), hobbies_of(Y,Z), X≠Y.

By applying this rule to the *EDB* we can deduce the new facts

same_hobbies(kurt,robert) and *same_hobbies(robert,kurt)*.

□

The next example shows how the LDL set-grouping construct " $<\ldots>$ " can be used in rule heads in order to define a set by the properties of its elements.

Example 11.22. Recall the *car EDB* predicate as defined in Example 11.20. We may define a new predicate

allcars(PERSON,VEHICLESET)

which associates the set of owned cars to each person:

allcars(PERSON, <VEHICLE>) :- car(VEHICLE, owner(PERSON,_))

If we apply this rule to the *EDB* of Example 11.20 then we can generate several new complex *IDB* facts, for instance the following one:

allcars(name(arthur,keller),{ vehicle(ny-858778, chevy-vega, 1983),
vehicle(ca-baby53, oldsmobile, 1953)),
vehicle(ny-358735, volkswagen, 1956)}).

Now recall the *possessed(PERSON,VEHICLE)* predicate from Example 11.20. By using this predicate we can define a new predicate:

owners(VEHICLE,PERSONSET)

whose first argument stands for a car and whose second argument consists of the set of all persons who ever possessed this car:

owners(VEHICLE, <PERSON>):- possessed(PERSON,VEHICLE) □

LDL offers several built-in predicates and functions for handling sets. The most important are:

- *member(E,S)*, a built-in predicate for expressing the fact that E is an element of the set S. Notice that E can be instantiated to a complex term containing sets as components.
- *union(A,B,S)*, a built-in predicate for expressing the fact that $S = A \cup B$.

Furthermore the equality and the unequality predicates and several arithmetic predicates are available in LDL.

Formally, enumerated sets can be represented by use of a special function symbol "*scons*" (set constructor). This function symbol is of arity 2; its first argument is a term T and its second argument is a set S. The expression $scons(T, S)$ represents the set $S \cup \{T\}$. For instance, the set $\{a, b, c\}$ can be represented as $scons(a, scons(b, scons(c, \{\})))$ or as
$scons(b, scons(c, scons(a, \{\})))$ and so on. Thus, enumerated sets in set notation can be viewed as a convenient shorthand for a corresponding nested *scons* term. The only set which needs a special representation is the empty set. This set is denoted by the special constant $\{\}$.

An important issue is the unification of sets. The set constructor *scons* is not a freely interpretable (i.e., uninterpreted) function symbol, but an interpreted symbol with a precise meaning. Two ground sets are unifiable iff they contain the same elements. For instance, the sets $scons(a, scons(b, scons(c, \{\})))$ and $scons(b, scons(c, scons(a, \{\})))$ are unifiable. This unifiability cannot be detected by any ordinary unification algorithm for terms with uninterpreted function symbols. Indeed, such an algorithm would fail, because the first arguments of the two terms are different constants (a and b) which are of course not unifiable. What we need is a more powerful concept of unification which takes into consideration the particular properties of the special function symbol *scons*. The characteristic properties of *scons* are the following:

COMMUTATIVITY: $\quad scons(X, scons(Y, S)) = scons(Y, scons(X, S))$
IDEMPOTENCE: $\quad scons(X, scons(X, S)) = scons(X, S)$

A unification method which respects the properties of special function symbols and constants which are interpreted according to a certain theory is called *theory unification*. Set unification is a particular type of theory unification because the function symbol *scons* and the symbol {} are interpreted as set theoretical entities.

Set unification (also called set matching) can be correctly defined by using the *scons* notation. However, for the sake of simplicity, we will use the standard set notation here. We consider finite sets of the form $\{t_1, \ldots, t_n\}$ where each t_i is a term, i.e., either a constant, or a variable, or a set of terms, or a nested term (which in turn may contain sets and variables as subterms). Recall that any constant is a ground term. A set is ground iff all of its elements are ground. An LDL term $f(t_1, \ldots, t_n)$ is ground iff each of its arguments is ground. Let us first define the unifiability of ground sets and terms and then proceed by considering general sets and terms. Once we have defined the concept of LDL term unification, an extension of this concept to literal unification is straightforward.

Two constants match iff they are equal. Two enumerated LDL ground sets A and B match iff each element of A matches with at least one element of B and each element of B matches with at least one element of A. Two ground terms $\psi(t_1, \ldots, t_n)$ and $\psi'(t'_1, \ldots, t'_n)$ match iff ψ and ψ' are the same function symbol and t_i matches with t'_i for $1 \leq i \leq n$. For example, the two terms $f(\{a, \{a, b, c\}\}, b)$ and $f(\{\{c, a, b\}, a\}, b)$ match.

Two general LDL terms $T1$ and $T2$ (either constants or variables or sets or function-terms possibly containing variables) are *equivalent* iff, after substituting a new and distinct constant symbol for each variable, these terms match. For example, the terms $f(\{X, \{X, Y, c\}\}, Y)$ and $f(\{\{c, X, Y\}, X\}, Y)$ are equivalent, because after substituting the constants x and y for X and Y respectively we yield matching ground terms. On the other hand, it is easy to see that the terms $f(\{X, \{X, Y, c\}\}, Y)$ and $f(\{\{c, Z, Y\}, Z\}, Y)$ are not equivalent. Note that, in particular, two ground terms which match are also equivalent.

Two general LDL terms t_1 and t_2 are *unifiable* iff there exists a substitution θ replacing zero or more variables by LDL terms such that $t_1\theta$ and $t_2\theta$ are equivalent. In this case, θ is called a unifier for t_1 and t_2.

A unifier θ for t_1 and t_2 is more general than a unifier γ iff there exists a substitution λ such that $\theta\lambda = \gamma$.

In the contexts of pure *Datalog* (see Sect. 6.1.3) and of general logic programming, a unifier θ for two literals L_1 and L_2 is a most general unifier (**mgu**) iff it is more general than any other unifier. As we noted in Sect. 6.1.3, there may exist several **mgus** for a couple of *Datalog* literals. However it can be seen that whenever L_1 and L_2 are unifiable, they have at least one **mgu**. Also, it follows from this definition of **mgu** that each **mgu** γ for L_1 and L_2 can be obtained from each other **mgu** δ for L_1 and L_2 by composing it with an appropriate variable renaming substitution ξ, i.e., $\gamma = \delta\xi$. Thus, to a pair of literals, up to variable renaming, there exists a unique **mgu**.

Unfortunately, such nice properties do not hold in the context of theory resolution. If we use the above definition of **mgu** for LDL terms, then we can easily find examples t_1 and t_2, such that t_1 and t_2 are unifiable but do not admit a **mgu**.

Example 11.23. Consider the terms $t_1 : \{X, Y\}$ and $t_2 : \{a, b\}$. There exist two essential unifiers $\theta = \{X/a, Y/b\}$ and $\sigma = \{X/b, Y/a\}$. All other unifiers are supersets of either θ or σ and are thus less general than θ or, respectively, σ. However, neither θ nor σ is a **mgu** in the traditional sense of the definition. Indeed, θ is not more general than σ because no substitution λ exists such that $\sigma = \theta\lambda$ and, for similar reasons, σ is not more general than θ. Thus t_1 and t_2 do not have an **mgu** even though they are unifiable. Both θ and σ are independent unifiers of maximal generality. □

As illustrated by Example 11.23, two unifiable LDL terms t_1 and t_2 (which are either sets or nested structures with sets as subterms) do not necessarily have a **mgu** but may have several maximally general unifiers which, in a certain sense, have an equal right of being. Thus, instead of considering a single **mgu**, we will deal with sets of maximally general unifiers.

More formally, a unifier θ of t_1 and t_2 is maximally general (or simply *maximal*) iff for each unifier λ which is more general than θ it also holds that θ is more general than λ.

A *base of maximal unifiers* (**bmu**) for a pair t_1, t_2 of LDL terms is a set Θ of unifiers for t_1 and t_2 which satisfies the following conditions:

1) *Completeness:* For each unifier λ of t_1 and t_2, there exists at least one unifier $\theta \in \Theta$ such that θ is more general than λ.
2) *Minimality:* No unifier in Θ is more general than any other unifier in Θ.

Example 11.24. The set of unifiers $\{\theta, \sigma\}$ of Example 11.23 is a **bmu**. □

It is very easy to see that each element of a **bmu** is indeed a maximal unifier. Furthermore, it can be seen that the **bmu** of two LDL terms is unique up to variable-renaming (as, similarly, the **mgu** of two pure *Datalog* literals is unique up to variable-renaming).

Our notion of unification for LDL terms extends naturally to LDL literals, i.e., to expressions of the form $\psi(t_1, \ldots, t_n)$, where ψ is a predicate symbol and $t_1, \ldots, t_n$ are (possibly complex) LDL terms.

Forward chaining and backward chaining in LDL or LDL-like languages can be defined in a similar way as in the context of pure *Datalog*. However, when unifying two literals containing sets, instead of using a **mgu**, one uses the set of unifications of a **bmu**. Each element of the **bmu** represents a possible alternative unification. Thus, an LDL rule of the form $L_0 : -L_1 \ldots L_n$ applied to a set of complex ground literals $K_1, \ldots, K_n$ such that each K_i matches with the corresponding L_i but with no other literal of the rule may produce several new inferred literals of the form $L_0\theta$ due to different choices of maximal unifiers.

Example 11.25. Consider an LDL rule $R : q(X) : -p(\{X, Y\})$ and an LDL ground fact $F : p(\{a, b\})$. R applied to F produces two new facts, namely $q(a)$ and $q(b)$ because the *bmu* of $p(\{X, Y\})$ and $p(\{a, b\})$ contains two maximal unifiers θ and σ (see Examples 11.23 and 11.25). □

It is easy to conceive algorithms for computing the **bmu** for pairs of LDL terms. Such algorithms can be built by using a standard unification algorithm as primitive. Note that the size of the output of such algorithms is often exponential in the size of their input because there may exist exponentially many independent maximal unifiers. Furthermore, the problem of testing whether two LDL terms are unifiable is NP-hard, so it is quite unlikely that a very efficient algorithm can be found for finding even just a single maximal unifier. Some information on the implementation of set unification in the LDL system is given in Sect. 12.1.

Special predicates *choice* and *partition*, allowing, respectively, to make nondeterministic choices and to divide a set into two disjoint parts are also available in LDL. These predicates, as well as some other particular features of LDL such as updating of relations, will be discussed in Sect. 12.1. In the next subsection we deal with some more fundamental problems concerning the meaning and well-definedness of simple LDL programs.

11.3.2 Semantics of Admissible LDL Programs

As for pure *Datalog* programs, it is possible to define a model-theoretic semantics for LDL programs. Such a semantics has been defined for a representative subset of LDL by Beeri, Naqvi, Ramakrishnan, Shmueli, and Tsur. The main issues of their approach are explained here in a somewhat simplified form.

The syntax of the LDL kernel we are dealing with in this subsection corresponds almost to the syntax we have considered (informally) in the last subsection. However, here we make the assumption that all enumerated sets (with or without variables as elements) are represented by use of the *scons* special functor. An LDL rule is either an *ordinary* rule or a *grouping* rule containing the set-constructor symbols "$<,>$" in its head. A grouping rule is subject to the following restrictions: (a) the body contains no set constructor symbols; (b) the head contains at most one occurrence of the form $< X >$; (c) the body contains no negated literal.

The main characteristics of the model-theoretic semantics of LDL are outlined in the sequel.

While for describing pure *Datalog* we use a Herbrand universe consisting merely of the set of all constant symbols (see Chap. 6), the semantics of LDL requires a generalized Herbrand universe U which, in addition, contains variable-free complex structures with a finite depth of nesting and finite sets of such structures. Thus any element of the LDL Universe U is either a constant or a complex ground term built from constants, uninterpreted function symbols (the symbol *scons* is not allowed to appear in any term of U), and sets, or a finite set of terms, possibly the empty set $\emptyset$. (Note that the word "set" is used here in its mathematical sense, i.e., independently of any representation.) The LDL Herbrand base consists of all facts of the form $\psi(t_1, \ldots, t_n)$ such that ψ is an n-ary predicate symbol and $t_1, \ldots t_n \in U$.

Based on the LDL universe U, the notion of *LDL interpretation* is defined in analogy to the concept of Herbrand interpretation. Constants are assigned to "themselves" in U, except the empty set constant $\{\}$ which is assigned the empty set $\emptyset$. Each nonbuilt-in n-ary function symbol f is assigned a mapping from U^n to U such that the tuple $(t_1, \ldots, t_n)$ is mapped by f to the term

$f(t_1,\ldots,t_n)$ for all $t_1,\ldots,t_n \in U$. Particular restrictions hold on the interpretation of the built-in functions and predicates: $scons(t,s)$ is interpreted as $\{t\} \cup s$; $member(t,s)$ has truth value true only if $t \in s$, etc. Each nonbuilt-in n-ary predicate symbol is assigned a particular predicate over U^n.

A *U-fact* is an object of the form $\psi(e_1,\ldots,e_n)$ where ψ denotes an n-ary predicate symbol (either an ordinary one or a built-in) and each e_i is an element of U. Similarly to classical Herbrand interpretations, each LDL interpretation can be identified with a set of U-facts.

Let I be an LDL interpretation. A U-fact F is true under (or satisfied by) I iff $F \in I$; otherwise F is false under I. A negated U-fact $\neg F$ is true under I iff $F \notin I$; otherwise F is false under I. Additional special conditions must be observed for U-facts with built-in predicate symbols. For instance, $union(e_1, e_2, e_3)$ is true under I only if e_1, e_2, and e_3 are sets and $e_1 \cup e_2 = e_3$.

A conjunction of U-facts and/or negated U-facts is true under I iff each of its members is true under I.

Let R be an LDL rule. We understand by a *ground substitution* for R a substitution which replaces each variable of R by an element of U. Note that if θ is a ground substitution it does not necessarily follow that all literals occurring in $R\theta$ are U-facts or negated U-facts. Indeed, these literals may still contain the symbol $\{\}$ and some *scons* functors which do not belong to the vocabulary of the universe U. Therefore we denote by $\overline{R\theta}$ the formula obtained from $R\theta$ by replacing each occurrence of $\{\}$ with $\emptyset$ and by transforming each *scons*-expression to the corresponding U-set. Note that all literals of $\overline{R\theta}$ are either U-facts or negated U-facts.

The truth value of an ordinary rule R under an interpretation I can be defined in analogy to pure *Datalog*: R is true under I iff for each ground substitution θ for R, either the body of $\overline{R\theta}$ evaluates to *false* or the head of $\overline{R\theta}$ evaluates to *true*. (Note that the body of $\overline{R\theta}$ is a conjunction of U-facts and/or negated U-facts.)

Let R be a grouping rule of the form $\psi(<Y>):-BODY$ where ψ stands for a predicate symbol and let I be an LDL interpretation. Let Ω be the set of all substitutions θ such that the body of $\overline{R\theta}$ is true under I and let $\Omega_Y = \{t \mid Y/t \in \Omega\}$.

R is true under I iff one of the following conditions hold

a) Ω is empty;
b) Ω_Y is finite and nonempty and $\psi(\Omega_Y) \in I$.

If Ω_Y is infinite, then the truth-value of R is undefined. In all other cases R is false under I.

Note that whenever the body of R is true under I for an infinite number of Y-values, the truth-value of the entire rule is undefined. Since the LDL universe U does not contain infinite sets as elements, we are not able to interpret the head of R as a U-fact.

A more general model theory of LDL which allows infinite sets up to any fixed transfinite cardinality $\aleph_\alpha$ has been developed by Beeri, Naqvi, Schmueli,

Tsur[†]. However, also in this more general theory a rule R may be of undefined truth-value w.r.t a particular interpretation I. This happens precisely when the body of R is true under I for a set of head-instances whose cardinality is greater than the allowed maximum cardinality $\aleph_\alpha$ for U-terms.

In a similar way, one can define the truth value under an LDL interpretation of a grouping rule whose head also contains some other variables in addition to the special variable in angle brackets. In this case, a set of grouping-variable values corresponds to each different combination of nongrouping variable values in the rule head. Since the formalization of this definition is somewhat more complicated and since it will not be used in the rest of this section, we omit it here.[†]

By an LDL *program* we mean a set of LDL rules and/or facts.

Let S be an LDL program and let I be an interpretation. I is a *model* of S iff each element of S is true under I.

Example 11.26. Consider an LDL program S consisting of the rule

$$allbadboys(<X>) \text{ :- } badboy(X)$$

and of the facts $badboy(jim)$ and $badboy(joe)$. The following sets of U-facts are models of S:

$M_1 = \{badboy(jim), badboy(joe), allbadboys(\{jim,joe\})\}$
$M_2 = \{badboy(jim), badboy(joe), badboy(jack), allbadboys(\{jim,joe,jack\})\}$

□

Note that the model intersection property does not hold in the context of LDL, i.e., the intersection of two models of a set S of LDL clauses is not necessarily a model of S. In particular, this means that a given set S of LDL clauses does not necessarily have a unique minimal model. Note that the failure of the model intersection property is not only due to the possible existence of negated literals in LDL rule bodies. The following example shows that there exist sets of purely positive LDL clauses for which the model intersection property fails to hold.

Example 11.27. Reconsider the set S and the models M_1 and M_2 of Example 11.26. Let $M = M_1 \cap M_2 = \{badboy(jim), badboy(joe)\}$. Evidently, M is not a model of S. Note that no subset of either M_1 or M_2 is a model of S; hence, M_1 and M_2 are both minimal models. □

The next example shows that sets of LDL clauses exist which do not have any model at all.

Example 11.28. Consider a set S' containing the following LDL clauses R and F:

$$R: \ p(<X>) \text{ :- } p(X). \qquad F: \ p(a).$$

[†] see [Beer 89] and [Naqv 89] for more details

Assume S' has a model M. Let $Z = \{t|p(t) \in M\}$. Note that Z must be finite, otherwise R would not be satisfied by M. Z is also nonempty for $a \in Z$. It follows that $p(Z) \in M$. By definition of Z it then follows that $Z \in Z$. This means that Z is a term with an infinite depth of nesting. But, by definition of U, all elements of U have a finite depth of nesting. This is a contradiction. It follows that S' has no model. □

Note that the set definition in the rule R of the above example is a typical case of cyclic self-reference. By trying to define the set $<X>$ we automatically define a new element of this set, namely, the set itself. Hence, the set which is apparently defined by rule R and fact F of Example 11.28 is defined in terms of itself – and thus is not correctly defined.

This type of self-reference is not fatal only in the context of LDL. Even in the context of general set theory such definitions are not allowed. A famous example of an incorrect set definition is *Russell's Paradox*. Consider the set W containing each set V such that $V \notin V$. Clearly this definition leads to a contradiction because if $W \in W$ then, by definition of W, $W \notin W$ and if $W \notin W$ then $W \in W$.

In order to avoid such cyclic self-referential definitions we have to impose certain restrictions on the syntax of LDL programs. On the other hand, LDL allows negative literals in rule bodies and this might lead to semantic ambiguities, as shown in Sect. 11.2.2. Recall that these ambiguities are also due to a kind of cyclic self-reference: predicates which are defined in terms of their own negation. This type of self-reference should also be avoided (for instance, by requiring stratification).

Instead of treating sets and negation separately, we will define one single condition, *admissibility*, guaranteeing both the absence of self-referential set definitions and the absence of predicates which are defined in terms of their own negation. Admissibility is a generalization of stratifiability. Before giving a formal definition of admissibility, we define some auxiliary concepts.

The *Augmented Extended Dependency Graph* $AEDG(P)$ of an LDL program P is obtained from $EDG(P)$ as follows: each edge $<p,q>$ of $EDG(P)$ is labeled with the sign @ iff a rule in P exists in which p is the head predicate symbol, there is an occurence of the form $<X>$ in the head, and q occurs in the body of the rule.

A *layering* of an LDL program P is a partition of the set of IDB predicate symbols of P into subsets $P_1, \ldots, P_n$ such that the following conditions a) and b) are satisfied:

a) if $p \in P_i$ and $q \in P_j$ and $<p,q>$ is an edge of $AEDG(P)$ then $i \geq j$.
b) if $p \in P_i$ and $q \in P_j$ and $<p,q>$ is an edge of $EDG(P)$ labeled with either "¬" or "@", then $i > j$.

Obviously, the notion of *layering* is a generalization of the concept of *stratification*. Indeed, for programs without grouping rules these two concepts coincide.

an LDL program P is *admissible* iff it has a layering.

It is easy to see that an LDL program P is admissible iff there is no cycle in $AEDG(P)$ containing an edge labeled $\neg$ or @. An algorithm similar to STRATIFY (see Sect. 11.2.2) can be used for computing a layering for an admissible LDL program.

It follows that admissible LDL programs contain neither self-referential set definitions nor predicates defined in terms of their own negation. Thus we would expect that all problems of semantic ambiguities are solved for this class of LDL programs, and that a) each positive LDL program has exactly one minimal model   and b) each LDL program with some negative literals has one model which is preferable to all other models (see Sect. 11.2.3).

Unfortunately, things are not so easy! Consider, for instance the LDL program S and the two models M_1 and M_2 of Example 11.26. This program is obviously admissible. However both M_1 and M_2 are minimal models, because neither has nontrivial subsets which are models of S. On the other hand, it is intuitively obvious that model M_1 of S has to be considered, in some sense, as the "smallest" model; in particular, M_1 is "smaller" than M_2.

The problem is that our traditional definition of a model M being smaller than a model M', namely $M \subseteq M'$, is not subtle enough to compare models with set-terms. A more adequate comparison criterion for LDL models is given by the partial ordering $\leq$ as defined in the sequel.

A U-fact $F : p(e_1, \ldots, e_n)$ is *dominated* by a U-fact $G : p(e'_1, \ldots, e'_n)$ iff for each i with $1 \leq i \leq n$ either $e_i = e'_i$ or, if e_i is a set, $e_i \subseteq e'_i$. If F is dominated by G, we write $F \leq G$. A mapping ρ from a set of U-facts to a set of U-facts is *preserving* iff each fact F is mapped to a fact $\rho(F)$ such that $\rho(F) \leq F$. If A is a set of U-facts, then $\rho(A)$ denotes the set $\{\rho(e) | e \in A\}$.

If A and B are sets of U-facts then $A \leq B$ iff there is a set $C \subseteq B$ and a preserving mapping ρ such that $\rho(C) = A$. In particular, if M and M' are models of an LDL program P and if $M \leq M'$ then M is called a *submodel* of M'.

Example 11.29. Consider the LDL program S and the two models M_1 and M_2 of Example 11.26. Consider the following subset C of M_2: $C = \{badboy(jim),$ $badboy(joe),\ allbadboys(\{jim,joe,jack\})\}$. Let us define a mapping ρ from C to M_1 as follows:

$$\rho(badboy(jim)) = badboy(jim);$$
$$\rho(badboy(joe)) = badboy(joe);$$
$$\rho(allbadboys(\{jim,joe,jack\})) = allbadboys(\{jim,joe\}).$$

Obviously ρ is preserving and $\rho(C) = M_1$. Thus $M_1 \leq M_2$, i.e., M_1 is a submodel of M_2. □

Let us now state our new definition of model minimality. A model M of an LDL program P is *minimal* iff there is no model M' of P such that $M' \neq M$ and $(M' - M) \leq (M - M')$.

Example 11.30. Model M_1 of Example 11.26 is minimal. This can be shown as follows.

Each other model M' of S must contain at least the facts F_1 : $badboy(jim)$ and F_2 : $badboy(joe)$ and thus also a fact F_3 of the form $allbadboys(\{jim,joe,\ldots\})$. Assume that $F_3 \in M_1$, i.e., $F_3 = allbadboys(\{jim,joe\})$. In this case, $M_1 - M' = \emptyset$ and thus $M' - M_1 \leq M_1 - M'$ is impossible, i.e., M_1 is minimal. Now assume $F_3 \notin M_1$. Thus $F_3 \in M' - M_1$ and F_3 is of the form $allbadboys(\{jim,joe,t\ldots\})$ where t is a U-term different from jim and joe, and $M_1 - M'$ contains the single fact G : $allbadboys(\{jim,joe\})$. Any mapping ρ from $M' - M_1$ to $M_1 - M'$ maps F_3 to G and is therefore not preserving. In other words, there is no preserving mapping from $M' - M_1$ to $M_1 - M'$, i.e., M_1 is minimal. $\square$

Note that the new definition of "minimal" is a generalization of the classical one. In the absence of set-terms, both definitions coincide. With the new definition, the following results can be proved:

- Each admissible LDL program has at least one minimal model.
- Each positive admissible LDL program has exactly one minimal model.
- Let P be an admissible LDL program. Given a layering of P, a minimal model of P can be computed bottom-up, layer by layer, using LFP-iteration, from the lowest to the highest layer, much in the way stratified *Datalog* programs are computed. This model, called the *standard model*, has properties similar to those of the perfect model of a *Datalog*¬ program. In particular, the standard model is independent of the chosen layering.
- The above results can be generalized to the setting where an LDL program P acts on an EDB of ground U-facts.

11.3.3 Data Models for Complex Objects

A logic programming language for data manipulation such as LDL should be conceived in accordance with an appropriate data model which formalizes the storage and retrieval principles and the manipulation primitives that a DBMS offers for the objects referenced by the language. Pure *Datalog*, for instance, can be based on the relational model in first normal form (nowadays often called the *flat* relational model) because the concept of a *literal* nicely matches the concept of a *tuple* in a data relation and because the single evaluation steps of a *Datalog* program can be translated into appropriate sequences of relational operations (see Chap. 8). On the other hand, logic programming languages such as LDL, dealing with structured objects, require more complex data models.

Quite a number of extensions of the relational model have been developed within the last few years, as attempts to capture the semantics of complex objects. The most famous ones are the NF^2 model by Jaeschke and Schek, the model of nested relations by Fischer and Thomas, the model of Abiteboul and Beeri (which is more general than the former two models), the "Franco-Armenian Data model" FAD by Bancilhon, Briggs, Khoshafian and Valduriez (based on a calculus for complex objects by Bancilhon and Khoshafian). References to papers describing these data models are given in the Bibliographic Notes of this chapter.

The FAD model has been successfully implemented at MCC and serves as the underlying data model in the LDL project. Some important issues of FAD are discussed in Sect. 12.1.

ALGRES, a quite powerful data model for complex objects incorporating all relevant features of the NF^2 model and some additional ones has been developed and implemented in Milan, Italy, in the context of the ESPRIT project Meteor. This data model uses an extended relational algebra as data manipulation language and is well suited as a base for implementing interpreters or compilers for logic data languages with complex objects. ALGRES is described in Chap. 12.

11.4 Conclusions

In this chapter we have studied several useful extensions of pure *Datalog*. In particular, it has been shown how built-in predicates can be used, how negation can be incorporated into *Datalog*, and how complex objects consisting of sets and nested terms can be added to the language.

All these additions result in an enhancement of expressive power and make the *Datalog* language more amenable to practical use as a query language for large databases or as a language for coding the rules of an expert database system. We expect that the intensive research activity centered around *Datalog* and its possible extensions will lead to the development and distribution of new products in the next decade. A number of interesting research prototypes have already been developed. Some are described in Chap. 12.

However, in spite of all existing prototypes, all topics treated in this chapter, as well as the ones mentioned in the introduction to this chapter, still belong to an active research area, and we expect that several new results will be published in the next years.

Many interesting open problems and intriguing alternative approaches to the ones outlined in this chapter can be found in the research papers quoted in the following Bibliographic Notes.

11.5 Bibliographic Notes

A general overview of different extensions of *Datalog* is given in [Gard 87a].

Safety has been originally defined in the context of Relational Calculus [Ullm 82]. The safety problem for nonrecursive Horn clauses is discussed extensively in [Zani86]. Safety of general *Datalog* queries is treated in [Banc 86b] where programs fulfilling our sufficient conditions for the safety of *Datalog* program (Sect. 11.1) are referred to as *strongly safe* programs. Safety can be considered w.r.t. particular evaluation procedures. This is done for instance in [Araf 86] where *monotonicity constraints* on predicates are used as additional semantic knowledge. Safety problems for particular extensions of *Datalog* are investigated in [Rama 87], in [Shmu 87a], and in [Kris 88b]. A somewhat weaker property than safety is *domain independence*. A *Datalog* query is domain independent iff the result of the query depends solely on the underlying *EDB* instance and not on the domain. Domain independence is investigated (in a general context) in [Nico 82].

An important principle used in both Artificial Intelligence and Database Theory is the *Full Information Principle* [Gelf 86]. This principle states that the positive information existing in a given theory (or in a given database) is *full*, i.e., no other positive information is available. There are several different formalizations of this principle. In this chapter we dealt with two of these formalizations: the Closed World Assumption (CWA) and Perfect Models Semantics (stratification).

The Closed World Assumption in the context of databases was introduced by Reiter [Reit 78]. Further developments of this idea can be found in [Reit 80], [Mink 82], and [Reit 84].

The idea of stratification was originally introduced by Chandra and Harel in [Chan 85]. A thorough investigation of the semantics of stratified (general) logic programs is carried out in [Apt 86] where it is also shown how stratified logic programs can be processed in a top-down fashion. In [Kemp 88] it is shown how the QSQR method (see Sect. 9.2.1) can be extended to stratified $Datalog^\neg$ programs. Other interesting papers concerning stratification are [VanG 86], [Naqv 86], [Przy 86], [Lifs 86], [Apt 87], and [Imie 88].

Local stratification is introduced in [Przy 86]; in this paper, local stratification is also compared to a particular type of Circumscription. Cholak [Chol 88] presents some undecidability results concerning local stratification.

The papers [Dahl 87], [Kola 87], [Kola 88], and [Abit 88a] deal with the expressive power of stratified *Datalog* and inflationary *Datalog*. In particular, in [Kola 88] and [Abit 88a] it is shown that inflationary $Datalog^\neg$ has the same expressive power as Fixpoint Logic (on finite structures). The fact that stratified $Datalog^\neg$ is less expressive than inflationary $Datalog^\neg$ is shown in [Dahl 87] and [Kola 87]. As background literature for Fixpoint Logic, query language expressiveness, query complexity, and related aspects, we suggest [Acze 77], [Chan 82a, 82b], [Fagi 74], and [Imme 86].

A different approach for deriving negated facts from logic programs has been proposed much earlier by Clark [Clar 78] and was extended by Lloyd [Lloy 87]. They consider the *completion* of a logic program by viewing the definitions of derived predicates as logical equivalences rather than as logical implications. The semantics of a logic program can then be defined as the set of all consequences (positive or negated) of its completion. The corresponding proof principle is called *negation as failure* and can be formalized as *SLDNF-resolution*, a particular extension of SLD-resolution. The use of program completions was further discussed by Sheperdson [Shep 85] [Shep 88]. Fitting [Fitt 85] and Kunen [Kune 87] refined this approach by using three valued logic: in their setting, an interpretation consists of a mapping which assigns to each atom of the Herbrand Base a truth value from the set $\{true, false, undefined\}$. The semantics of a $Datalog^\neg$ program P is then defined as the minimum three-valued model of

P's completion (according to the partial ordering *undefined* < *true* and *undefined* < *false*. In the context of *Datalog*¬, these approaches are not completely satisfactory.

A recently introduced and very promising approach, also based on three valued logic, is the *well-founded semantics* by Van Gelder, Ross, and Schlipf [VanG 88]. Their method nicely extends the stratified approach to arbitrary logic programs with negation. In particular, a superset of the locally stratified programs is semantically characterized by *total* models, i.e., models such that each fact of the Herbrand Base has either truth value "*true*" of "*false*". These models coincide with the perfect models mentioned in Sect. 6.3. Other programs, on the other hand, can be characterized by *partial* models where single facts may assume truth value "*undefined*". A fixpoint method for computing the well-founded partial model is given in [VanG 89], while a resolution-based procedural semantics for well-founded negation is provided in [Ross 89]. Further important papers related to well-founded semantics are [Bry 89c], where the relationship to logical constructivism is investigated and [Przy 89]. In both of these papers, several other relevant references can be found.

Other formalizations of the Full Information Principle not treated in this book are *Circumscription* [McCa 80], [McCa 86], [Gelf 86], [Gene 87], and *Positivism* [Bido 86].

The principles of representing and manipulating complex objects in a deductive framework are discussed in [Zani 85]. Some basic features of the LDL language are described in [Tsur 86]. A first formalization of a subset of LDL (called LDL1 there) is presented in [Beer 87a]. This paper also shows how LDL programs can be evaluated according to the Magic Set method. A more precise formalization of the semantics of LDL and a revision of some key-issues concerning grouping rules is given in [Beer 89]. Our Sect. 11.3.2 is based on parts of this paper. Some interesting model-theoretic considerations on LDL programs are made in [Shmu 87b]. The essential result of this paper is that there exist non-admissible grouping programs with a unique minimal model which do not have equivalent admissible programs (without any additional predicate symbol).

Another (somewhat less expressive) approach to incorporate sets into the *Datalog* formalism has been undertaken by Kuper [Kupe 86]. Theory unification with particular regard to sets is studied in [Stic 81].

The possibility of updating EDB relations through LDL programs has been briefly mentioned in this chapter. An example and more comments will be given in Sec. 12.1 of the next chapter. The underlying theory for performing EDB updates was carried-out by Naqvi and Krishnamurthy in [Naqv 88] and is very well explained in [Naqv 89]. In [Kris 89] it is shown how transaction processing primitives can be incorporated into LDL.

The *partition* and *choice* constructs of LDL (which will be discussed in more detail in Sect. 12.1) were introduced in [Kris 88c]. They are also carefully discussed in [Naqv 89].

An interesting problem, not treated in this book, is the consistency problem for monovalued data functions. A monovalued data function f is an evaluable function symbol with a particular fixed interpretation. Such functions can be defined by the rules of a logic program. However, the unicity of the function value must be ensured. The papers [Abit 88a], [Lamb 88a], and [Lamb 88b] deal with this problem.

An excellent overview and comparison of data models for complex objects is given in [Abit 87a, 87b]. The NF^2 model is introduced in [Jaes 82]. The Model of Fischer and Thomas is described in [Fisc 83]. The model of Abiteboul and Beeri is described in [Abit 87a] and [Abit 87b]. The FAD Model and its underlying calculus are explained in [Banc 87c] and [Banc 86e]. References to ALGRES are given in the next chapter. Several other important references to data models for complex objects can be found in [Abit 87a]. The *Token Object Model*, an interesting approach for compiling nested *Datalog* into flat *Datalog* by Chen and Gardarin is described in [Chen 88].

A good overview of uncertainty issues and fuzzy reasoning is given in [Gene 87]. A theoretical basis for incorporating certainty factors into Logic Programming is presented in [VanE 86].

11.6 Exercises

11.1 Show that the safety condition for *Datalog* programs with built-ins given in Sect. 11.1 is not a necessary condition: find a program P which does not satisfy this condition but which nevertheless is safe.

11.2 Write a program which for each given *Datalog* program P tests whether P satisfies our safety condition.

11.3 Show that it is possible to express universal quantification by use of negation signs in rule bodies: assume that p and q are unary EDB predicates and show that one can write a $Datalog^\neg$ program such that the fact $r(a)$ is true iff the formula $\forall X(p(X) \Rightarrow q(X))$ is true.

11.4 Write an interpreter for stratified $Datalog^\neg$ programs. The interpreter should test its input program for stratifiability and, in the positive case, compute the output of the program w.r.t. an EDB. You may use *Prolog* or *Lisp* and assume, just as in Exercise 7.2, that both the $Datalog^\neg$ program and the EDB are available as lists of clauses in main memory.

11.5 Compute $M_{P \cup E}$ for the "railroad" program P of Example 11.10 and for the following extensional database E:

link(milano,genova)
link(milano,torino)
link(torino,genova)
link(milano,venezia)

Use different stratifications (for instance, those given in Example 11.11) and show that they lead to the same result.

11.6 Show that local stratification is not a necessary condition for the existence of a perfect model by indicating a $Datalog^{\neg}$ program P which is not locally stratified but which has a perfect model.

11.7 Write an interpreter for $Datalog^{\neg}$ according to the inflationary semantics.

11.8 Let us represent by $set([t_1,\ldots,t_n])$ in *Prolog* the set $\{t_1,\ldots,t_n\}$ where $t_1\ldots t_n$ are (possibly complex) terms. Write a *Prolog* program defining a binary predicate $match$ such that $match(X,Y)$ is true iff X and Y are set-theoretically unifiable.

11.9 Are the following complex terms t_1 and t_2 unifiable ? If they are, indicate a **bmu**.

$t_1: f(\{Z,\{Y,g(Z)\},b\},a);\ t_2: f(\{\{g(c)\},V,b\},W).$

11.10 Let us represent sets by use of the *scons* notation as in basic LDL. Write a *Prolog* program defining a predicate $matching$ such that $matching(X,Y)$ is true iff X and X are terms that match.

11.11 Write a program in *Prolog* or in *Lisp* for testing the admissibility of LDL programs.

Chapter 12
Overview of Research Prototypes for Integrating Relational Databases and Logic Programming

This chapter presents an overview of some of the research prototypes which are under development for the integration of relational databases and logic programming. We present:

a) The LDL project, under development at Microelectronics and Computer Technology Corporation (MCC) at Austin, Texas. The project's goal is to implement an integrated system for processing queries in Logic Data Language (LDL), a language which extends Datalog.
b) The NAIL! project (Not Another Implementation of Logic!), under development at Stanford University with the support of NSF and IBM. NAIL! processes queries in Datalog, but interfaces a conventional SQL database system (running on IBM PC/RT).
c) The POSTGRES system, under development at Berkeley University. POSTGRES is a large project for developing a new generation database system, extending relational technology to support complex objects, data types, and rules (called alerts or triggers); in this chapter, we focus on rule management.
d) The FIFTH GENERATION Project, under development at the Institute for New Generation Computer Technology (ICOT) in Tokio, Japan. The guiding principle of the project is to select *Prolog* as unifying element for the development of computer architecture, programming style, and database approach. In this chapter, we illustrate features related to the development of knowledge bases.
e) The Advanced Database Interface (ADE) of the KIWI Esprit project, sponsored by the EEC. KIWI is a joint effort for the development of knowledge bases, programmed through an object-oriented language (OOPS), and interfaced to an existing relational database. ADE is joint effort of the University of Calabria, of CRAI, and of ENIDATA (Italy).
f) The ALGRES Project, under development in the frame of the METEOR Esprit project, sponsored by the EEC. The ALGRES project extends the relational model to support nonnormalized relations and a fixpoint operator, and supports *Datalog* as programming language. ALGRES is a joint effort of the Politecnico di Milano and of TXT-Techint (Italy).
g) The PRISMA project, under development at the University of Twente in Enschede, the Centre for Mathematics and Computer Science in Amsterdam, and Philips Research Laboratories in Eindhoven, sponsored by the Dutch research fund SPIN. PRISMA is a large project for the development of a

multiprocessor, main-memory, relational database machine that supports a logic interface, called PRISMAlog.

The above systems are rather inhomogeneous; note that the placement of NAIL! and ADE within this section might be considered as inappropriate, because these systems are coupled to existing relational systems, rather than being fully integrated with them. However, they support compilation techniques specifically designed for *Datalog*. Among systems reviewed in this section, the only one which supports *Prolog* is the FIFTH GENERATION Project; this approach, however, is fully innovative, and certainly can be classified among integrated systems.

The unifying element of systems presented in this section is that they can all be considered fairly advanced research prototypes, still under development. This section presents an overview of the *frontiers of logic approaches to databases*.

12.1 The LDL Project

The LDL project is directed towards two significant goals. The first one is the design of Logic Data Language (LDL), a declarative language for data-intensive applications which extends pure *Datalog* with sets, negation, and updates. The second goal is the development of a system supporting LDL, which integrates rule base programming with efficient secondary memory access, transaction management, recovery and integrity control. The LDL system belongs properly to the class of integrated systems; the underlying *database engine* is based on relational algebra and was developed specifically within the LDL project. Note that some of the features of LDL we mention here have been discussed in detail in Sect. 11.3.

The LDL language supports complex terms within facts and rules. Examples of facts with complex terms are:

$champion(mcEnroe, tennis).$

$champion(prost, car - racing, drives(mc - laren, turbo, honda)).$

$champion(gullit, soccer, team(milan, winner(equipe - prize, 1988))).$

$champion(berger, car - racing, drives(ferrari, turbo, ferrari)).$

$champion(maradona, soccer, team(napoli, winner(world - cup, 1986))).$

In the above examples, the complex terms *drives*, *team*, and *winner* enable the flexible description of sport champions. For car-racing champions, the car driven is indicated, with the type and brand of the motor; for soccer champions, the team and special prizes won are indicated, with the prizes' name and year. An example of a rule with complex terms is:

$soccer - winners(N, Y) \; :- \; champion(N, soccer, team(\_, winner(\_, Y))).$

The LDL language supports set terms, as discussed in Chap. 11. For instance, it is possible to state the fact

$$play(tennis, \{lendl, wilander, becker, mcEnroe\}).$$

Set support requires specialized unification algorithms, as well as the availability of specialized *primitives* for set-oriented processing. For instance, given a database including the above fact, the goal:

$$? - play(tennis, \{lendl, becker, mcEnroe, wilander\}).$$

should be answered *yes* for any permutation of the elements within the set. The *set generation* is expressed in LDL in rule form, by denoting the set $s = \{x|p(x)\}$ through the rule: $s(< X >) : - p(X).$, where $p(X)$ is a predicate on X. For instance, assuming a binary relationship *belongs* between players and their teams, the following rule creates for each team the set of its players:

$$team - set(Team, < Player >) : - belongs(Player, Team).$$

In addition to the builtin predicates *member(E,S)* and *union(A,B,S)*, whose meaning has been defined in Sect. 11.3.1, there are some other useful builtins for manipulating sets. The predicates *partition* and *choice*, for instance, are particulaly suited for defining defining aggregate functions on sets.

The *partition* predicate $partition(P1, P2, S)$ applies to a set S having cardinality greater than one, and produces partitions of S into two disjoint subsets $P1$ and $P2$.

The following predicate uses the *partition* predicate to evaluate the cardinality of a set:

$$card(\{\}, 0).$$

$$card(\{X\}, 1).$$

$$card(S, V) : - partition(P1, P2, S),$$

$$card(P1, V1),$$

$$card(P2, V2),$$

$$V = V1 + V2.$$

Note that the first rule succeeds with empty sets, the second rule with singleton sets, and the third rule succeeds only with sets which have cardinality greater than one.

Note also that a computation of the cardinality of a set according to the above rules will not be very efficient in general. The reason is that to a given set S, there are a several possible partitions $\{P1, P2\}$, all of which are considered by the LDL system. However, in order to compute $card(S, V)$, just one partition $\{Pa, Pb\}$ of S would suffice, since all other partitions lead to the same result. In the above program, the same result fact $card(S, V)$ is computed several times.

Considering just one partition instead of all possible partitions would eliminate a considerable amount of duplicate work.

The builtin predicate *choice* allows to suppress multiple solutions by choosing nondeterministically one particular solution. Assume that the variables $X_1, \ldots, X_n, Y_1, \ldots, Y_m$ occur in a rule body (maybe among others). If the expression $choice((X_1, \ldots, X_n), (Y_1, \ldots, Y_m))$ occurs in the rule body, then, for each satisfying instantiation of the X-variables, each Y-variable is constrained to assume only one value. In other words, the relation of chosen values for the variables $X_1, \ldots, X_n, Y_1, \ldots, Y_m$ satisfies the functional dependency $X_1 \ldots X_n \rightarrow Y_1 \ldots Y_m$.

To illustrate the use of the *choice* predicate, consider an EDB predicate $employee(Name, Sex, Age, Income, Dept)$ storing data about each employee of a company, where each employee is uniquely identified by his name, and assume that a party should be arranged such that from each department only one female and one male employee are invited, regardless of the employee's age and income. The embarrassing task of selecting the party guests can be delegated to the following LDL program:

$$party\_guest(Name) :- employee(Name, Sex, Age, Income, Dept)$$

$$choice((Dept, Sex), (Name)).$$

The *choice* predicate in the above program makes sure that the *Name* variable does not get more than one instantiation for each possible combination of instantiations of the *Dept* and *Sex* variables.

The first argument of the *choice* predicate may consist of an empty list of variables. In this case, each variable occurring in the second argument is constrained to assume a single value. We shall use the *choice* predicate with an empty list as first argument to improve our $card(S, V)$ program:

$$card(\{\}, 0).$$

$$card(\{X\}, 1).$$

$$card(S, V) :- partition(P1, P2, S),$$

$$choice((), (P1)),$$

$$card(P1, V1),$$

$$card(P2, V2),$$

$$V = V1 + V2.$$

The *choice* predicate in the body of the third clause enforces that only one partition $\{P1, P2\}$ of the set S is selected and that all other possible partitions of S are suppressed. The unique partition is chosen nondeterministically; this means that the programmer has no control over its exact structure.

Several other aggregate functions can be defined and efficiently implemented by use of the *choice* predicate. It is interesting to notice that the *choice* predicate is a generalization of *Prolog*'s cut construct in a nonprocedural environment.

LDL supports stratified negation, as defined in Chap. 11; programs which incorporate negation but are not stratified are regarded as *inadmissible programs*, and can be detected by the LDL compiler.

Finally, LDL supports updates through special rules. The body of an update rule incorporates two parts, a *query part* and a *procedure part*. The former part identifies the tuples to be updated, and should be executed first; as an effect of the query part, all variables of the procedure part should get bound. Then, the procedure part is applied to the underlying database; the procedure part contains *insert* operations (denoted by the "+" functor) and *delete* operations (denoted by the "-" functor). The order of execution of the procedure is left-to-right, unless the system detects that the procedure has the *Church-Rosser* property (namely, the procedure produces the same answer for every permutation of its components); in this case, the most efficient order of execution is decided by the system. For example, the following rule increases the salary of employees $e1$, $e2$, $e3$, and $e4$.

$$increase - salary() : - member(X, \{e1, e2, e3, e4\}),$$

$$emp(X, Sal), Newsal = Sal * 1.1,$$

$$- emp(X, Sal),$$

$$+ emp(X, Newsal).$$

Notice that the *Church-Rosser* property holds if $e1$, $e2$, $e3$, and $e4$ are distinct employees.

We now consider how rules are compiled, optimized, and executed. The compiler uses the notion of *dynamic adornments*, by knowing which places will be bound to constant values at goal execution; these places will have a *deferred constant*. This is quite similar to the optimization of parametric queries which takes place in conventional database systems. With nonrecursive rules, the main problem considered by the optimizer is determining the order and method for each join. With recursive rules, the compiler tries to *push selection constants*, by applying a variation of the method regarded in Chap. 9 as *variable reduction*. This entails searching for recursive rules equivalent to the original ones; the search considers some special, simple changes of the rule structure, thus producing programs which are semantically equivalent to the given one. If these methods fail, then the LDL optimizer uses the methods of *magic set* and *counting* for rule rewriting.

Special attention is dedicated to rules containing set terms. We recall that in set matching the following two set terms match: $friend\{john, X, paul\}$ and $friend\{mary, paul, Z\}$ with substitutions $\{X/mary, Z/john\}$. Each rule containing a set term is replaced at compile time with a set of rules, each containing ordinary terms; in this way, set matching is reduced to ordinary matching. However, several ordinary matching patterns are required for dealing with each set matching. In order to reduce them, elements of sets within facts are stored in acending ascii order. Thus, the set term $friend-s\{john,tom,Y\}$ corresponds to ordinary terms $friend-o\{Y,john,tom\}$, $friend-o\{john,Y,tom\}$, and $friend-o\{john,tom,Y\}$, where variable Y can occupy any of the positions of the ordered set, but $john$ precedes tom in ascii order.

The improved LDL program, produced after the application of various rewriting methods, is translated into an expression of relational algebra extended by the fixpoint operation; this espression is given as input to the optimizer. Relational expressions are internally represented through *processing graphs*, directed graphs whose nodes correspond either to database relations or to operations; edges indicate relationships between operations and their operands. A particular transformation, called *contraction*, turns processing graphs into trees, by substituting cycles with special nodes corresponding to fixpoint operations.

In this latter representation, leaf nodes correspond to database relations, nonleaf nodes correspond to algebraic operations; these are restricted to joins, unions, differences, and fixpoints. All selections and projections that are directly applicable to database relations are included in the final processing strategy and not further considered in the optimization. In spite of these simplifications, there is an extremely high number of *processing trees* equivalent to the given one, obtained by permuting the order of execution of operations, by permuting the order of evaluation of operands of joins and fixpoints, and by selecting the appropriate technique for computing joins and fixpoints among various available methods. The optimizer uses a cost model (which considers CPU, disk I/O, and communication costs) to associate a cost to each of these alternatives. If the number of alternatives is not prohibitive, the search strategy among alternatives is exhaustive; otherwise, heuristic approaches are used.

The execution environment for LDL is an algebraic machine, which is capable of performing retrieval and manipulation of complex terms, as well as efficient joins and unions. Join methods include *pipelined* joins, in which tuples are transmitted along the *processing tree* while they are produced, and *materialized* joins, in which the entire join result is produced before starting the evaluation of the next operation in the *processing tree*. The target language for the algebraic machine, called FAD, supports these basic operations together with updates, aggregate operations, and general control structures (required for implementing fixpoints).

12.2 The NAIL! Project

The NAIL! (Not Another Implementation of Logic!) project at Stanford University aims at supporting the optimal execution of *Datalog* goals over a relational database system. The fundamental assumption of the project is that *no single strategy is appropriate for all logic programs*, hence much attention is devoted to the development of an *extensible* architecture, which can be enhanced through progressive additions of novel strategies.

The language supported by NAIL! is *Datalog* extended by function symbols, negation (through the *not* operator) and sets (through the *findall* operator); rules must be stratified with respect to the *not* and *findall* operators.

We exemplify the use of the *findall* operator. Consider the following database of facts (with no complex terms supported):

$$champion(mcEnroe, tennis).$$

$$champion(prost, car-racing).$$

$$champion(gullit, soccer).$$

$$champion(berger, car-racing).$$

$$champion(maradona, soccer).$$

The *findall* operator is used as follows:

$$champion-list(L):-findall(X,(champion(X,soccer)),L).$$

The intuitive interpretation of *findall* is as follows: L in the third place denotes the list of terms X such that the condition specified as the second argument of *findall* holds for X. In the above example, we obtain:

$$champion-list([gullit, maradona]).$$

Note that NAIL! does not support set terms or set unification, as LDL does; the *findall* predicate really produces a list, though the language does not control the order of elements within the list. The *findall* operator can be used to produce the same effect as the grouping in SQL; for instance, the goal:

$$champion-list(S,L):-findall(X,(champion(X,S)),L).$$

corresponds to the following result:

$$champion-list(tennis, [mcEnroe]).$$

$$champion-list(soccer, [gullit, maradona]).$$

$$champion-list(car-racing, [prost, berger]).$$

The NAIL! system considers a *Datalog* program and a user's goal as input. Each predicate or each group of mutually recursive predicates is considered separately; these constitute the *strong components* of the dependency graph illustrated in Chap. 9. From the user's goal, the system generates the *adorned goal*, a synthetic representation consisting of the name of the goal predicate with a superscript string; letters b and f in the superscript denote places which are, respectively, *bound* and *free*. Thus, the goal $p(a, X, b)$ over predicate p corresponds to the adorned goal p^{bfb}. It should be noted that the computation of an adorned goal p might require the computation of other adorned goals over other predicates q, or even of different adornments for p.

The NAIL! system uses *capture rules* as basic strategies. Each capture rule applies to particular *adorned goals*, and produces the corresponding result relations. Each capture rule consists of a pair of algorithms; the first one concerns the *applicability* of a capture rule, the second one concerns the *substantiation* of the capture rule, namely, the evaluation of the result relation. A capture rule is applicable to an adorned goal if the rule can evaluate the result relation with a finite number of iterations; thus, applicability algorithms express sufficient conditions for convergence of substantiation algorithms. Applicability is based on the adorned goal only, while substantiation requires the notion of the particular goal constants.

The architecture of NAIL! is illustrated in Fig. 12.1.

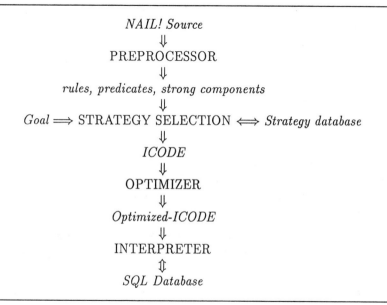

Fig. 12.1. The NAIL! Architecture

The *preprocessor* rewrites the source NAIL! program by isolating *not* and *findall* operators, and by treating disjunction through several conjunctive rules. After preprocessing, the NAIL! program is represented through its predicates and rules; stored rules are of three standard types:

a) Rules whose body is a conjunction of literals;
b) Rules whose body is a single negated literal;
c) Rules whose body is a single *findall* subgoal.

Predicates are partitioned into *strongly connected components*, as defined in the dependency graph.

The *strategy selection* module takes as input the user's goal (in particular, the *adorned goal*) and consults a database of successful strategies; it produces as output the best execution strategies for solving the user's goal and all other goals related to it, expressed using the internal language ICODE. ICODE includes statements of relational algebra together with control statements, procedures, and instructions for managing main-memory stacks.

During the strategy selection process, an internal structure called *rule-goal graph* is generated. The rule-goal graph indicates adornments for goals, and the order in which subgoals of a rule have to be processed. When two subgoals G_1 and G_2 share some variables and G_1 is executed before G_2, there is *sideways information passing* from G_1 to G_2, because common variables are bound to values determined in the solution of G_1 and can in fact be considered *bound* from G_2's viewpoint. Thus, determining the order of subgoals corresponds to determining the direction of sideways information passing.

Capture rules apply to adorned goals; the applicability algorithm of a capture rule tries to build a rule-goal graph for the goal such that each of its nodes is *annotated*. A node is annotated when there exists a strategy for capturing it; thus, this process is inherently recursive.

Capture rules are classified according to the form of the rules whom they apply to; these are:

a) Database predicates;
b) Built-in arithmetic predicates;
c) Negated predicates;
d) Nonrecursive rules without function symbols;
e) Nonrecursive rules with function symbols;
f) Linear recursive rules without function symbols;
g) Nonlinear recursive rules without function symbols;
h) Recursive rules with function symbols.

Note that the class of rules with the *findall* predicate is not yet included. Classes of capture rules as well as instances of capture rules can be easily extended.

There are basically 4 kinds of substantiation rules:

1) *basic* rules, to be used to "capture" nodes which represent nonrecursive predicates.
2) *sideways* rules, which use *sideways information passing* to transmit bindings among goals.
3) *top-down* rules, which use resolution.
4) *bottom-up* rules, which use forward chaining.

This classification includes a variety of evaluation methods; capture rules can be considered a flexible and extensible *control mechanism* for the execution of logic goals.

The strategy selection process is helped by the *Strategy Database*; this database stores the outcome of previous attempts to solve adorned goals together with the corresponding annotated rule-goal graphs. Adorned goals are stored as *capturable* (if the attempt succeeded) or *uncapturable* (if the attempt failed). In inspecting the database, the *bound-is-easier* assumption is used: if an adorned goal g^{bff} is capturable, then any adorned goal g with additional positions bound is also capturable (e.g., g^{bbf} or g^{bfb} or g^{bbb} are capturable). Vice versa, if an adorned goal g^{bff} is uncapturable, then any adorned goal g with additional positions free is uncapturable (e.g., g^{fff} is uncapturable). The use of the *bound-is-easier* assumption is similar to the use of subsumption between predicates for simplifying the retrieval of cached data in **CPR** systems, as discussed in Chap. 4.

Thanks to the *bound-is-easier* assumption, the search of one annotated rule-goal graph for a given user's goal can be performed efficiently (in polynomial time in the size of the rules). However, there might be several annotated rule-goal graphs corresponding to the same user's adorned goal; one of them corresponds to the most efficient execution. NAIL! produces just one of the possible annotated rule-goal graphs; however it uses a heuristic reordering of goals so that the ones with *the most bound arguments* are considered first. This heuristic corresponds to ordering relations involved in a n-ary join according to the number of elementary conditions in the selections which are applied to them, and then proceeding with binary joins according to this order; this is a classic heuristic of database systems that ensures very bad execution strategies are avoided, though no claim of optimality can be made.

The ICODE statements produced as a result of the strategy selection process are optimized and then executed through an interpreter, which translates ICODE retrieval statements to SQL when needed.

12.3 The POSTGRES Project

POSTGRES is a large project for developing a new generation database system, extending relational technology to support complex objects, data types, rules, versions, historic data, and new storage media; many of the features of POSTGRES are outside the scope of this book.

POSTGRES is the successor of the INGRES project, and in particular provides extensions to the QUEL query language of INGRES. New POSTQUEL commands are used for specifying iterative execution of queries, alerts, triggers, and rules. "Iterative" queries support transitive closures; linguistically, this is achieved by appending an asterisk to each command that should be iteratively executed. For instance, the following query builds the ancestors of "Maradona" from the binary relation $PARENT(Parent, Child)$:

retrieve * *into ANCESTOR(P.parent, P.child)*

from P in PARENT, A in ANCESTOR

where P.child = *"Maradona"*

or P.parent = *A.child*

The retrieval command is iterated until there are no changes in the ANCESTOR relation. Note that the retrieval command includes a disjunction in the selection predicate; the condition $P.child = "Maradona"$ generates the first tuples, while the subsequent tuples are generated by the join condition $P.parent = A.child$.

Attribute declarations of the ANCESTOR relation descend from attribute declarations of the PARENT relation.

The ANCESTOR relation can also be defined through views, using the following commands:

$$range\ of\ P\ is\ PARENT$$

$$range\ of\ A\ is\ ANCESTOR$$

$$define\ view\ ANCESTOR(P.all)$$

$$define\ view * ANCESTOR(A.parent, P.child)$$

$$where\ A.child = P.parent$$

This example shows that views can be defined through multiple commands, and in particular some of them may include an iterative command. Thus, the first command generates the set of all parent pairs, and the second command generates recursively all subsequent generations. These views can be queried just as all other relations. For instance,

$$retrieve\ (ANCESTOR.parent)$$

$$where\ ANCESTOR.child = "Maradona"$$

This retrieval command needs to be composed with the view definition in order to produce an expression directly evaluable on the EDB relations. Note that the view mechanism described above permits the definition of conflicting view definitions, e.g. definitions producing various answers for the same query. In this case, conflict resolution is based on assignment of priority to definitions; this feature is however quite procedural, and resembles the use of ordering in *Prolog* rules.

Alerts and triggers are commands which are active indefinitely, and may be "awakened" as a consequence of database changes. They are both specified by adding the keyword "always" to POSTQUEL statements; alerts are added to retrieval commands, while triggers are added to update (append, replace, or delete) commands. Alerters and triggers are permanently active; the system supports their efficient monitoring. An example of an alerter is:

$$retrieve\ always\ (PART.all)$$

$$where\ PART.name = "belt"$$

This alerter is issued by an application program; the effect is to return data from the database system to the application program whenever a PART record satisfying the selection is changed, thus "alerting" the application program.

An example of a trigger is:

$$delete\ always\ ORDER$$

$$where\ count\ (PART.name$$

$$where\ PART.name = ORDER.part) = 0$$

This trigger deletes ORDER tuples when they make reference to parts which do not exist in the PART relation, hence providing a partial implementation of referential integrity constraints.

Another advanced feature of POSTQUEL concerns "demand" (or "virtual") columns, e.g., columns which are not extensionally stored, but rather are evaluated as needed. An example of demand column definition is:

$$replace\ demand\ PERSON\ (class = \text{``child''})\ where\ PERSON.age < 12$$

$$replace\ demand\ PERSON\ (class = \text{``normal''})\ where\ 12 \leq PERSON.age$$

$$and\ PERSON.age < 60$$

$$replace\ demand\ PERSON\ (class = \text{``senior''})\ where\ PERSON.age \geq 60$$

A query accessing a person's class causes the "demand" column to be evaluated.

The implementation of alerters, triggers, and demand columns is done through special locks, which are placed on the data; alerters and triggers use T-locks, demand columns use D-locks. Whenever a transaction writes a data object which is locked with a T-lock, the corresponding dormant "always" commands are activated. The lock manager maintains information about dormant commands, stored into system's tables. Whenever a command is issued on a demand column locked by a D-lock, the demand column is substituted into the command being executed, with a "composition" mechanism. This "composition" proceeds recursively, with a *Prolog*-like style of unification and backward chaining, depth-first search strategy.

POSTQUEL uses a process-per-user architecture that can be built upon the conventional Unix operating system (rather than a server-based architecture, which would require a specialized operating system). There is one POSTMASTER process in each machine, permanently active, which contains the lock manager and all other "common" database services. Each application program then has its own version of the POSTGRES run-time system, which can be executing several commands at the same time on behalf of that application.

12.4 The FIFTH GENERATION Project

The FIFTH GENERATION Project, under development at the Institute for New Generation Computer Technology (ICOT) in Tokyo, Japan, has selected logic programming as the unifying element in the development of computer architecture, programming style, and database approach. In this section, we illustrate features related to the development of knowledge bases.

The architecture of knowledge-base systems in the FIFTH GENERATION Project has evolved in two stages.

a) During the first stage, a *three-layer architecture* was developed, based on the use of three components: the user's application, a knowledge management system supporting the KAISER language, and DELTA, a relational database management system capable of executing operations of relational algebra. KAISER, developed in the early years of the Project, supports induction, deduction, and abduction, derived through meta-programming techniques using *Prolog*. Horn clause queries supported by KAISER are translated into relational algebraic queries that can be executed on DELTA. Within this architecture, the user's application and KAISER stay on a user workstation, while DELTA stays on a server workstation; the two workstations are connected through a local area network (LAN). This approach corresponds to a low-level interface that would be classified as "tightly coupled" according to definitions given in Chap. 4; problems in overall performance were detected, because "a large number of commands and responses had to be transferred through a physically narrow channel". Another problem was the "poor extensibility of the relational database manager" [Itoh 86].

b) Subsequently, a *four-layer architecture* was developed, based on the addition of a knowledge base management system. This system is placed on the server workstation, being fully integrated with the database component. Thus, the server really is a "knowledge-base machine"; in contrast, the user workstation is a pure "inference machine". A specialized, Horn clause interface, called "Horn Clause Transfer" (HCT), separates the two machines. Communication is based on Horn clauses, rather than relational commands; this reduces the communication requirements. The knowledge management system is KAPPA, an evolution of KAISER which supports natural language processing and a theorem prover.

In the four-layer architecture, an essential role is played by the Horn Clause Transfer (HCT), which performs both clause rewriting and the translation into algebraic equations. For example, consider the following Horn clauses over a database storing two relations FATHER and MOTHER, with the obvious meaning:

$$?- ancestor(taro, Y).$$
$$ancestor(X, Y) :- parent(X, Y).$$
$$ancestor(X, Y) :- parent(X, Z), ancestor(Z, Y).$$
$$parent(X, Y) :- edb(father(X, Y)).$$
$$parent(X, Y) :- edb(mother(X, Y)).$$

The Horn clauses are rewritten as follows:

$$ancestor(X,Y) :- edb(father(X,Y)).$$

$$ancestor(X,Y) :- edb(mother(X,Y)).$$

$$ancestor(X,Y) :- edb(father(X,Z)), ancestor(Z,Y).$$

$$ancestor(X,Y) :- edb(mother(X,Z)), ancestor(Z,Y).$$

Then, the Horn clauses are translated into an optimized algebraic equation which incorporates the notion of the query constants:

$$ANCESTOR = \Pi_2(\sigma_{1=taro}FATHER$$
$$\cup\, \sigma_{1=taro}MOTHER$$
$$\cup\, \sigma_{1=taro}\Pi_{1,4}(FATHER \underset{2=1}{\bowtie} ANCESTOR)$$
$$\cup\, \sigma_{1=taro}\Pi_{1,4}(MOTHER \underset{2=1}{\bowtie} ANCESTOR)$$

Note that the equations produced incorporate the "reduction of variables" discussed in Chap. 10.

An interesting feature of the approach proposed by the FIFTH GENERATION Project is the possibility of also storing rules within the database. For that purpose, term relations have been defined, and a unification relational operation has been introduced. All operations of relational algebra, such as selections and joins, have a suitable extension to support predicates which are built through unification operations between terms, rather than arithmetic comparisons between columns or constant values.

The most peculiar feature of the FIFTH GENERATION Project is the high amount of effort placed on specialized hardware architectures. The knowledge base machine has an advanced, parallel architecture where logic processors and memory devices are linked through a high-speed network. A processor controls multiple dedicated hardware systems and distributes to them the computation of simple repeated processing loads over massive amounts of data, generating suitable data flows between them. Specialized functions supported by dedicated hardware include a sort/merge algorithm and the processing of algebraic operations, such as selections and joins; these functions are implemented in VLSI. A "unification engine" will also support unification of variable length terms through dedicated hardware. Hardware development makes the FIFTH GENERATION Project really distinguished; a possible success of this effort might in fact advance tremendously the state of the art in the overall area of logic programming interaction with databases.

12.5 The KIWI Project

KIWI is an Esprit Project, sponsored by the EEC, for the development of knowledge bases. The KIWI Project includes the design of an Object-Oriented Programming Language (OOPS) and a knowledge-base environment based on three loosely connected layers:

1) The *User Interface* (UI) that assists the user in the interaction with KIWI, mainly through a graphic interface.
2) The *Knowledge Handler* (KH) that implements the various features of OOPS except from management of facts.
3) The *Advanced Database Interface* (ADE) for connecting to an existing relational database.

In this section we describe ADE, a joint effort of the University of Calabria, CRAI, and ENIDATA (Italy). ADE supports a simple extension of pure *Datalog* which deals with negation and objects; an object is constituted by multiple tuples spread over several relations, all sharing the same object identifier.

ADE uses a *Prolog* system coupled with a relational database; *Datalog* programs are translated into *Prolog* programs which enforce the fixpoint semantics of *Datalog*, ensuring a safe implementation which terminates by producing the set of all answers. The compilation method used by ADE is the *Mini-Magic* method, a variation of the *Magic Set* method; the *Magic Counting* method is also being considered.

In the optimization of one query, each strong component is considered separately; the order of computation is that induced by the dependency graph of strong components. The computation is performed with a query-subquery approach: if a particular strong component contains a derived predicate defined in an inferior strong component, then the computation of the superior strong component is suspended, and the computation of the inferior strong component is fired.

We exemplify the compilation of *Datalog* rules by means of *Prolog* rules. Consider the following *Datalog* rules:

$$p(X,Y) :- db_1(X,Z), q(Z,Y).$$
$$p(X,Y) :- db_2(X,Y).$$
$$q(X,Y) :- db_3(\_,X), p(Y,\_).$$

Note that p and q are mutually recursive and form a strong component. Let $?- p(X,Y)$ be the query, without binding. Then, the *Prolog* rules generated would be:

$$?- go.$$
$$go :- db_1(X,Z), q(Z,Y), not(p(X,Y)), assert(p(X,Y)), go.$$
$$go :- db_2(X,Y), not(p(X,Y)), assert(p(X,Y)), fail.$$
$$go :- db_3(\_,X), p(Y,\_), not(q(X,Y)), assert(q(X,Y)), go.$$

Note that:

1) The facts for derived relations p and q are progressively asserted into the memory-resident *Prolog* database; at the end of the computation, the rule *go* fails when the fixpoint is reached.
2) Checking for existence of facts in the *Prolog* database before asserting them preserves tuple unicity and forces the final failure when the fixpoint is reached.
3) Nonrecursive *Datalog* rules are dealt with by forcing a failure which causes the *Prolog* system to enumerate all cases within the rule activation; recursive *Datalog* rules are dealt with by including a recursive call to the *go Prolog* predicate.

The query-subquery relationship between strong components is modeled by defining a distinct predicate for each strong component, and by inserting an instance of the predicate of the dependent subcomponent in each rule of the superior strong component, as exemplified by the following:

$$p(X,Y) :- db_1(X,Z), q(Z,Y).$$
$$q(X,Y) :- db_2(X,Y).$$
$$q(X,Y) :- db_3(\_,X), q(Y,\_).$$

The above *Datalog* rules correspond to two strong components for predicates p and q; q is used in the definition of p, hence the strong component for q depends on that for p. The translation of these rules into *Prolog* rules is as follows:

$$?- go_1.$$
$$go_1 :- db_1(X,Z), not(go_2), go_2, q(Z,Y), not(p(X,Y)), assert(p(X,Y)), go_1.$$
$$go_2 :- db_2(X,Y), not(q(X,Y)), assert(q(X,Y)), fail.$$
$$go_2 :- db_3(\_,X), q(Y,\_), not(q(X,Y)), assert(q(X,Y)), go_2.$$

Note that this schema introduces a rather useless repeated execution of the go_2 goal, which fails at the first invocation by producing all the tuples for the q predicate, and at subsequent interactions fails without introducing new tuples; but this schema can be improved to avoid this problem.

Finally, the important case of binding propagation from the user goal is considered. In that case, a predicate invoking the magic set method over bound elements is inserted, as exemplified by the following:

$$?- p(a, X).$$

$$p(X, Y) :- db_1(X, Z), p(Z, Y).$$

$$p(X, Y) :- db_2(X, Y).$$

The corresponding *Prolog* rules with the magic set predicate inserted are:

$$?- go_1.$$

$$go_1 :- magic(X), db_1(X, Z), p(Z, Y), not(p(X, Y)), assert(p(X, Y)), go_1.$$

$$go_1 :- db_2(X, Y), not(p(X, Y)), assert(p(X, Y)), fail.$$

$$magic(a).$$

Additional rules are obviously required to implement the behavior of the magic set method within *Prolog*. Binding propagation can be achieved only if every recursive rule of the program is *safe w.r.t. the binding*; conditions for safety are omitted here. Further, note that the KIWI system uses, in fact, a variation of the *magic set method*, called the *mini-magic* method. Finally, note that the KIWI implementation uses an optimized schema which resembles *semi-naive* evaluation, whereas the schema presented here resembles the *naive* evaluation.

The proposed method is quite general and can be used with any *Prolog* interface to a relational database system, as discussed in Part I of this book; in particular, KIWI uses C-Prolog and Ingres within a Unix environment.

12.6 The ALGRES Project

The ALGRES Project is part of the Esprit project "METEOR", sponsored by the EEC, for supporting the specification and development of complex software systems. ALGRES is a vehicle for the integration of two research areas: the extension of the relational model to deal with complex objects and operations, and the integration of databases with logic programming. In particular, ALGRES extends the relational model to support nonnormalized relations and a fixpoint operator, and supports *Datalog* as programming language. ALGRES is a joint effort of the Politecnico di Milano and of TXT-Techint (Italy).

The data model incorporates the standard elementary types (character, string, integer, real, boolean) and the type constructors RECORD, SET, MULTISET, and SEQUENCE. The schema of an ALGRES object consists of a hierarchical structure of arbitrary (but finite) depth built through the above type constructors. The algebraic language ALGRES-PREFIX includes:

1) The classical algebraic operations of selection, projection, cartesian product, union, difference (and derived operators, e.g. join) suitably extended to deal with new type constructors and with multiple-level objects.

2) Restructuring operations (nesting and unnesting) which modify the structure of objects.
3) Operations for the evaluation of tuple and aggregate functions over objects and sub-objects.
4) Operations for type transformations, which enable the coercion of any type constructor into a different one.
5) A *closure* operator which applies to an algebraic expression and iterates the evaluation of the expression until the result reaches a fixpoint.

Due to the complexity of ALGRES-PREFIX, efforts were invested in formally defining the semantics of the language by using two different formalisms, based on attribute grammars and on abstract data types, respectively.

ALGRES-PREFIX is not a user-friendly language; it can be considered as the assembler of the extended algebraic machine. Two programming paradigms are supported on top of ALGRES-PREFIX:

1) ALGRESQL is an extension of the SQL language which enables a *structured, English-like* query and manipulation language, comparable to other recent proposals; programs in ALGRESQL are translated into ALGRES-PREFIX before execution.
2) Pure *Datalog* is also supported; computationally, the closure operation of ALGRES-PREFIX provides enough expressive power for the support of recursive queries in *Datalog*. Extensions to include sets and negations are under consideration.

Finally, ALGRESQL is interfaced with the "C" programming language in order to exploit existing libraries, and to access the facilities of ALGRES from conventional programs.

ALGRES is coupled to a commercial relational database system, but is not implemented on top of it. This is for the sake of efficiency: execution of operations would be expanded into a long chain of database operations, causing very long response times. This danger is particularly serious for operations involving nested loops, or nested relations, or computation of fixpoints for transitive closures. It is assumed that the ALGRES system will be typically applied to complex data structures of medium size, rather than to very large databases: examples are CAD and engineering systems, knowledge bases, software engineering systems, and some office automation systems. As a consequence of these assumptions, ALGRES objects are stored as main-memory data structures, which are initially loaded from an external database, and returned to mass memory after manipulation.

The abstract machine, named RA (*Relational Algebra*), reflects these assumptions: only a fraction of the database is supposed to be present in main memory at any time, and speed (rather than memory) is the limiting factor. ALGRES runtime efficiency will not compete with traditional imperative Algol-like solutions, but it is intended to compare favourably with other existing productivity tools or high-level languages which have been successfully used for rapid prototyping, such as SETL and *Prolog*.

The core of the ALGRES environment consists of the ALGRES-PREFIX to RA translator and the RA interpreter. The RA instruction set provides traditional algebraic operations on normalized relations as well as special features; it has been designed so as to provide instructions that can be efficiently executed over the objects stored in the main-memory algebraic machine.

The ALGRES-PREFIX to RA translator and RA interpreter are being implemented on a SUN workstation, with the INFORMIX Database Management System providing permanent storage of objects.

12.7 The PRISMA Project

The PRISMA project aims at the development of a multiprocessor, main-memory, relational database machine. This machine is developed as a tightly-coupled distributed database, which aims at achieving high performance by data fragmentation and generation of parallel query-schedules. The PRISMA database machine is developed to be easily extendible, and will consist of 100 data-processors, each with 16 Mbyte of memory; it supports two different programming paradigms: standard SQL, and PRISMAlog, which is an extension of *Datalog*. SQL is used for data definition and updates, whereas PRISMAlog is a pure data retrieval language that enables easy definition of rules, including recursive ones.

PRISMAlog is an extension of pure *Datalog* that incorporates negation, arithmetic operations, aggregates, and grouping. Because the relations in the database are defined through the SQL-interface, the attributes of relations have an associated type. This typing is used by the PRISMAlog parser as an additional check on the correctness of programs. Negation is allowed in stratified programs, where it has the strict interpretation of set-difference. This means that negation requires two predicates in the body of a rule that are union-compatible.

An example of aggregates and grouping in PRISMAlog is given by the following program:

mean_sal_per_dep(Dep, avg(Sal)) ← employee(Name, Dep, Mgr, Sal).
? mean_sal_per_dep(D, S).

This program results in the average salary per department, given the base relation EMPLOYEE.

PRISMAlog programs are translated into the underlying target language for the PRISMA machine, which is an *extended Relational Algebra* (XRA). The meaning of a PRISMAlog program is given by its translated version in XRA, accounting for a standard fixpoint semantics and a strict set-oriented model of computation. Thereby, PRISMAlog programs are completely declarative in nature, and the order of rules and predicates within rules, is not relevant. The underlying target language, XRA, includes a fixpoint operator (called μ-calculus expression) that is used to express recursion.

In the PRISMA project, query optimization is done at the algebraic level by the relational query optimizer, which uses specific join algorithms and needs to know information about processor availability in such a distributed environment.

Optimization of the class of *regular recursive queries* (equivalent to systems of linear equations as introduced in Chap. 8) is done by an algebraic rewriting method. The output of this algorithm is a sequence of standard *Relational Algebra* statements, and *transitive closure* operations on relational expressions. Query constants are pushed into the transitive closure operations by choosing appropriate algorithms for their computation. For instance, by starting from the query constant and gradually joining the result with the original relation, only the 'magic cone' is created, thereby avoiding useless computations.

The PRISMAlog architecture takes advantage of the fragmentation of the base relations, by reformulating general transitive closure queries into several subqueries, each one only accessing a relevant subportion of the data. Thereby the processing of transitive closure queries on fragmented databases is speeded up considerably, leading to a faster computation of the original recursive query.

12.8 Bibliographic Notes

The main reference to the LDL project is the recent book of Naqvi and Tsur [Naqv 89]; a short overview of the LDL project can be found in [Chim 87]. Features for dealing with complex terms in LDL are presented by Zaniolo [Zani 85]; a language overview is given by Tsur and Zaniolo [Tsur 86]; the treatment of sets and negation is formally presented by Beeri et al. [Beer 87a]; the FAD intermediate language is described by Bancilhon et al. [Banc 87c]; query optimization is presented by Krishnamurthy and Zaniolo [Kris 88a].

A general overview of the NAIL! project is presented by Morris, Ullman, and Van Gelder in [Morr 86]; an update can be found in [Morr 87]. An algorithm for ordering subgoals in NAIL! is discussed by Morris [Morr 88]; the algorithm produces as output a special ("annotated") rule-goal graph if that exists. The testing of applicability of a particular class of capture rules, called top-down rules, is discussed by Ullman and Van Gelder [Ullm 85b].

A general overview of the POSTGRES project can be found in the collection of articles edited by Stonebraker and Rowe [Ston 87]. The report contains five papers, on the design, data model, rule management, storage system, and object management in POSTGRES. The first and third of these papers discuss, respectively, the general design of POSTGRES (Stonebraker and Rowe), and rule management through iterative commands, alerters, triggers, and demand data (Stonebraker, Hanson, and Hong).

A general reference for the FIFTH GENERATION Project can be found in the paper by Fuchi [Fuch 84]; the role of logic programming is described in the paper by Fuchi and Furukawa [Fuch 86]. The knowledge base part of the FIFTH GENERATION Project is described by Itoh [Itoh 86]; the relational database machine DELTA is described by Murakami et al. [Mura 83], term unification is described by Morita et al. [Mori 86].

A general overview of KIWI and ADE can be found in the paper by Del Gracco et al. [DelG 87]; an update is in [Sacc 87b]. The mini-magic variation to the magic set approach, used in KIWI, is described by Sacca' and Zaniolo [Sacc 86a]; the conditions for safety of a rule w.r.t. a binding are presented in [Sacc 87b]; the method of magic counting is also due to Sacca' and Zaniolo [Sacc 87a].

A general overview of the ALGRES Project can be found in papers by Ceri, Crespi-Reghizzi et al. [Ceri 88a, 88d]; a short version is [Ceri 88b]. The translation from *Datalog* to equations of relational algebra is described by Ceri, Gottlob, and Lavazza [Ceri 86a], and the optimization of algebraic equations is discussed by Ceri and Tanca [Ceri 87a]. The formal specification of operations in algebra through abstract data types is presented by Lavazza, Crespi-Reghizzi, and Geser [Lava 87]; the treatment given to null values is presented by Ferrario and Zicari [Ferr 87].

A general overview of the PRISMA project can be found in [KAHV 86] and in [Aper 88]. The rewriting mechanism for regular recursive queries as used in the PRISMA project is described in [Aper 86a]; a parallel strategy for solving general recursive queries can be found in [Aper 86b]. The PRISMAlog language and its optimization is described in [Hout 88a]. A strategy to solve transitive closure queries on fragmented databases and make efficient use of parallelism is described in [Hout 88b].

Recent overviews of the projects on *databases and logic* were presented by Zaniolo [Zani 87] in a dedicated issue of IEEE-Data Engineering and by Gardarin and Simon [Gard 87] in TSI.

Bibliography

[Abit 86] Abiteboul, S. and S. Grumbach; "Une approche logique de la manipulation d'objets complexes", *INRIA, Int. Rep. 1986*.

[Abit 87] Abiteboul, S. and S. Grumbach; "Bases de donnees et objets complexes", *Techniques et Sciences Informatiques*, 6:5, 1987.

[Abit 87b] Abiteboul, S. and C. Beeri, "On the Power of Languages for the Manipulation of Complex Objects", *Manuscript*, abstract in: *Proc. International Workshop on Theory and Applications of Nested Relations and Complex Objects*, Darmstadt, West Germany, 1987.

[Abit 88a] Abiteboul,S. and R. Hull; "Data Functions, Datalog and Negation" *Proc. ACM-SIGMOD Conference, 1988*.

[Abit 88b] Abiteboul,S. and V. Vianu; "Procedural and Declarative Database Update Languages" *Proc. ACM SIGMOD-SIGACT Symp. on Principles of Database Systems, Austin, 1988*, pp.240-250.

[Acze 77] Aczel, P.; "An Introduction to Inductive Definitions", *The Handbook of Mathematical Logic* (ed. J. Barwise), North Holland, 1977, pp.739-782.

[Aho 79a] Aho, A.V. and J.D. Ullman; "Universality of data retrieval languages", *Sixth ACM Symp. on Principles of Programming Languages*, San Antonio, January 1979.

[Aho 79b] Aho, A.V., Y. Sagiv, and J.D. Ullman; "Equivalences among Relational Expressions", *SIAM Journal of Computing*, 8:2, 1979.

[Aper 86a] Apers, P.M.G., M.A.W. Houtsma, F. Brandse; "Extending a Relational Interface with Recursion", *Proc. of the 6th Advanced Database Symposium*, Tokyo, Japan, 1986, pp. 159-166.

[Aper 86b] Apers, P.M.G., M.A.W. Houtsma, F. Brandse; "Processing Recursive Queries in Relational Algebra", *Data and Knowledge (DS-2), Proc. of the Second IFIP 2.6 Working Conf. on Database Semantics, Albufeira, Portugal, November 1986*, R.A. Meersman and A.C. Sernadas (eds.), North Holland, 1988, pp. 17-39.

[Aper 88] Apers, P.M.G., M.L. Kersten, H. Oerlemans; "PRISMA Database Machine: A Distributed Main-Memory Approach", *Proc. International Conference Extending Database Technology (EDBT'88)*, Venice, 1988; and Springer LNCS, No. 303, 1988.

[Apt 82] Apt, K.R., and M.H. VanEmden; "Contributions to the theory of logic programming", *Journal of the ACM*, 29:3, 1982.

[Apt 86] Apt, K.R., H. Blair, and A. Walker; "Towards a theory of declarative knowledge", *IBM Res. Report RC 11681*, April 1986.

[Apt 87] Apt, K.R., and J.M. Pugin; " Maintenance of Stratified Databases viewed as Belief Revision Systems", *Proc. Sixth ACM SIGMOD-SIGACT Symp. on Principles of Database Systems*, San Diego, CA, 1987, pp.136-145.

[Araf 86] Arafati, F., et al.; "Convergence of Sideways Query Evaluation", *Proc. Fifth ACM SIGMOD-SIGACT Symp. on Principles of Database Systems*, Cambridge, MA, March 1986, pp. 24-30.

[Banc 85] Bancilhon, F.;"Naive Evaluation of Recursively Defined Relations", *On Knowledge Based Management Systems – Integrating Database and AI Systems*, Brodie and Mylopoulos eds. Springer-Verlag, Berlin, 1985.

[Banc 86a] Bancilhon, F., D. Maier, Y. Sagiv, and J.D. Ullman; "Magic sets and other strange ways to implement logic programs", *Proc. ACM SIGMOD-SIGACT Symp. on Principles of Database Systems*, Cambridge (MA), March 1986.

[Banc 86b] Bancilhon, F. and R. Ramakrishnan; "An amateur's introduction to recursive query processing", *Proc. of the ACM-SIGMOD Conference*, Washington D.C., May 1986.

[Banc 86c] Bancilhon, F. and R. Ramakrishnan; "Performance evaluation of data intensive logic programs", *Foundations of Deductive Databases and Logic Programming*, ed. J. Minker, Washington D.C., 1986, and Morgan-Kaufman, 1988.

[Banc 86d] Bancilhon, F., D. Maier, Y. Sagiv, and J.D. Ullman; "Magic sets: algorithms and examples", *Manuscript*, 1986.

[Banc 86e] Bancilhon F. and S. Khoshafian ;" A Calculus for Complex Objects", *Proc. SIGMOD 86*, 1986.

[Banc 87] F. Bancilhon, T. Briggs, S. Khoshafian, and P. Valduriez; "FAD, a Powerful and Simple Database Language", *Proc. 13th Int. Conf. on Very Large Data Bases*, Brighton, September 1987.

[Baye 85] Bayer, R.; "Query Evaluation and Recursion in Deductive Database Systems" *Manuscript*, 1985.

[Beer 87a] Beeri, C. et al.; "Sets and negation in a logical database language (LDL1)", *Proc. Sixth ACM SIGMOD-SIGACT Symp. on Principles of Database Systems*, San Diego, CA, March 1987.

[Beer 87b] Beeri, C. and R. Ramakrishnan; "On the power of magic", *Proc. Sixth ACM SIGMOD-SIGACT Symp. on Principles of Database Systems*, San Diego, CA, March 1987.

[Beer 89] Beeri, C., S. Naqvi, O. Shmueli, and S. Tsur; "Set Constructors in a Logic Database Language", manuscript, 1989, submitted to *The Journal of Logic Programming*.

[Bido 86] Bidoit, N. and R. Hull; "Positivism vs. Minimalism in Deductive Databases", *Proc. Fifth ACM SIGMOD-SIGACT Symp. on Principles of Database Systems*, Cambridge, MA, *March 1986* pp.123-132.

[Bocc 86a] Bocca J.; "On the Evaluation Strategy of EDUCE"; *Proc.e ACM-SIGMOD Conference*, Washington D.C., May 1986.

[Bocc 86b] Bocca, J., H. Decker, J.-M. Nicolas, L. Vielle, and M. Wallace; "Some Steps Toward a DBMS-Based KBMS", *Proc. IFIP World Conference*, Dublin, 1986.

[Brod 84] Brodie, M. and M. Jarke; "On Integrating Logic Programming And Databases", *Proc. First Workshop on Expert Database Systems*, Kiawah Island, SC, October 1984; *Expert Database Systems*, ed. L. Kerschberg, Benjamin/Cummings, 1986.

[Brod 86] Brodie, M.L. and J. Mylopoulos (eds); *On Knowledge–Base Management Systems*, Topics in Information Systems, Springer-Verlag, 1986. [Brod 88] Brodie, M.L.; "Future Intelligent Information Systems: AI and Database Technologies Working Together", *Readings in Artificial Intelligence and Databases*, 1988, Morgan Kaufman, San Mateo.

[Bry 86] Bry, F. and Manthey; "Checking Consistency of Database constraints: a Logical Basis", *Proc. 12th Int. Conf. on Very Large Data Bases*, Kyoto, August 1986.

[Bry 88] Bry, F., H. Decker and R. Manthey; " A Uniform Approach to Constraint Satisfaction and Constraint Satisfiability in Deductive Databases", *Proc. International Conference Extending Database Technology (EDBT'88)*, Venice, 1988; and Springer LNCS, No. 303, 1988.

[Bry 89a] F. Bry; "Towards an Efficient Evaluation of General Queries: Quantifiers and Disjunction Processing Revisited", *Proc. ACM-Sigmod*, Portland, 1989.

[Bry 89b] F. Bry; "Query Evaluation in Recursive Databases: Bottom-up and Top-down Reconciled", *Manuscript*, 1989.

[Bry 89c] Bry, F.; "Logic Programming as Constructivism: A Formalization and its Application to Databases", in: *Eighth ACM Symposium on Principles of Database Systems (PODS)*, March 1989, pp.34-50.

[Ceri 84] Ceri, S., and G. Pelagatti; *Distributed Databases: Principles and Systems*, McGraw-Hill, 1984.

[Ceri 85] Ceri, S. and G. Gottlob; "Translating SQL into Relational Algebra: Semantics, Optimization, and Equivalence of SQL Queries", *IEEE-Transactions on Software Engineering*, SE 11:4, April 1985.
[Ceri 86a] Ceri, S., G. Gottlob, and L. Lavazza; "Translation and optimization of logic queries: the algebraic approach", *Proc. 12th Int. Conference on Very Large Data Bases*, Kyoto, Aug. 1986.
[Ceri 86b] Ceri, S., G. Gottlob, and G. Wiederhold; "Interfacing relational databases and Prolog efficiently", *Proc. First Intl. Conf. on Expert Database Systems*, Charleston, 1986; and *Expert Database Systems*, ed. L. Kerschberg, Benjamin-Cummings, 1987.
[Ceri 87] Ceri, S. and L. Tanca; "Optimization of systems of algebraic equations for evaluating Datalog queries", *Proc. 13th Int. Conf. on Very large Data Bases*, Brighton, September 1987.
[Ceri 88a] Ceri, S., S. Crespi Reghizzi, F. Lamperti, L.Lavazza, and R. Zicari; " ALGRES: an Advanced Database System for Complex Applications", *IEEE Software* (to appear).
[Ceri 88b] Ceri, S., S. Crespi Reghizzi, G. Gottlob, F. Lamperti, L.Lavazza, L.Tanca, and R. Zicari; "The ALGRES project" *Proc. Int. Conf. Extending Database Technology (EDBT88)*, Venice, 1988; and Springer LNCS, No 303, 1988.
[Ceri 88c] Ceri, S., F. Gozzi, and M. Lugli; "An Overview of PRIMO: a Portable Interface between Prolog and Relational Databases", *Int. Report, Univ. Modena, CS School*, March 1988.
[Ceri 88d] S. Ceri, S. Crespi-Reghizzi, L. Lavazza, and R. Zicari: "ALGRES: a system for the specification and prototyping of complex databases", *Int. Rep. 87-018, Dip. Elettronica, Politecnico di Milano*.
[Ceri 88e] S. Ceri, F. Garzotto; "Specification and Management of Database Integrity Constraints through Logic Programming", *Int. Rep. 88-025, Dip. Elettronica, Politecnico di Milano*.
[Ceri 89] Ceri, S., G. Gottlob, and G. Wiederhold; "Efficient database access through Prolog", *IEEE-Transactions on Software Engineering*, Feb. 1989.
[Chak 84] Chakravarthy, U.S., D. Fishman, and J. Minker; "Semantic Query Optimization in Expert Systems and in Database Systems", *Proc. First Workshop on Expert Database Systems*, Kiawah Island (SC), October 1984; ed. L. Kerschberg, Charleston, 1984 and *Expert Database Systems*, ed. L. Kerschberg, Benjamin-Cummings, 1986.
[Chak 86] Chakravarthy, U.S., J. Minker, and J. Grant; "Semantic Query Optimization: Additional Constraints and Control Strategies", *Proc. First Int. Conference on Expert Database Systems*, ed. L. Kerschberg, Charleston, 1986, and *Expert Database Systems*, ed. L. Kerschberg, Benjamin-Cummings, 1987.
[Chak 87] Chakravarthy, U.S., J. Grant, and J. Minker; "Foundations of Semantic Query Optimization for Deductive Databases", *Proc. Int. Workshop on the Foundations of Deductive Databases and Logic Programming*, ed. J. Minker, Washington D.C., August 1986, and Morgan-Kaufman, 1988.
[Cham 76] Chamberlin, D.D. et al.; "Sequel 2: A Unified Approach to Data Definition, Manipulation, and Control", *IBM Journal of Research and Development*, 20:6, 1976.
[Chan 82a] Chandra, A.K. and D. Harel; "Horn clauses and the fixpoint hierarchy", *Proc. ACM SIGMOD-SIGACT Symp. on Principles of Database Systems*, Los Angeles, 1982.
[Chan 82b] Chandra, A. and D. Harel; "Structure and Complexity of Relational Queries", *Journal of Computer and Systems Sciences* 25, 1982, pp.99-128.
[Chan 85] Chandra, A. and D. Harel; "Horn Clause Queries and Generalizations", *Journal of Logic Programming*, 1, 1985, pp.1-15.
[Chan 73] Chang, C.L. and R. C. Lee, *Symbolic Logic and Mechanical Theorem Proving*, Academic Press, 1973.
[Chan 81] Chang, C.; "On the Evaluation of Queries Containing Derived Relations in Relational Databases", *Advances in Database Theory*, vol.1, H. Gallaire, J. Minker and J.M. Nicholas, Plenum Press, N.Y., 1981.

[Chan 84] Chang, C.L. and A. Walker; "PROSQL: A Prolog Programming Interface with SQL/DS", *Proc. First Workshop on Expert Database Systems*, Kiawah Island, SC, October 1984; *Expert Database Systems*, L. Kerschberg (editor), Benjamin-Cummings, 1986.

[Chen 88] Chen, Q. and G. Gardarin; "An Implementation Model for Reasoning with Complex Objects", *SIGMOD 88*, pp.164-172.

[Chim 87] Chimenti, D., T. O'Hare, R. Krishnamurthy, S. Naqvi, S. Tsur, C. West,and C. Zaniolo; "An Overview of the LDL System", *Special Issue on Databases and Logic*, IEEE − Data Engineering, 10:4, ed. C. Zaniolo, December 1987.

[Chol 88] Cholak P. ; "Post Correspondence Problem and Prolog Programs", Technical Report, Univ. of Wisc., Madison, *Manuscript*, 1988.

[Clar 78] Clark, K.L.; "Negation as Failure", in *Logic and Databases*, H. Gallaire and J. Minker (eds), Plenum Press, 1978, pp.293-322.

[Cloc 81] Clocksin, W.F., and C.S. Mellish; *Programming in Prolog*, Springer-Verlag, 1981.

[Codd 70] Codd, E.F.; "A relational model of data for large shared data banks", *Communications of the ACM*, 13:6, June 1970.

[Colm 85] Colmerauer, A.; "Prolog in 10 Figures", *Communications of the ACM*, 28:12, 1985.

[Cupp 86] Cuppens, F. and R. Demolombe; "A PROLOG-Relational DBMS interface using delayed evaluation", Workshop on Integration of Logic Programming and Databases, Venice, December 1986.

[Dahl 87] Dahlhaus, E.; "Skolem Normal Forms Concerning the Least Fixpoint", *Computation Theory and Logic*, (E. Börger Ed.), Springer Lecture Notes in Computer Science 270, 1987, pp.101-106.

[Date 83] Date, C.J.; *An Introduction to Database Systems, Vol. II*, Addison-Wesley, Reading, Massachusetts, 1984.

[Date 86] Date, C.J.; *An Introduction to Database Systems, Vol. I*, Fourth Edition, Addison-Wesley, Reading, Massachusetts, 1984.

[DelG 87] Del Gracco, C., M. Dispinzieri, A. Mecchia, P. Naggard, C. Pizzuti, D. Sacca'; "Design Documentation of ADE", Report C3, Esprit Project P1117.

[Deno 86] D. Denoel, D. Roelants, and M. Vauclair; "Query Translation for Coupling Prolog with a Relational Database Management System"; Workshop on Integration of Logic Programming and Databases, Venice, December 1986.

[Deck 86] Decker H.; "Integrity Enforcement on Deductive Databases", *Proc. First Int. Conference on Expert Database Systems*, Charleston, 1986, and *Expert Database Systems*, Benjamin-Cummings, 1987.

[Deva 86] Devanbu, P. and R. Agrawal; "Moving selections into fixpoint queries", *Manuscript*, Bell Labs, Murray Hill, 1986

[Epsi 86] Esprit Project 530 Epsilon (organizer); *Workshop on Integration of Logic Programming and Databases*, Venice, December 1986.

[Fagi 74] Fagin, R.; "Generalized first-order Spectra and Polynomial Time Recognizable Sets", *Complexity of Computations* (R. Karp Ed.), SIAM-AMS Proc. 7, 1974, pp.43-73.

[Ferr 87] Ferrario, M. and R. Zicari;"Extending the ALGRES Datamodel and Algebra to Handle Null Values", *Int. Rep. 88-001, Dip. Elettronica, Politecnico di Milano*.

[Fisc 83] Fischer, P. and S. Thomas;"Operators for Non-First-Normal-Form Relations ",*Proc. 7th int. Computer Software Applications Conf.*, Chicago, 1983.

[Fuch 84] Fuchi, K.; "Revisiting Original Philosophy of Fifth Generation Computer Project", *International Conference on Fifth Generation Computer Systems*, 1984.

[Fuch 86] Fuchi, K. and K. Furukawa; "The role of Logic Programming in the Fifth Generation Computer Project", *Int. Logic Programming Conference, July 1986*.

[Fitt 85] Fitting, M.; "A Kripke-Kleene Semantics for Logic Programs", *Journal of Logic Programming*, 2:4, 1985, pp.295-312.

[Gall 78] Gallaire, H. and J. Minker (eds); *Logic and Databases*, Plenum Press, 1978.

[Gall 81] Gallaire, H., J. Minker, and J.M. Nicholas; *Advances in Database Theory, Vol I*, Plenum Press, 1981.

[Gall 84a] Gallaire, H., J. Minker, and J.M. Nicholas; *Advances in Database Theory, Vol II*, Plenum Press, 1984.
[Gall 84b] Gallaire, H., J. Minker, and J-M. Nicolas; "Logic and Databases: A Deductive Approach", *ACM Computing Surveys*, 16:2, June 1984.
[Gall 87] Gallaire, H. and J.-M. Nicholas; "Logic Approaches to Knowledge and Databases at ECRC", *Special Issue on Databases and Logic*, IEEE – Data Engineering, 10:4, Zaniolo, C. (ed.), December 1987.
[Gall 88] H. Gallaire; "Logic and Databases", Tutorial Notes, *Proc. International Conf. Extending Database Technology (EDBT88)*, Venice, 1988.
[Gard 86] Gardarin, G. and C. De Maindreville; "Evaluation of database recursive logic programs as recurrent function series", *Proc. of the ACM-SIGMOD Conference*, Washington D.C., May 1986.
[Gard 87a] Gardarin, G. and E. Simon; "Les systemes de gestion de bases se donnees deductives", *Technique et Science Informatiques*, 6:5, 1987.
[Gard 87b] Gardarin, G.; "Magic Functions: a Technique to Optimize Extended Datalog Recursive Programs", *Proc. 13th Conference on Very Large Databases*, Brighton, UK, 1987.
[Gelf 86] Gelfond, M., H. Przymusinska, and T. Przymuzinski; "The Extended Closed World Assumption and its Relationship to Parallel Circumscription", *Proc. ACM SIGMOD-SIGACT Symp. on Principles of Database Systems*, Cambridge,MA, March 1986.
[Gene 87] Genesereth, M. and N.J. Nilsson; *Logical Foundations of Artificial Intelligence*, Morgan-Kaufmann, 1987.
[Gott 85] Gottlob, G. and A. Leitsch; "On the Efficiency of Subsumption Algorithms", *Journal of the ACM*, 32:2 (1985), pp.280-295.
[Gott 85a] Gottlob, G. and A. Leitsch; "Fast Subsumption Algorithms" in: *Lecture Notes in Computer Science*, Vol. 204, II (Springer Verlag, 1985), pp.64-77.
[Gott 87] Gottlob G.; "Subsumption and Implication", *Information Processing Letters* 24 (1987), pp.109-111.
[Gozz 87] Gozzi, F. and M. Lugli; "Design and Development of Efficient Interfaces between Prolog Environments and Relational Databases", *Diploma thesis, Un. Modena, Comp. Sc. School*, December 1987.
[Guess 87] Guessarian, I.; "Some fixpoint techniques in algebraic structures and application to computer science", *INRIA-MCC Intern. Workshop*, Austin, Texas, 1987.
[Gure 86] Gurevich, Y. and S. Shelah; "Fixed Point Extensions of First Order Logic", *Annals of Pure and Applied Logic*, vol. 32, 1986, pp.265-280.
[Hens 84] Henschen, L.J. and S.A. Naqvi; "On compiling queries in recursive first order databases", *Journal of the ACM*, 31:1, 1984.
[Hout 88a] Houtsma, M.A.W., H.J.A. van Kuijk, F. Flokstra, P.M.C. Apers, M. L. Kersten; "A Logic Query Language and its Algebraic Optimization for a Multiprocessor Database Machine",*Technical Report INF-88-52, University of Twente*, 1988.
[Hout 88b] Houtsma, M.A.W., P.M.G. Apers, S. Ceri; "Parallel Computation of Transitive Closure Queries on Fragmented Databases", *Technical Report INF-88-56, University of Twente*, 1988.
[Imie 86] Imielinski T. and S. Naqvi; "Explicit Control of Logic Programs through Rule Algebra", *Proceedings of the 7th Annual ACM Symposium on Principles of Database Systems (PODS 88)*, Austin, TX, 1988.
[Imme 86] Immermann, N.; "Relational Queries Computable in Polynomial Time" *Information and Control* 68, 1986, pp.86-104.
[Ioan 85] Ioannidis, Y. E. and E. Wong; "An algebraic approach to recursive inference" *Univ. of Cal. at Berkeley, Electronics Res. Lab. Int. Rep. n. UCB/ERL M85/92*, 1985.
[Ioan 87a] Ioannidis, Y. E. and E. Wong "Trasforming non-linear recursion into linear recursion", *Manuscript*, 1987.
[Ioan 87b] Ioannidis, Y.E., J. Chen, M.A. Friedman, and M.M. Tsangaris; "BERMUDA – An Architectural Perspective on Interfacing Prolog to a Database Machine"; *University of Winsconsin, CS Dept.*, Tech. Rep. 723, October 1987.

[Itoh 86] Itoh, H.; "Research and Development on Knowledge Base Systems at ICOT", *Proc. 12th Int. Conference on Very Large Data Bases*, Kyoto, Aug. 1986.

[Jaes 82] Jaeschke, B. and H.J. Schek;"Remarks on the Algebra of Non First Normal Form Relations", *Proc. ACM SIGMOD-SIGACT Symp. on Principles of Database Systems*, Los Angeles, 1982, pp.124-138.

[Jaga 87] Jagadish, H.V., R. Agrawal and L. Ness; "A study of transitive closure as a recursion mechanism", *Proc. of the ACM-SIGMOD Conference*, S. Francisco, May 1987.

[Kemp 88] Kemp, D.B. and R.W. Topor; "Completeness of a Top-Down Query Evaluation Procedure for Stratified Databases", in: *Proc. 5th Int. Connf. and Symp. on Logic Programming*, Seattle, USA, Aug.88.

[Kers 84] Kerschberg, L. (ed.); *Proc. First Workshop on Expert Database Systems*, Kiawah Island, 1984, and *Expert Database Systems*, Benjamin-Cummings, 1986.

[Kers 86] Kerschberg, L. (ed.); *Proc. First Int. Conference on Expert Database Systems*, Charleston, 1986, and *Expert Database Systems*, Benjamin-Cummings, 1987.

[Kers 88] Kerschberg, L. (ed); *Proc. Second Int. Conference on Expert Database Systems*, Tyson Corner (Virginia), 1988.

[KAHV 86] Kersten, M.L., P.M.G. Apers,M.A.W. Houtsma, H.J.A. van Kuijk, R.L.W. van de Weg; "A Distributed Main-Memory Database Machine", *Proc. of the 5th International Workshop on Database Machines*, Karuizawa, Japan, 1987; and *Database Machines and Knowledge Base Machines*, M. Kitsuregawa and H. Tanaka (eds.), Kluwer Science Publishers, 1988, pp. 353–369.

[King 81] King, J.; "Quist: A System for Semantic Query Optimization in Relational Databases", *Proc. 7th Int. Conference on Very large Data Bases*, Cannes, 1981.

[Klee 67] Kleene, S.C.; *Mathematical Logic*, John Wiley & Sons, New York, 1967.

[Kife 86] Kifer, M. and E.L. Lozinskii; "Filtering data flow in deductive databases", *Proc. 1st International Conference on Database Theory*, Roma, September 1986.

[Kola 87] Kolaitis, Ph.G.; "On the Expressive Power of Stratified Datalog Programs", *Preprint*, Stanford University, November 1987.

[Kola 88] Kolaitis, Ph.G. and Ch.H. Papadimitriou; "Why not Negation by Fixpoint?", *Proc. ACM SIGMOD-SIGACT Symp. on Principles of Database Systems 1988*, pp.231-239.

[Kort 86] Korth, H.F. and A. Silberschatz; *Database Systems Concepts*, McGraw-Hill, Computer Science Series, 1986.

[Kowa 86] Kowalski, R. and F. Sadri; "An Application of General-Purpose Theorem Proving to Database Integrity" *Foundations of Deductive Databases and Logic Programming*, J. Minker ed., Washington, 1986, and Morgan-Kaufman, 1988.

[Kowa 87] Kowalski R. A., F. Sadri, and P. Soper; "Integrity Checking in Deductive Databases", *Proc. Int. Conf. Very Large Data Bases*, Brighton, Sept. 1987.

[Kris 88a] Krishnamurthy, R. and C. Zaniolo; "Optimization in a Logic Based Language for Knowledge and Data Intensive Applications", *Proc. International Conf. Extending Database Technology (EDBT88)*, Venice, 1988; and Springer LNCS, No 303, 1988.

[Kris 88b] Krishnamurthy, R., R. Ramakrishnan, and O. Shmueli; "A Framework for Testing Safety and Effective Computability of Extended Datalog" *ACM SIGMOD Int. Conf. on Management of Data*, Chicago, IL, June 1988, pp.154-163.

[Kris 88c] Krishnamurthy R. and S. Naqvi; "Non-deterministic Choice in Datalog", *Proceedings of the 3rd International Conference on Data ans Knowledge Bases*, Jerusalem, 1988.

[Kris 89] Krishnamurthy R., S. Naqvi, and C. Zaniolo; "Database Updates and Transactions in LDL", *manuscript, MCC*, submitted for publication.

[Kune 87] Kunen, K.; "Negation in Logic Programming", *Journal of Logic Programming*, 4:4, 1987, pp.289-308.

[Kuni 82] Kunifuji, S. and H. Yokota; "Prolog and Relational Databases for Fifth Generation Computer Systems", *Proc. Workshop on Logical Bases for Databases*, Toulose, December 1982.

[Kupe 86] Kuper, G.M.; "Logic Programming with Sets ", *Proc. ACM SIGMOD-SIGACT Symp. on Principles of Database Systems 1987*, pp.11-20.

[Lamb 88a] Lambrichts, E., P. Nees, J. Paredaens, P. Peelman, L. Tanca; "MilAnt: an extension of Datalog with Complex Objects, Functions and Negation", - Internal Report, 1988, University of Antwerp (Dept. of Computer Science).
[Lamb 88b] Lambrichts, E., P. Nees, J. Paredaens, P. Peelman, L. Tanca; "Integration of Functions in the Fixpoint Semantics of Rule Based Systems", *Proc. 2nd Symposium on Mathematical Fundamentals of Database Thoery*, Visegrag (Hungary), June 1989, and *LNCS*, Springer Verlag, 1989.
[Lass 82] Lassez, J.L., V.L. Nguyen, and E.A.Sonenberg; "Fixed Point Theorems and Semantics: a Folk Tale", *Information Processing Letters*, 14:3, May 1982.
[Lava 87] Lavazza, L., S. Crespi Reghizzi, and A. Geser; "Algebraic ADT Specification of an Extended Relational Algebra and their Conversion into a Running Prototype", *Workshop on Algebraic Methods, Theory, Tools and Applications*, Passau, West Germany, June 1987, and Springer Verlag.
[Lifs 86] Lifschitz, V.; "On the Declarative Semantics of Logic Programs with Negation", *Proc. of the Workshop on Foundations of Deductive Databases and Logic Programming*, Washington D.C. (J.Minker Ed), August 1986, pp. 420-432.
[Lloy 86] Lloyd, J., E.A. Sonenberg and R. Topor; "Integrity constraint checking in Stratified Databases", *Tech. Rep., University of Melbourne, CS Dept.*, 1986.
[Lloy 87] Lloyd, J., *Foundations of Logic Programming*, Second, Extended Edition, Springer Verlag, 1987.
[Love 78] Loveland, D.W.; *Automated Theorem Proving: A Logical Basis* North Holland, New York, 1978.
[Lozi 85] Lozinskii, E.; "Evaluating queries in deductive databases by generating" *Proc. Int. Joint conference on Artificial Intelligence*, 1985.
[Mall 86] Malley, V. and B. Zdonik; "A Knowledge-based Approach to Query Optimization", *Proc. First Intl. Conf. on Expert Database Systems*, Charleston, 1986; and *Expert Database Systems*, L. Kerschberg (ed.), Benjamin-Cummings, 1987.
[Maie 83] Maier, D.; *The Theory Of Relational Databases*, Computer Science Press, Rockville, Md., 1983.
[Maie 84] Maier, D.; "Databases and the Fifth Generation Project: is Prolog a Database Language?", *Proc. of the ACM-SIGMOD Conference*, 1984.
[Maie 88] Maier, D. and D.S.Warren; "Computing With Logic", Benjamin/Cummings, Menlo Park, CA, 1988.
[Marq 83] Marque-Pucheu, G. "Algebraic Structure of Answers in a Recursive Logic Database", *Acta Informatica*, 1983.
[Marq 84] Marque-Pucheu, G., J. Martin Gallausiaux and G. Jomier; "Interfacing Prolog and Relational Database Management Systems", *New Applications of databases*, Gardarin and Gelenbe eds. Academic Press, London, 1984.
[McCa 80] McCarthy, J.; "Circumscription – A Form of Non–Monotonic Reasoning", *Artificial Intelligence* 13, 1980, pp.27-39.
[McCa 86] McCarthy, J.; "Applications of Circumscription in Formalizing Common Sense Knowledge", *Artificial Intelligence* 28, 1986, pp.89-116.
[McKa 81] McKay, D. and S. Shapiro; "Using Active Connection Graphs for Reasoning with Recursive Rules", *Proc. 7th International Joint Conference on Artificial Intelligence*, 1981.
[Mend 64] Mendelson, E.; *Introduction to Matematical Logic,*, Van Nostrand-Reinhold, Princeton, New Jersey, 1964.
[Mend 85] Mendelzon, A.; "Functional Dependencies in Logic Programs", *Proc. 11th Int. Conf. Very Large Data Bases*, Stockolm, 1985.
[Mink 82] Minker,J.; "On indefinite Databases and the Closed World Assumption" *Proc. of the Sixth Conference on Automated Deduction*, Springer LNCS, No 138, 1982, pp. 292-308.
[Mink 86] Minker, J. (ed.); *Foundations of Deductive Databases and Logic Programming*, Washington, 1986, and Morgan-Kaufman, 1988.

[Miss 84] Missikoff, M. and G. Wiederhold; "Towards a Unified Approach for Expert and Database Systems", *Proc. First Workshop on Expert Database Systems*, L. Kerschberg ed., Kiawah Island, 1984, and *Expert Database Systems*, Benjamin-Cummings, 1986.

[Moff 86] Moffat, D.S., and P.M.D. Gray; "Interfacing Prolog to a Persistent Data Store", *Proc. Int. Conf. on Logic Programming*, London, 1986.

[Mori 86] Morita, Y., H. Yokota, K. Nishida, and H. Itoh; "Retrieval-by-Unification Operation on a Relational Knowledge Base" *Proc. 12th Int. Conf. on Very Large Data Bases*, Kyoto, August 1986.

[Morr 86] Morris, K., J. D. Ullman, and A. Van Gelder; "Design overview of the Nail! system", *Proc. Int. Conf. on Logic Programming*, London, 1986.

[Morr 87] Morris, K., J. Naughton, Y. Saraiya, J. Ullman, and A. Van Gelder; "YAWN! (Yet Another Window on NAIL!)", *Special Issue on Databases and Logic*, IEEE – Data Engineering, 10:4, Zaniolo, C. (ed.), December 1987.

[Morr 88] Morris, K.; "An Algorithm for Ordering Subgoals in Nail!", *Proc. ACM SIGMOD-SIGACT Symp. on Principles of Database Systems*, Austin, 1988.

[Mura 83] Murakami, K., T. Kakuta, N. Miyazaki, S. Shibayama, H. Yokota; "A Relational Database Machine, First Step to Knowledge Base Machine", *Proc. 10th Symp. on Computer Architecture*, June 1983.

[Naqv 84] Naqvi, S.A.; "Prolog and Relational Databases: A Road to Data-Intensive Expert Systems", *Proc. First Workshop on Expert Database Systems*, Kiawah Island, SC, Oct.1984; and *Expert Database Systems*, L. Kerschberg (editor), Benjamin-Cummings, 1986.

[Naqv 86] Naqvi, S.A. ; "A logic for negation in database systems", *Workshop on Deductive Databases*, Univ. of Maryland, August 1986.

[Naqv 88] Naqvi S. and R. Krishnamurthy; "Database Updates in Logic Programming", *Proceedings of the 7th Annual ACM Symposium on Principles of Database Systems (PODS 88)*, Austin, TX, 1988.

[Naqv 89] Naqvi S. and S. Tsur; "A Logical Language for Data and Knowledge Bases", Computer Science Press, New York, 1989.

[Nejd 87] Nejdl, W.; "Recursive strategies for answering recursive queries – the RQA/FQI strategy", *Proc. 13th Int. Conference on Very large Data Bases*, Brighton, September 1987.

[Nico 82] Nicolas J.M. and R. Demolombe; " On the Stability of Relational Queries", *Proc. of int. Workshop on Formal Bases for Data Bases*, Toulouse, France, December 1982.

[Pars 83] Parsaye, K.; "Logic Programming and Relational Databases", *IEEE Computer Society DataBase Engineering Bulletin*, 6:4, Dec.1983.

[Przy 86] Przymusinski T.; "On the Semantics of Stratified Deductive Databases", *Proc. Workshop on Foundations of Deductive Databases and Logic Programming*, Washington D.C., J. Minker ed., August 1986, pp.433-443, and Morgan-Kaufman, 1988.

[Przy 89] Przymusinski, T.C.; "Every Logic Program Has a Natural Stratification and an Iterated Least Fixed Point Model", in: *Eighth ACM Symposium on Principles of Database Systems (PODS)*, March 1989, pp.11-21.

[Qian 87] Qian, X. and D.R. Smith; "Integrity Constraint Reformulation for Efficient Validation ", *Proc. Int. Conf. Very Large Data Bases*, Brighton, Sept. 1987.

[Quin 87] Quintus Computer Systems Inc., Mountain View, California; "Quintus Prolog Data Base Interface Manual", Version 1, June 29, 1987.

[Rama 87] Ramakrishnan, R., F. Bancilhon, and A. Silberschatz; "Safety of recursive Horn clauses with infinite relations", *Proc. ACM SIGMOD-SIGACT Symp. on Principles of Database Systems*, S. Diego, California, March 1987.

[Rama 88a] Ramakrishnan, R., C. Beeri and R. Krishnamurty; "Optimizing existential Datalog queries", *Proc. ACM SIGMOD-SIGACT Symp. on Principles of Database Systems*, Austin, Texas, March 1988.

[Rama 88b] Ramakrishnan, R.; "Magic templates, a spellbinding approach to logic evaluation", *Proc. of the Logic Programming Conference*, August 1988.

[Reit 78] Reiter, R.; "On closed world databases"; *Logic and Databases*, H. Gallaire and J. Minker (eds), Plenum Press, New York, 1978.

[Reit 80] Reiter, R.; "Equality and Domain Closure in First-Order Databases", *Journal of the ACM*, 27:2, 1980, pp.235-249.

[Reit 84] Reiter, R.; "Towards a Logical Reconstruction of Relational Database Theory", *On Conceptual Modelling: Perspectives from Artificial Intelligence, Databases, and Programming Languages*, M.L.Brodie and J.W. Schmidt Eds. Springer-Verlag, New York, 1984, pp.191-233.

[Robi 65] Robinson, J.A.; "A machine oriented logic based on the resolution principle", *Journal of the ACM*, **12**, 1965.

[Robi 68] Robinson, J.A.; "The generalized resolution principle", *Machine Intelligence*, vol.3, (D.Michie ed.) American Elsevier, N.Y., 1968.

[Roel 87] Roelants, D.; "Recursive rules in logic databases", *Report R513 Philips Research Laboratories, Bruxelles, March 1987*, submitted for publication.

[Rohm 86] Rohmer, J., R.Lescoeur and J.M.Kerisit; "The Alexander method: a technique for the processing of recursive axioms in deductive databases", *New Generation Computing*, 4, Springer-Verlag, 1986.

[Ross 89] Ross, A.; "A Procedural Semantics for Well Founded Negation in Logic Programs", in: *Eighth ACM Symposium on Principles of Database Systems (PODS)*, March 1989, pp.22-32.

[Sacc 86a] Sacca', D. and C. Zaniolo; "On the implementation of a simple class of logic queries for databases", *Proc. ACM 1986 SIGMOD-SIGACT Symp. on Principles of Database Systems*, Cambridge (MA), March 1986.

[Sacc 86b] Sacca', D., and C. Zaniolo; "Implementing recursive logic queries with function symbols", *MCC Technical Report DB-401-86*, December 1986.

[Sacc 87a] Sacca', D. and C. Zaniolo; "Magic counting methods", *Proc. of the ACM-SIGMOD Conference*, S. Francisco, May 1987.

[Sacc 87b] Sacca', D., M. Dispinzieri, A. Mecchia, C. Pizzuti, C. Del Gracco, and P. Naggar; "The Advanced Database Environment of the KIWI System", *Special Issue on Databases and Logic*, IEEE – Data Engineering, 10:4, Zaniolo, C. (ed.), December 1987.

[Sagi 87] Sagiv, Y.; "Optimizing Datalog programs", *Proc. ACM 1987 SIGMOD-SIGACT Symp. on Principles of Database Systems*, S.Diego (CA), March 1987.

[Sche 86] Scheck, H.J. and M.H. Scholl; "The relational model with relation-valued attributes", *Information Systems* 1986.

[Scio 84] Sciore, E., and D. S. Warren; "Towards an Integrated Database-Prolog System", *Proc. First Workshop on Expert Database Systems*, Kiawah Island, SC, Oct.1984; *Expert Database Systems*, L. Kerschberg (editor), Benjamin-Cummings, 1986.

[Shea 86] Sheard, T. and D. Steample; "Automatic Verification of Database Transaction Safety", Coins Tech. Rep. 86-30, Univ. Massachusetts, Amherst.

[Shep 85] Shepherdson, J.C., "Negation as Failure II", *Journal of Logic Programming*, 2:3, 1985, pp.185-202.

[Shep 88] Shepherdson, J.C., "Negation in Logic Programming", in: *Foundations of Deductive Databases and Logic Programming*, J. Minker Editor, Los Altos, CA, 1988, pp.19-88.

[Shmu 87a] Shmueli, O.; "Decidability and Expressiveness Aspects of Logic Queries", *Proc. ACM 1987 SIGMOD-SIGACT Symp. on Principles of Database Systems*, S.Diego (CA), March 1987, pp.237-249.

[Shmu 87b] Shmueli, O. and Sh. Naqvi; "Set Grouping and Layering in Horn Clause Programs", *Proc. International Conference on Logic Programming*, 1987, pp.152-177.

[Smit 86] Smith, D.E., M.R. Genesereth, M.L. Ginsberg; "Controlling Recursive Inference", *Artificial Intelligence*, 30:3, 1986.

[Ster 86] Sterling, L. and E. Shapiro; *The Art of Prolog*, MIT-Press, 1986.

[Stic 81] Stickel, M.E.; "A Unification Algorithm for Associative Commutative Functions", *JACM* 28:3, July 1981.

[Ston 87] M. Stonebraker and L.A. Rowe (editors); "The Postgres Papers", Berkeley University, *Memorandum* No. UCB/ERL M86/86, June 87 (revised version).

[Tanc 88] Tanca L.; "Optimization of Recursive Logic Queries to Relational Databases", (in italian) *Ph.D. Thesis, Politecnico di Milano and Universita' di Napoli*, 1988.

[Tars 55] Tarski, A.; "A lattice theoretical fixpoint theorem and its applications", *Pacific Journal of Mathematics*, n.5, 1955.

[Tsur 86] S. Tsur and C. Zaniolo; "LDL: A Logic-based Query Language", *Proc. 12th Int. Conf. Very Large Data Bases, Kyoto, 1986*.

[Ullm 82] Ullman, J.D., *Principles of Database Systems*, Second Edition, Computer Science Press, Rockville, Md., 1982.

[Ullm 85a] Ullman, J.D., "Implementation of logical query languages for databases", *ACM Transactions on Database Systems*, 10:3, 1985.

[Ullm 85b] Ullman, J.D. and A. Van Gelder; "Testing Applicability of Top-Down Capture Rules", Int. Rep. STAN-CS-85-1046, Stanford University, to appear in *ACM-Journal*.

[VanE 76] Van Emden, M.H., and R. Kowalski; "The semantics of predicate logic as a programming language", *Journal of the ACM*, 4, October 1976.

[VanE 86] Van Emden M.H. "Quantitative Deduction and its Fixpoint Theory" *Journal of Logic Programming* 1986:1 pp37-53.

[VanG 86] Van Gelder, A.; "Negation as Failure Using Tight Derivations for General Logic Programs", *Proc. Third IEEE Symp. on Logic Programming*, 1986, pp.137-146.

[VanG 88] Van Gelder, A., A. Ross, and J.S. Schlipf; "The Well-Founded Semantics for General Logic Programs", in: *Seventh ACM Symposium on Principles of Database Systems (PODS)*, March 1988, pp.221-230.

[VanG 89] Van Gelder, A.; "The Alternating Fixpoint of Logic Programs with Negation", in: *Eighth ACM Symposium on Principles of Database Systems (PODS)*, March 1989, pp.1-10.

[Viei 86a] Vieille, L.; "Recursive axioms in deductive databases: the Query-Subquery approach", *Proc. First Int. Conf. on Expert Database Systems*, L. Kerschberg ed., Charleston, 1986.

[Viei 86b] Vieille, L.; "A database complete proof procedure based on SLD resolution", *Proc. 4th Int. Conf. on Logic Programming ICLP '87*, Melbourne, Australia, May 1987.

[Viei 87] Vieille, L.; "From QSQ to QoSaQ: global optimization of recursive queries", *Proc. Second Int. Conference on Expert Database Systems*, L. Kerschberg ed., Tyson Corner, 1988.

[Wagn 86] Wagner Dietrich, S. and D.S. Warren; "Extension Tables: Memo Relations in Logic Programming", *Tech. Rep. 86/18*, Computer Science Department, SUNY at Stony Brook, 1986.

[Warr 81] Warren, D.H.D.;"Efficient Processing of Interactive Relational Database Queries Expressed in Logic", *Proc. 7th Int. Conf. Very Large Data Bases*, Cannes, 1981.

[Wied 87] Wiederhold, G.; *File Organization for Database Design*, McGraw-Hill, Computer Science Series, 1987.

[Zani 84] Zaniolo, C.; "Prolog: A Database Query Language For All Seasons", *Proc. First Workshop on Expert Database Systems*, Kiawah Island, SC, Oct.1984; *Expert Database Systems*, L. Kerschberg (editor), Benjamin-Cummings, 1986.

[Zani 85] Zaniolo, C.; "The representation and deductive retrieval of complex objects", *Proc. 11th Int. Conference on Very large Data Bases*, August 1985.

[Zani 86] Zaniolo, C.; "Safety and compilation of non recursive Horn cluses", *MCC Technical Report DB-088-85*, 1986.

[Zani 87a] Zaniolo, C. (ed.); *Special Issue on Databases and Logic*, IEEE – Data Engineering, 10:4, December 1987.

[Zani 87b] Zaniolo, C., and D. Sacca'; "Rule Rewriting Methods for Efficient Implementation of Horn Logic", *MCC Tech. Rep. DB-084-87*, March 1987, Austin, Texas.

Author Index

Abiteboul, S. 240, 244, 267
Aczel, P. 267
Agrawal, R. 205, 270, 272
Aho, A.V. 26, 206, 267
Apers, P.M.G. 267, 271, 272
Apt, K.R. 267
Arafati, F. 267

Bancilhon, F. 15, 161, 202, 206, 240, 265, 267, 268, 274
Bayer, R. 268
Beeri, C. 205, 235, 236, 244, 265, 267, 268, 274
Bidoit, N. 268
Blair, H. 267
Bocca, J. 62, 74, 268
Brandse, F. 267
Briggs, T. 240, 268
Brodie, M. 38, 268
Bry, F. 62, 143, 268

Ceri, S. 26, 62, 74, 143, 161, 266, 268, 269, 271
Chakravarthy, U.S. 143, 269
Chamberlin, D.D. 26, 269
Chandra, A.K. 242, 269
Chang, C.L. 26, 74, 269, 270
Chen, J. 63, 74, 244, 271
Chen, Q. 270
Chimenti, D. 270
Cholak, P. 242, 270
Clark, K.L. 242, 270
Clocksin, W.F. 26, 270
Codd, E.F. 26, 270
Colmerauer, A. 26, 270
Crespi, Reghizzi S. 266, 269, 273
Cuppens, F. 62, 74, 270

Dahlhaus, E. 270
Date, CJ. 26, 270
De Maindreville, C. 161, 271
Decker, H. 143, 268, 270
Del Gracco, C. 266, 270, 275
Demolombe, R. 62, 74, 270, 274
Denoel, D. 62, 74, 270
Devanbu, P. 205, 270
Dispinzieri, M. 270, 275

Fagin, R. 270
Ferrario, M. 266, 270
Fischer, P. 240, 244, 270
Fishman, D. 269
Fitting, M. 242, 270
Flokstra, F. 271
Friedman, M.A. 63, 74, 271
Fuchi, K. 265, 270
Furukawa, K. 265, 270

Gallaire, H. 14, 15, 270, 271
Gardarin, G. 15, 62, 121, 161, 244, 266, 270, 271
Garzotto, F. 143, 269
Gauss 128, 129, 147, 161
Gelfond, M. 271
Genesereth, M.R. 121, 271, 275
Geser, A. 266, 273
Ginsberg, M.L. 121, 275
Gottlob, G. 26, 62, 74, 161, 266, 269, 271
Gozzi, F. 38, 74, 269, 271
Grant, J. 143, 269
Gray, P.M.D. 274
Grumbach, S. 267
Guessarian, I. 271
Gurevich, Y. 271

Hanson: 265,
Harel, D. 242, 269
Henschen. 128,, 154, 161, 203, 271
Herbrand, J. 88, 89
Hong. 265,
Houtsma, M.A.W. 267, 271, 272
Hull, R. 267, 268

Imielinski, T. 271
Immermann, N. 271
Ioannidis, Y.E. 63, 74, 271
Itoh, H. 265, 272, 274

Jacobi. 128,, 145, 146, 161
Jaeschke, B. 240, 272
Jagadish, H.V. 272
Jarke, M. 38, 268
Jomier, G. 273

Kakuta, T. 274
Kemp, D.B. 272
Kerisit, J.M. 275
Kerschberg, L. 15, 272
Kersten, M.L. 267, 271, 272
Khoshafian, S. 240, 268
Kifer, M. 205, 272
King, J. 143, 272
Kleene, S.C. 26, 272
Kolaitis, Ph.G. 272
Korth, H.F. 26, 272
Kowalski, R. 26, 121, 143, 272, 276
Krishnamurthy, R. 243, 265, 270, 272, 274
Kunen, K. 242, 272
Kunifuji, S. 272
Kuper, G.M. 243, 272

Author Index

Lambrichts, E. 273
Lamperti, F. 269
Lassez, J.L. 273
Lavazza, L. 161, 266, 269, 273
Lee, R.C. 26, 269
Leitsch, A. 271
Lescoeur, R. 275
Lifschitz, V. 273
Lloyd, J. 26, 242, 273
Loveland, D.W. 273
Lozinskii, E.L. 205, 272, 273
Lugli, M. 38, 74, 269, 271

Maier, D. 26, 268, 273
Malley, V. 143, 273
Manthey, R. 143, 268
Marque-Pucheu, G. 273
Martin Gallausiaux, J. 273
McCarthy, J. 273
McKay, D. 273
Mecchia, A. 270, 275
Mellish, C.S. 26, 270
Mendelson, E. 26, 273
Mendelzon, A. 273
Minker, J. 14, 15, 143, 269, 270, 271, 273
Missikoff, M. 62, 274
Miyazaki, N. 274
Moffat, D.S. 274
Morita, Y. 266, 274
Morris, K. 265, 274
Murakami, K. 266, 274
Mylopoulos, J. 268

Naggard, P. 270, 275
Naqvi, S. 38, 128, 154, 161, 203, 235, 236, 243, 265, 268, 270, 271, 272, 274, 275
Naughton, J. 274
Nees, P. 273
Nejdl, W. 274
Ness, L. 272
Nguyen, V.L. 273
Nicolas, J.M. 14, 15, 268, 270, 271, 274
Nilsson, N.J. 271
Nishida, K. 274

O'Hare, T. 270
Oerlemans, H. 267

Papadimitriou, Ch.H. 272
Paredaens, J. 273
Parsaye, K. 38, 274
Peelman, P. 273
Pelagatti, G. 268
Pizzuti, C. 270, 275
Przymusinska, H. 271
Przymusinski, T. 271, 274
Pugin, J.M. 267

Qian, X. 143, 274
Quintus Computer Systems Inc.. 274

Ramakrishnan, R. 15, 161, 202, 205, 206, 235, 268, 272, 274
Reiter, R. 242, 274, 275
Robinson, J.A. 26, 121, 275
Roelants, D. 15, 62, 74, 270, 275
Rohmer, J. 275
Ross, A. 243, 275, 276
Rowe, L.A. 265, 275

Sacca', D. 205, 266, 270, 275, 276
Sadri, F. 143, 272
Sagiv, Y. 26, 206, 267, 268, 275
Saraiya, Y. 274
Schek, H.J. 240, 272, 275
Schlipf, J.S. 243, 276
Scholl, M.H. 275
Sciore, E. 38, 275
Seidel 128, 129, 147, 161
Shapiro, E. 26, 275
Shapiro, S. 273
Sheard, T. 143, 275
Shelah, S. 271
Shepherdson, J.C. 242, 275
Shibayama, S. 274
Shmueli, O. 235, 236, 268, 272, 275
Silberschatz, A. 26, 272, 274
Simon, E. 15, 62, 121, 266, 271
Skolem, T. 88, 89
Smith, D.E. 121, 275
Smith, D.R. 143, 274
Sonenberg, E.A. 273
Soper, P. 143, 272

Steample, D. 143, 275
Sterling, L. 26, 275
Stickel, M.E. 275
Stonebraker, M. 265, 275

Tanca, L. 266, 269, 273, 276
Tarski, A. 121, 276
Thomas, S. 240, 244, 270
Topor, R.W. 272, 273
Tsangaris, M.M. 63, 74, 271
Tsur, S. 235, 236, 265, 268, 270, 274, 276

Ullman, J.D. 15, 26, 206, 265, 267, 268, 274, 276

Valduriez, P. 240, 268
van de Weg, R.L.W. 272
Van Emden, M.H. 26, 121, 267, 276
Van Gelder, A. 243, 265, 274, 276
van Kuijk, H.J.A. 271, 272
Vauclair, M. 62, 74, 270
Vianu, V. 267
Vielle, L. 121, 161, 268, 276

Wagner-Dietrich, S. 62, 276
Walker, A. 74, 267, 270
Wallace, M. 268
Warren, D.H.D. 276
Warren, D.S. 38, 62, 273, 275, 276
West, C. 270
Wiederhold, G. 26, 62, 74, 269, 274, 276
Wong, E. 271

Yokota, H. 272, 274

Zaniolo, C. 15, 38, 205, 265, 266, 270, 272, 275, 276
Zdonik, B. 143, 273
Zicari, R. 266, 269, 270

Subject Index

AC-notation 177
Active instances 42
ADE 260, 266
Admissibility 238–240
Adornment 41, 166
Aggregate functions 209
Alerter 256
Algebraic
   equation 133
   naive evaluation 145
   queries 133
ALGRES 241, 244, 262, 266
Alphabet 77
Ancestor 137, 163, 171, 182, 186, 202
Answer substitution 119
Anti-trust control problem 30
Apex method 205
Arithmetic built-in predicates 211
Arity 16, 23, 78
Arity-PROLOG 71
Assert 25
Atom 23, 78
Attributes 16
Axiom nodes 177

Backtracking 24, 25
Backward chaining 94, 107, 121
   completeness 110
Base
   conjunction 47
      connected 47
   of maximal unifiers 234
BERMUDA 69
Bill of materials problem 34
Binding 83
Bmu 234
Bottom element 102
Bottom-up 125, 127
   computation 100
   evaluation 145
Bound-is-easier assumption 254, 255
Breadth-first search 112, 125, 127
Built-in predicate 132, 175, 208, 210

C-Prolog 262
Caching of data and queries 58
Caching of data 58
Caching of queries 60
Capture rules 253
Cardinality 16
Cartesian product 19, 135
Certainty factor 244
CGW 70
Chain base conjunction 47
Chaotic method 147
Choice predicate 235, 249, 250
Church-Rosser property 250
Circumscription 242, 243
Classification of optimization methods 124
Clause 78
   definite 113
   goal 80
   ground 79
   horn 79, 92
   positive 79
   Prolog 23
   formulas 80
   World Assumption (CWA) 208, 211, 212, 242
Common subexpressions 187, 200
Complementary rule 169
Completeness 98, 115, 121
   of backward chaining 110
   of EP 99
   of SLD-resolution 119
Completion 242
Complex objects 208, 228, 240
Components
   strong 189, 252
   strongly connected 189, 200
Composition 83, 134
Cone 171, 176, 201
   generalized 201
Conjunctive query 48
Connected base conjunction 47
Cons(S) 91, 99, 105
Consequence 88, 89, 91

under the CWA 212
Const 77
Constant 23, 77
  reduction 128, 129
  relation 130
Cost metrics 203
Counting method 128, 129, 174, 205
Coupling approach 12
CPR system 40, 42
Cut 25, 50
Cyclic databases 176

Data model 240, 244
Database
  deductive 1
  engine 44
  extensional 81
  formula 42
  intensional (IBD) 4, 29
  interface 43
  predicates 41
Datalog 10, 75, 77, 79, 86, 92, 124, 213–215
  clause 82
  expressive power 142
  extensions of pure 208
  goal 82
  inflationary 227
  program 82
  pure 208
  semantics 221, 223
DBMS 240
Declarative semantics 224
Definite clauses 113
DELTA 266
Denotational Semantics 101
Dependency graph 136, 147, 189, 194, 217, 238
  augmented extended 238
Depth-first search 112, 125, 127
Derivation tree 112
Derived relation 29
Difference 19
Differential 151, 153
Distinguished 166
Domain 16
  independence 242

EDG(P) 217
EDUCE 67
Efficiency 126
EHB 81
Elementary Production (EP) 95, 220
  completeness 99
EPred 81

Equation of relational algebra 124, 130
Equivalence transformations 18
Equivalent programs 164
ESPRIT 241
ESTEAM 68
Evaluable function symbol 229
Evaluation methods 128
Expert database system 1, 15
Expressive power 226, 142
Extended Dependency Graph 217
Extensional database (EDB) 81
Extensions of pure Datalog 208

Fact 24, 79
  inference 94, 95
FAD 240, 241, 244, 251, 265
Fail 25
FIFTH GENERATION Project 1, 15, 257, 259
Filter 179, 180
Findall 252
First-order logic 78
Fixpoint 103, 145
  Least 103, 221, 223
  Logic 227, 242
  Semantics 101, 121
  Theorem 103, 121
  Theory 101
Forward chaining 94, 100, 125
Function symbols 78, 229
  evaluable 229
Functional interpretation of predicates 154
Functor 23
Fuzzy reasoning 244

General semi-naive 152
Goal 24, 80, 86, 90, 223
  adorned 252
  clause 80
  derived 115
  projection 134
  selection 134
  structure 126, 127
  node 108
Graph representation of a binary relation 176
Ground 78
  clause 79
  expansion 224
  literal 78
  substitution
Grouping rule 235, 236
Growing transformation 221

Halting condition 112, 113, 115, 156
Herbrand
  Base (HB) 81, 242, 243
  interpretation 88, 89, 91, 214
  model 89, 105, 213, 214
  Universe 78, 235
    generalized 235
Horn clause 79, 92

IDM500 65
IHB 81
Incomplete knowledge 29
INFER 95, 99, 100
INFER1 96
Inference 97
  control 29
  engine 100
Inflationary 212, 221, 226, 227
INFORMIX 264
INGRES 65, 67, 68, 255, 262
Input ports 178
Integration approach 12
Integrity constraints 143
Intensional Database (IDB) 4, 29
Interpretation 87
IPred 81

J-equations 186
Join 19, 135
  natural 19

KAISER 258
KAPPA 258
Key 16
KIWI 260, 266
  Project 260
Knowledge base management system 1

Lattice 102
  complete 102
Layering 238
LCPR system 45
LDL (Logic Data Language) 228, 229, 231–237, 240, 243, 247, 265
  Herbrand base 235
  interpretation 235
  program 237, 238, 239
Least fixpoint (LFP) 103, 106, 221, 223
  iteration 94, 101, 103, 106, 221
  Herbrand Model 90, 91
Linear 139, 141, 194
  refutation tree 114
  refutation 114
  with respect to q_i 138

with respect to X_i 139
Linearity 126
Literal 23, 78
  ground 78
  negative 78
  ordinary 215
  positive 78
Local stratification 184, 224, 225, 242
Logic 127
  goal 114, 133
  mathematical 86
  Programming 77, 92, 113, 229
Logical rewriting methods 163
Loose coupling 12
Loosely coupled systems 45

Magic
  Counting method 205, 260
  rule 168
  set 128, 129, 165, 169, 201, 205, 243
Marking 191, 192
Matching 24
Mathematical
  linearity 140, 151
  Logic 86
MCC 228, 241
Meta-rule 95
MGU 84, 85, 95, 115, 116, 233, 234
Mini-Magic method 260
Minimal model 223, 225, 237, 239, 240
Model 88
  intersection property 92, 237
  minimality 221
  partial 243
  perfect 224–226, 242, 243
  Theory 86, 87
Modified rule 168
Monotonic
  constraints 241
  transformation 102, 104
Monovalued data function 243
Most
  general unifier (MGU) 84, 233
  instantiated goal 159
Mutually
  dependent 200
  recursive 138

NAIL! 251, 265
Naive 128, 129, 179
Natural
  join 19
  semijoin 19
Negation 52, 208, 211, 212, 215

as failure 242
Negative
    clause 79
    facts 213
    information 29, 211
    literal 78
Nested relations 240
NF^2 model 240, 241
Nondeterminism 209
Nonnormalized relations 262
Nonprocedural 18
NP-hard 235
Null
    (or empty) relation 16
    values 209

One-tuple-at-a-time 125
Optimization 57, 126, 127
Oracle 65, 71
Order sensitive 29
Output ports 178

Partial
    model 243
    order 101
Partially ordered 101
Partition 235
    predicate 248
Perfect model 224–226, 242, 243
Performance 203, 204
Positive Relational Algebra 18, 227
Possible worlds 86, 87
POSTGRES 255, 265
Pre-fetching 61
Predicate 23, 77
    built-in 132, 175, 208, 210
    partition 248
PRIMO 71
Principal variable 185
PRISMA 264, 266
PRO-SQL 65
Program
    completion 242
    datalog 82
    Prolog 23
    structure 126
    transformation 126
Projection 19, 135
Prolog 1, 9, 23–27, 40–43, 71, 80, 115
    clause 23
    engine 24, 42
    interface 43
    program 23
Proof 97

Theory 86, 87, 94
tree 24, 97, 100, 108, 110, 111
    depth 98
    full 110
Pure
    Datalog 208
    evaluation methods 126, 127
Push operation 180, 181

QSQ 113, 120, 121, 128, 155, 157
QSQI 157
QSQR 157, 242
QUEL 43
Query 18, 86, 156, 185
    complexity 242
    generalized 156
    language 18, 29
    subsetting 189, 200
QUINTUS-PROLOG 72

RA+ 18, 130
Reachable 167
Recursion 24, 29, 53, 101
Recursive 137, 138
    predicate 24
Reducible by substitution 191
Reduction
    of constants 193, 195, 200
    of variables 193, 200
Reflexivity 101
Refutation 114, 118
    procedure 125
    tree 117, 118
Relation 16
    derived 29
    node 177
    temporary 156
Relation-axiom graph 177
Relational
    algebra 18, 127
    calculus 18
    model 16, 240
Relationally complete 18
Resolution 94, 107, 113, 116, 120, 121
Resolvent 116
Retract 25
Rewriting
    methods 126, 128, 172
    of algebraic systems 185
RQA/FQI method 160
Rule 23, 79
    adorned 166
    composition 49
    grouping 235, 236

modified   168
  structure   127
Rule-goal graph   254
Russell's Paradox   238

Safety   210, 214, 241, 242
  condition   82, 210
Same-generation   149, 171, 174, 201, 202
Satisfaction   87
Schema   16
Scons   229
Search
  strategy   125
  technique   127
  tree   108–111
Selection   18, 135
  function   113, 115, 116, 118, 157
Self-reference   238
Semantic   127
  optimization   126, 143
Semantics   221, 223
  declarative   224
  inflationary   212, 221, 226
  of Datalog   213
  of logic programs   92
Semi-naive   128, 129, 151
  by rewriting   183, 184
  evaluation   150
Semijoin   19, 52
  natural   19
Set   101, 231
  matching   233, 250
  terms   248
  unification   233
Set-oriented   18, 125
Sideways information passing   165, 166, 201, 254
Simplistic evaluation   179
SLD
  refutation   116, 117, 119, 120
  resolution   94, 113, 118–121, 242
    completeness   119
SLDNF resolution   242
Soundness   98, 121
  of backward chaining   110
  of EP   98
  of SLD-resolution   119
  theorem   119
Special semi-naive   128
SQL   21, 26, 43, 71
SQL/DS   65, 67
Stability   141
  chain   141
Stable   194

Standard model   240
Static
  adornment   41
  filtering   128, 177, 181, 201, 205
  halting condition   112
Stratification   218–220, 223–226, 238
  algorithm   226
Stratified   219, 220
  Datalog   212, 215, 227
  program   215, 216, 219, 225
Stratum   218, 219, 220, 223
Structure of the goal   126, 127
Subgoal   108
Submodel   239
Subquery   156
Substantiation of a capture rule   253
Substitution   83, 84, 108
  ground
Subsume   84, 90, 113
Subsumption   59, 83, 92
Subtree factoring   113, 156
Success set   119
Syntactic optimization   126
Syntactic   127
Syntax-directed translation   130
System of algebraic equations   124, 130, 185

TCPR system   46
Term   78, 23
Termination   126, 129
Theorem proving   80
Theory unification   233, 243
Tight coupling   13
Tightly coupled systems   46, 258
Token Object Model   244
Top goal   108
Top-down   94, 125, 127, 155, 201
  computation   107, 108, 111, 155
Transformation, growing   221
Transitive closure   265
Transparency   43
Traversal order   127
Tree traversal   112, 115
Trigger   257
Truth   86
  value   88, 89
Tuple-oriented   18
Tuples   16
Typing of variables   209

U-equations   186
Uncertainty   209
Unifiable   84, 85
Unification   83

general   229
Unifier   84, 95
Uniform equivalence   206
UNIFY   65, 72
Union   19, 135
Union-join normal form   185
Universal quantifier   80
Unix   257, 262
Unsafe   210
Up-flat-down   202
Updates   209, 243, 250
Upper bound   101

Var   77
Variable   23, 77
  anonymous   23, 29
  principal   185
  relation   130, 185
Variables reduction   128
Variant   81, 116
View   22, 29

Well-founded semantics   243